INDUSTRIAL
MICROWAVE
SENSORS

The Artech House Microwave Library

Analysis, Design, and Applications of Fin Lines, Bharathi Bhat and Shiban K. Koul
E-Plane Integrated Circuits, P. Bhartia and P. Pramanick, eds.
Filters with Helical and Folded Helical Resonators, Peter Vizmuller
GaAs MESFET Circuit Design, Robert A. Soares, ed.
Gallium Arsenide Processing Techniques, Ralph Williams
Handbook of Microwave Integrated Circuits, Reinmut K. Hoffmann
Handbook for the Mechanical Tolerancing of Waveguide Components, W.B.W. Alison
Handbook of Microwave Testing, Thomas S. Laverghetta
High Power Microwave Sources, Victor Granatstien and Igor Alexeff, eds.
Introduction to Microwaves, Fred E. Gardiol
LOSLIN: Lossy Line Calculation Software and User's Manual, Fred. E. Gardiol
Lossy Transmission Lines, Fred E. Gardiol
Materials Handbook for Hybrid Microelectronics, J.A. King, ed.
Microstrip Antenna Design, K.C. Gupta and A. Benalla, eds.
Microstrip Lines and Slotlines, K.C. Gupta, R. Garg, and I.J. Bahl
Microwave Engineer's Handbook: 2 volume set, Theodore Saad, ed.
Microwave Filters, Impedance Matching Networks, and Coupling Structures, G.L. Matthaei, L. Young and E.M.T. Jones
Microwave Integrated Circuits, Jeffrey Frey and Kul Bhasin, eds.
Microwaves Made Simple: Principles and Applications, Stephen W. Cheung, Frederick H. Levien, et al.
Microwave and Millimeter Wave Heterostructure Transistors and Applications, F. Ali, ed.
Microwave Mixers, Stephen A. Maas
Microwave Transition Design, Jamal S. Izadian and Shahin M. Izadian
Microwave Transmission Line Filters, J.A.G. Malherbe
Microwave Transmission Line Couplers, J.A.G. Malherbe
Microwave Tubes, A.S. Gilmour, Jr.
MMIC Design: GaAs FETs and HEMTs, Peter H. Ladbrooke
Modern Spectrum Analyzer Theory and Applications, Morris Engelson
Monolithic Microwave Integrated Circuits: Technology and Design, Ravender Goyal, et al.
Nonlinear Microwave Circuits, Stephen A. Maas
Terrestrial Digital Microwave Communications, Ferdo Ivanek, et al.

INDUSTRIAL
MICROWAVE
SENSORS

Ebbe Nyfors
&
Pertti Vainikainen

Artech House

Library of Congress Cataloging-in-Publication Data

Nyfors, Ebbe G., 1956-
  Industrial measurement sensors / Ebbe G. Nyfors, Pertti
Vainikainen.
      p. cm.
  Includes bibliographical references.
  ISBN 0-89006-397-4
  1. Measuring instruments. 2. Detectors. I. Vainikainen, Pertti,
1957- . II. Title.
TA165.N94 1989   89-15197
681'.2--dc20   CIP

International Standard Book Number: 0-89006-397-4
Library of Congress Catalog Card Number: 89-15197

10 9 8 7 6 5 4 3 2 1

Contents

Preface   xi

Chapter 1   Microwave Sensors   1
    1.1   Introduction   1
    1.2   The Electromagnetic Spectrum   3
    1.3   Propagation of Electromagnetic Waves   4
        1.3.1   Maxwell's Equations and the Wave Equation   4
        1.3.2   Plane Waves and Complex Number Notation   6
        1.3.3   Propagation of Microwaves   9
        1.3.4   Guiding of Microwaves and Boundary
                Conditions for Conducting Surfaces   10
        1.3.5   Radiation and Antennas   16
    1.4   Interaction of Electromagnetic Waves with Matter   20
        1.4.1   Propagation Factor in Dielectric Medium   20
        1.4.2   Refraction and Reflection   23
    1.5   Microwave Sensors   28
        1.5.1   How Microwave Sensors Work   28
        1.5.2   The Measurement System   35
        1.5.3   How to Choose the Type of Sensor   36
        1.5.4   Interference (RFI or EMI) and Safety   36
    1.6   Advantages and Disadvantages of Microwave Sensors   38
    References   39

Chapter 2   Dielectric Properties of Materials   41
    2.1   Introduction   41
    2.2   Polarization Phenomena in Matter   41
        2.2.1   Relative Permittivity ε_r and Boundary Conditions   41
        2.2.2   Polarization in Dense Materials   47
        2.2.3   Typical Frequency Dependence of ε_r in
                Solids and Liquids   48

2.2.4 Electronic Polarization   49
2.2.5 Atomic Polarization   51
2.2.6 Orientation Polarization in Gases   52
2.2.7 Orientation Polarization in Liquids and Solids,
       the Debye Relation   55
2.2.8 Ion Conductivity and the Maxwell-Wagner Effect   58
2.3 Water   62
2.3.1 The Water Molecule   62
2.3.2 Liquid Water   63
2.3.3 Water Vapor   66
2.3.4 Bound Water   66
2.3.5 Ice   68
2.4 Mixing Formulas and Structure Dependence of Permittivity   69
2.4.1 General   69
2.4.2 Structure-Independent Formulas   70
2.4.3 Structure-Dependent Formulas   70
2.5 Multiparameter Measurement   79
2.5.1 The Measurement Situation   79
2.5.2 The Number of Measurements   80
2.5.3 Auxiliary Measurements   81
2.5.4 Microwave-Multiparameter Measurement   82
2.5.5 Short Cuts   86
2.6 Examples and References   87
2.6.1 Humid Air   87
2.6.2 Moist Wood   90
2.6.3 Short Summary and References   92
2.7 Permittivity Measurements in the Laboratory   93
2.7.1 Introduction   93
2.7.2 Permittivity Measurement Techniques   94
2.7.3 Measurement of Different Materials   108
2.7.4 Examples   111
References   121
Chapter 3 Resonator Sensors   127
3.1 Introduction   127
3.1.1 The Resonance Phenomenon   127
3.1.2 Microwave Resonators   130
3.2 Theory   133
3.2.1 Resonator Filled with Dielectric   133
3.2.2 Partially Filled Resonator, Perturbation Formulas   141
3.3 Resonator Structures   145
3.3.1 Coupling Devices   145

	3.3.2	Hollow Cavity Resonator	150
	3.3.3	Coaxial and Helical Resonators	154
	3.3.4	Strip Resonators	155
	3.3.5	Two-Conductor Line Resonator	160
	3.3.6	Slotline Resonator	161
	3.3.7	Open Quasioptical Resonator	162
	3.3.8	Dielectric, Ferromagnetic, and Other Resonators	163
3.4	Measurement of Resonant Frequency and Quality Factor		164
	3.4.1	Introduction	164
	3.4.2	The Basic Measurements	164
	3.4.3	Different Resonance Measurement Techniques	170
	3.4.4	Practical Aspects	184
3.5	Examples of Reported Applications of Resonators		186
	3.5.1	Hollow Cavities	186
	3.5.2	Coaxial and Helical Resonators	191
	3.5.3	Strip Resonators	193
	3.5.4	Two-Conductor Line and Slotline Resonators	194
	3.5.5	Open Quasioptical Resonators	196
	3.5.6	Dielectric and Ferromagnetic Resonators	196
	References		197
Chapter 4	Transmission Sensors		201
4.1	Introduction		201
4.2	Free-Space Transmission Sensors		202
	4.2.1	Sensor Structures	202
	4.2.2	Theory	202
	4.2.3	Practical Considerations	211
4.3	Guided Wave Transmission Sensors		213
	4.3.1	Introduction	213
	4.3.2	Sensor Structures	213
4.4	Measurement Equipment for Transmission Sensors		214
	4.4.1	Introduction	214
	4.4.2	Measurement Techniques	216
	4.4.3	Practical Aspects	221
4.5	Applications of Transmission Sensors		222
	4.5.1	Free-Space Systems	222
	4.5.2	Guided Wave Systems	225
	References		229
Chapter 5	Special and Hybrid Sensors		231
5.1	Introduction		231
5.2	Selected Examples		231

5.2.1   Measurement of Thickness and Permittivity of a Dielectric Layer on a Conducting Plane Using Surface Waves   231

5.2.2   Detection of Knots in Sawed Timber   234

5.2.3   Strength Grading of Timber by Combining Several Measurement Methods   234

5.2.4   Detection of Fatigue Cracks Based on Third-Order Nonlinearities   238

5.2.5   Gas Analysis by Microwave Spectrometry   238

5.2.6   Automatic Moisture Meter Based on Microwave Drying   239

5.2.7   Laboratory Meter for Fast and Simultaneous Determination of Permittivity and Density   241

References   243

Chapter 6   Reflection and Radar Sensors   245

6.1   Introduction   245

6.2   Reflection Sensors   245

6.2.1   Open-Ended Transmission Line Sensors   245

6.2.2   Shielded and Other Transmission Line Sensors   252

6.2.3   Free-Space Sensors   252

6.3   Radar Sensors   255

6.3.1   Introduction   255

6.3.2   Basic Principles of Operation   255

6.3.3   Applications of Radar Sensors   259

References   263

Chapter 7   Radiometer Sensors   267

7.1   Introduction   267

7.2   Basics of Radiometry   267

7.2.1   Black-Body Radiation   267

7.2.2   Brightness Temperature   269

7.2.3   Antenna Temperature   270

7.2.4   Radiometer   272

7.2.5   Receiver Noise Temperature and Sensitivity   274

7.3   Subsurface Radiometry and Thermography   276

7.3.1   Equation of Radiation Transfer   276

7.3.2   Visibility   276

7.3.3   Contacting Antennas   279

7.3.4   Correlation Radiometer   279

7.3.5   Aperture Synthesis Thermography   283

7.3.6   Multifrequency Radiometry   285

7.4   Applications of Radiometer Sensors   286

7.4.1   Medicine   286

	7.4.2	Industrial Applications	287
	7.4.3	Remote Sensing	287
References			288
Chapter 8	Active Imaging		291
8.1	Introduction		291
8.2	Holography		291
	8.2.1	True Holography	291
	8.2.2	Quasiholography	294
	8.2.3	Two-Dimensional Imaging of a Half-Space	300
	8.2.4	Imaging of a Rotating Sample	302
	8.2.5	Applications of Microwave Holographic Imaging	303
8.3	Tomography		304
	8.3.1	Introduction	304
	8.3.2	General Principles	304
	8.3.3	Medical Applications of Diffraction Tomography	305
References			307
Chapter 9	Appendices		309
9.1	Transmission Lines		309
	9.1.1	Basic Equations	309
	9.1.2	TEM Lines	310
	9.1.3	Waveguides	311
9.2	Resonant Cavities		313
	9.2.1	Rectangular Cavities	313
	9.2.2	Circular Cavities	314
9.3	Bessel Functions		316
9.4	Vector Relations		318
	9.4.1	Basic Relations	318
	9.4.2	Differential del Operator	319
	9.4.3	Vector Identities	320
9.5	Signal Flow Graphs		321
	9.5.1	Basic Elements	321
	9.5.2	Reduction of Flow Graphs	321
	9.5.3	Example: Measurement of Reflection Response of a Resonator	322
Reference			325
List of Symbols			327
Index			339

Preface

The purpose of this book is to provide the engineer, scientist, and university teacher with knowledge about the possibilities of the microwave technique to implement practical measurement sensors. Current applications, research results, some future ideas, and the underlying theories are described.

During the 1970s and 1980s, a rapid change has occurred in the controlling and steering of manufacturing processes. Formerly based on personal experience, observations, and some simple measurements, the job has now been given to a computer. This has caused a great need for sensors that can provide the control system with the necessary input information about the state of the process. Sensors are desired for measuring almost anything that comes to mind, ranging from moisture (in solids—surface, granules, sheets, bodies—liquids, or gases) and density to thickness, vibration, and temperature. The measurement sites may be conveyors or pipes and ovens, and the environmental conditions are often hostile. A great deal of research has been done on different physical phenomena to try to solve measurement problems. Today there are many different techniques in use, one of which is the microwave technique.

This book primarily concerns measurement applications in factories, but it also covers related areas in which the same kinds of sensors are used, such as nondestructive testing, collecting of ground truth data in remote sensing, close-range sensing and imaging of concealed objects, and medical applications.

We realized, in 1979, when we went into the field of microwave applications, the lack of good textbooks dealing with the subject. Since then, this has made difficult the research on new applications and especially the product development in the industry. When, in 1984, we had the opportunity to start a course on industrial applications of microwaves at the Helsinki University of Technology, our teaching was based primarily on scientific reports and publications.

This book attempts to satisfy the need for a textbook for those who have been exposed to microwave measurement applications. A short theoretical description of basic measurement principles and electrical properties of materials is given. Practical considerations related to measurement places, sensitivity requirements, and choice of sensor structure are treated, and design principles are presented where possible. Current applications are used throughout the book as illustrative examples, and references are made to many scientific reports. The collection of examples is chosen to be representative and broad, based on the information available to the authors.

The reader is assumed to know the fundamentals of physics, but not familiar with microwave technology. Knowing the basics of electronics will help to understand the sections dealing with measurement technology. The book is intended to be useful as a textbook for the university teacher at the undergraduate level as well as for process or product development engineers in industry.

The book is the result of a close and continuous cooperation between the authors. Matters were always first discussed, and the major part of the text was written by Nyfors, while Vainikainen wrote the sections dealing with measurement technology and electronics (Sections 2.7, 3.4, and 4.4) and the appendices. Later the various parts were revised to form a homogeneous text.

We want to direct our sincere thanks to all those who have worked with us in the research group for industrial applications; Prof. M. Tiuri, Dr. Tech. A. Sihvola, and Lic. Tech. P. Jakkula, who read the manuscript and made many valuable comments and suggestions for improvements; Ms. K. Sippola, who very patiently typed the manuscript; Mr. H. Frestadius, who skillfully drew the diagrams; our families and friends, who gave us their support; and Artech House, the publisher of the book.

EBBE NYFORS AND PERTTI VAINIKAINEN
ESPOO, FINLAND
MARCH 31, 1989

Chapter 1
Microwave Sensors

1.1 INTRODUCTION

The ability to measure has always been a central problem in technology and science. Today, when the control and steering of production processes are becoming computerized, measurement technology is increasing in importance. The requirements on the measurement sensors have also changed. Features enabling contactless, nondestructive, nondrifting, on-line, and real-time measurements are frequently in demand. In addition, the sensors should be able to measure not only the surface, but also inside a volume. They should be cheap so that they can be built in the form of arrays measuring moisture profiles, mass distributions, *et cetera*. The biggest change, however, has occurred in the range of nonelectrical quantities requiring measurement. In earlier days, these were merely temperature, weight, flow, time, and size, but today finding a quantity that nobody wants to measure is difficult.

In principle, a measurement sensor can be based on any physical phenomenon that is affected by the quantity to be measured. The sensors available today represent a wide range of phenomena, and a good amount of research is being done in new fields, some of which will bring solutions to specific problems and others will probably give birth to new families of sensors.

A large number of sensors are based on the interaction of electromagnetic energy with matter. Depending on the operating frequency, they are called resistive, capacitive, microwave, infrared (IR), optical, ultraviolet (UV), x-ray, and gamma-ray sensors. Because of the different frequencies, the applications and the technology involved are different as well. In this book, we will be only dealing with microwave sensors. We will try to give an idea of the variety of different applications, ranging from moisture measurement to tomographic imaging. The emphasis will be on the technology, fields of application, and measurement methods, but we will also study the properties of the materials to be measured.

Microwave sensor technology began to develop in the early 1960s, resulting in the first commercial sensors, but the initial enthusiasm soon vanished. The major reasons for this were, first, that the microwave technology was not yet suitable for industrial applications. The components were big, expensive, and in some cases unreliable. A second reason was the insufficient understanding of the material properties. Often there is more than one unknown (for example, moisture, density, or temperature) that calls for multiparameter or hybrid measurement systems and good mathematical models describing the material properties. The third reason was the lack of suitable calculating and data acquisition technology. Combining the signals from several measurement sensors to obtain the result was difficult. Today, however, the situation is different. Suitable microwave and data handling technologies are available, and we know a lot more of the materials we want to measure.

Designing a new application requires knowledge of both sensor technology and the material to be measured. This book is primarily about the technology, but one chapter is devoted to the materials. We have especially devoted Chapter 1 to the basics of electromagnetic fields and waves and their interaction with matter, which naturally leads to the possibility of measuring material properties with microwaves. The technical arrangements for measuring material properties (i.e., the types of sensor) are introduced in Chapter 1, and advice is given about choosing the type of sensor suitable for a specific application. The chapter ends with a study of the main advantages and disadvantages of measuring with microwaves as compared to other techniques.

Chapter 2 concerns the materials to be measured. Although this topic is treated in only one chapter, it plays a very important part in the process of designing a new application. The design process usually starts with a thorough study of the electrical properties of the material as a function of the quantity to be measured and all the other variables that may disturb the measurement. The designer should be fully aware of the occurring variations in temperature, composition, structure, density, for example, and their influence on the electrical properties of the material to be measured when planning the measurement system. In Chapter 2, we describe the physical phenomena related to the electrical properties of materials and discuss the practical consequences of the theory. This general discussion should provide appropriate tools for dealing with most of the materials the reader may need to measure.

In Chapters 3 through 8, we present the different types of sensors. For each sensor, we start with a theoretical description of the working principle and then describe how it can be used for measurement purposes. For the major types of sensor, we have tried to include enough rules and hints for designing it and the measurement electronics. We include reported examples of applications as a further attempt to link the theory to practice. Throughout the book we have also tried to present the basics of radio engineering to the extent that such is needed by an engineer familiar with lower frequencies for him or her to be able to accomplish a microwave sensor application.

1.2 THE ELECTROMAGNETIC SPECTRUM

Microwaves, light, and x-rays have one thing in common, they are *electromagnetic waves*. The difference lies in their wavelength and frequency. Figure 1.1 shows the electromagnetic spectrum on a logarithmic scale. We can see that short waves correspond to high frequencies and long waves to low frequencies. This is because the frequency and wavelength are related through the speed of propagation:

$$\lambda f = c \qquad (1.1)$$

where λ is the wavelength, f is the frequency, and c is the speed of propagation. In vacuum, all electromagnetic energy propagates with the same speed, $c_0 = 2.998 \times 10^8$ m/s, often called the speed of light. In other media, the speed is lower and the wavelength is shorter than in vacuum, but the frequency is unchanged.

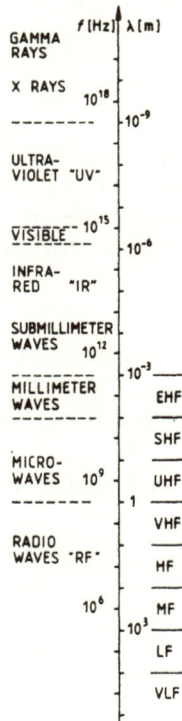

Figure 1.1 The electromagnetic spectrum.

All of the frequency limits for different kinds of radiation mentioned in Figure 1.1 are not exactly defined, except for the abbreviations listed on the right. For *microwaves,* the upper frequency limit is usually given as 30 GHz (= 3×10^{10} Hz), but the lower limit varies from 300 MHz to 1 GHz. Above microwaves come millimeter waves and submillimeter waves. The boundary of the infrared is decided more by technology than frequency. In this book, we will consider frequencies in the range below infrared and down to about 100 MHz. For the sake of brevity, we will refer to this range by the term "microwaves." Consequently, the wavelength varies from 3 m to less than 1 mm, but certainly not in the micrometer range, which the name erroneously implies. A characteristic of microwaves is that the wavelength is on the same order of size as the components and devices we use to guide and handle the waves.

Electromagnetic radiation has a dual nature. It acts as both a wave and a particle. The particulate behavior is explained by the quantum physics. Electromagnetic energy cannot exist in arbitrary amounts, but is always a multiple of a smallest possible amount, called a *quantum* (or *photon* for visible light). The energy of the quantum is related to the frequency:

$$E_q = hf \tag{1.2}$$

where E_q is the energy of the quantum and h is the Planck's constant ($h = 6.626 \times 10^{-34}$ Js). In the upper part of the spectrum, the particulate behavior is pronounced, and we therefore have spoken of rays. In the microwave range, the energy of the quantum is so small that we normally can disregard the quanta and speak only of waves.

1.3 PROPAGATION OF ELECTROMAGNETIC WAVES

1.3.1 Maxwell's Equations and the Wave Equation

The electromagnetic fields are completely described by *Maxwell's equations:*

$$\nabla \times \bar{\mathbf{E}} = -\frac{\partial \bar{\mathbf{B}}}{\partial t} \tag{1.3a}$$

$$\nabla \times \bar{\mathbf{H}} = \frac{\partial \bar{\mathbf{D}}}{\partial t} + \bar{\mathbf{J}} \tag{1.3b}$$

$$\nabla \cdot \bar{\mathbf{D}} = \rho \tag{1.3c}$$

$$\nabla \cdot \bar{\mathbf{B}} = 0 \tag{1.3d}$$

where ∇ is the vectorial *del* operator (for the rules of vector calculation, see Section 9.4), $\times$ denotes the vector product, $\cdot$ denotes the scalar product, ∂ denotes the partial derivative, $\bar{\mathbf{E}}$ is the electric field strength, $\bar{\mathbf{D}}$ is the electric flux density, $\bar{\mathbf{H}}$ is the magnetic field strength, $\bar{\mathbf{B}}$ is the magnetic flux density, $\bar{\mathbf{J}}$ is the current density, ρ is the charge density, and t is time. Boldface letters are used for vectors. The electric and the magnetic quantities are further related by the constitutive relations:

$$\bar{\mathbf{D}} = \varepsilon \bar{\mathbf{E}} \tag{1.4a}$$

$$\bar{\mathbf{B}} = \mu \bar{\mathbf{H}} \tag{1.4b}$$

where ε is the *electric permittivity* (also called the *dielectric constant*) of the medium and μ is the *magnetic permeability*. In isotropic media ε and μ are plain constants, but in anisotropic media either or both are tensors. In an area free of electric charge, (1.3) and (1.4) can be combined to give

$$\nabla^2 \bar{\mathbf{E}} - \mu\varepsilon \frac{\partial^2 \bar{\mathbf{E}}}{\partial t^2} = 0 \tag{1.5}$$

which is the three-dimensional *wave equation*. This means that the ability to propagate as waves is a very basic feature of the electromagnetic fields. Equation (1.5) can also be derived in the same form for $\bar{\mathbf{H}}$, and a wave will always have both magnetic and electric fields, which, in most practical cases, are perpendicular to each other in space. The wave equation can be solved for different situations, for example, leading to plane and spherical waves in space and guided waves in cables, waveguides, and other transmission lines.

In the wave equation (1.5), we can identify the constant preceding the second term as the inverse of the square of the speed of propagation:

$$c = \frac{1}{\sqrt{\mu\varepsilon}} \tag{1.6}$$

The values for the permittivity and the permeability in vacuum are

$$\varepsilon_0 = 8.854 \times 10^{-12} \quad \text{F/m} \tag{1.7a}$$

$$\mu_0 = 4\pi \times 10^{-7} \quad \text{H/m} \tag{1.7b}$$

giving, for the speed of propagation,

$$c_0 = 2.998 \times 10^8 \quad \text{m/s}$$

which is the same as the speed of light given in Section 1.2. In other media, the constants obtain higher values and they are usually expressed relative to the values in vacuum:

$$\varepsilon = \varepsilon_r \varepsilon_0 \qquad (1.8a)$$

$$\mu = \mu_r \mu_0 \qquad (1.8b)$$

where the subscript r stands for "relative." In media where the constants ε_r, μ_r, or both are greater than one, the speed of propagation is lower than in vacuum.

1.3.2 Plane Waves and Complex Number Notation

The wave equation (1.5) is in the general form. We know, however, that the time dependence of the steady-state wave solutions is sinusoidal. Other waveforms and transients can be expressed as spectra of sinusoidal waves by using the Fourier series expansion or the Fourier transform. Limiting the discussion to sinusoidal or so-called *time-harmonic solutions* is therefore justified.

In free space, a simple solution of the wave equation is that of a *plane wave* traveling in the direction of the x-axis:

$$\bar{E} = \bar{E}_0 \cos[\phi_0 + \omega(t - x\sqrt{\mu\varepsilon})] = \bar{E}_0 \cos\left[\phi_0 + \omega\left(t - \frac{x}{c}\right)\right]$$

$$= \bar{E}_0 \cos(\phi_0 + \omega t - kx) \qquad (1.9)$$

where $\bar{E}_0$ is the *amplitude* (peak value) of the wave (Figure 1.2), ϕ_0 is an arbitrary constant (angle of the cosine function at origin, at $t = 0$), ω is the angular frequency ($\omega = 2\pi f$), and k is the *propagation factor* or wave number. From (1.1) and (1.9), we can express the propagation factor in several ways:

$$k = \frac{\omega}{c} = \omega\sqrt{\mu\varepsilon} = \frac{2\pi}{\lambda} \qquad (1.10)$$

The wave has also a magnetic field (Figure 1.3) because, as we can see from Maxwell's equations, a time-dependent electric field creates a magnetic field and *vice versa*. For plane waves, the magnetic field is perpendicular to the electric field. The ratio between the amplitudes of the electric field and the magnetic field is determined by the medium and is called the *wave impedance* (see (1.45)).

In physics and electrical engineering, common practice is to use complex numbers instead of trigonometric functions to express the time dependence of time-har-

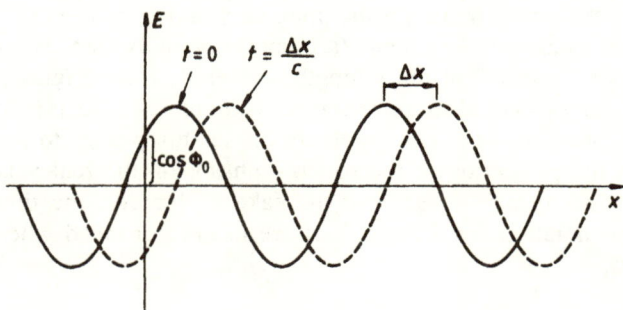

Figure 1.2 The electric field of a plane wave traveling along the x-axis. The magnetic field vector is perpendicular to the electric field vector and simultaneous in phase (see Figure 1.3).

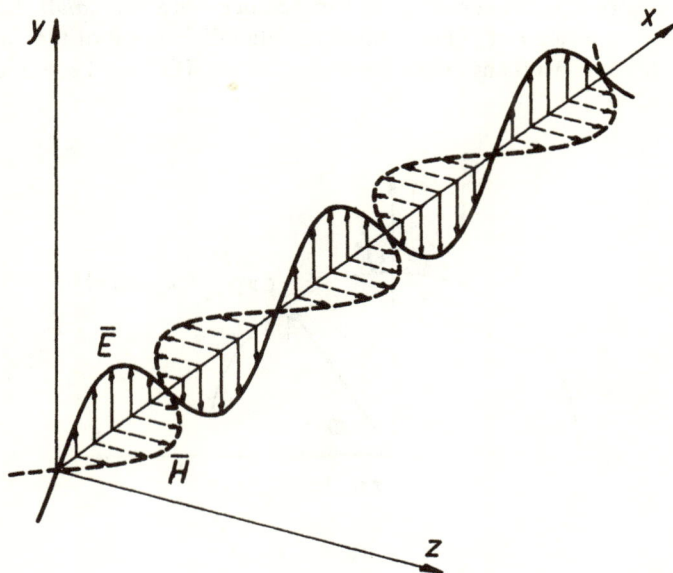

Figure 1.3 The electric and magnetic field vectors of a plane wave traveling along the x-axis at $t = 0$.

monic fields and signals. This is simply because complex numbers "behave" in the same way as the fields, and they are easier with which to calculate than the trigonometric functions. Expressed as a complex number, the plane wave in (1.9) is

$$\bar{E} = \bar{E}_0 \, \text{Re}\{\exp[j(\phi_0 + \omega t - kx)]\}$$
$$= \bar{E}_0 \, \text{Re}\{\exp[j(\phi_0 + \omega t)] \cdot \exp(-jkx)\} \qquad (1.11)$$

where Re means the real part and j is the imaginary unit (j = $\sqrt{-1}$). The exponential function with an imaginary exponent (argument) actually depicts, in the complex plane, a vector or "pointer" of unity length, centered at the origin, thus making an angle with the real axis equal to the argument (Figure 1.4). Such pointers are called *phasors*. The expression ωt in the argument causes the phasor to rotate with time. The real part is the projection of the rotating phasor on the real axis, which is the same as the cosine term in (1.9). For the sake of brevity, the time dependence, constant ϕ_0, and notation for the real part are usually dropped where they are not necessary. Thus,

$$\bar{\mathbf{E}} = \bar{\mathbf{E}}_0 \exp(-jkx) \tag{1.12}$$

is the same as (1.9) and (1.11). The above form is that which we will be using in this book for waves and other time-harmonic signals such as currents and voltages. The same convention is universally used in science with the small difference that physicists write $-i$ instead of j. For a more detailed discussion of the complex number notation, please consult any of the following books: [Collin, 1966; Gardiol, 1984; or Harrington, 1961].

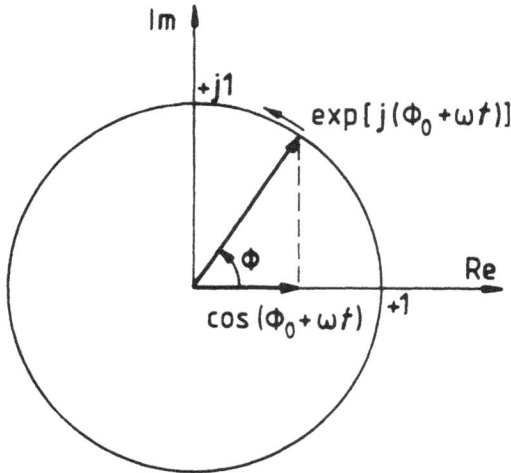

Figure 1.4 The complex number notation depicts a "pointer" rotating in the complex plane. The pointer is called a *phasor*. The actual signal is the real part of the phasor.

Written in a more general form, independent of the coordinate system (1.12) will be

$$\mathbf{E} = \bar{\mathbf{E}}_0 \exp(-j\bar{\mathbf{k}} \cdot \bar{\mathbf{r}}) \tag{1.13}$$

where $\bar{k}$ is the *wave vector* of magnitude k (propagation factor), pointing in the direction of travel of the wavefront, and $\bar{r}$ is the position vector.

The exponent in (1.12) or (1.13) gives the angle of the complex number or the cosine function at any point in space, which is called the phase angle or simply the *phase* of the wave. We can see that the phase is constant in the plane perpendicular to $\bar{k}$, which is the reason for calling them plane waves.

If the wave propagates in a lossy medium, the amplitude will be attenuated. By using the complex number notation, this can be taken into account by defining k as complex: $k = k' - jk''$. Now, k'' will cause part of the exponent in (1.11), (1.12), or (1.13) to be real and negative, thus describing the attenuation. In the most general case, $\bar{k}'$ and $\bar{k}''$ are vectors pointing in different directions. Such waves are called *inhomogeneous waves* (see Section 1.4.2).

We have frequently used the very obvious word, "wavelength," but we have still not defined it. The definition is *the wavelength is the distance between two consecutive surfaces with the same phase*.

1.3.3 Propagation of Microwaves

In the previous section we saw that plane waves propagate in free space in a given direction without changing. This is approximately true for waves that have traveled a long distance compared to the dimensions of the volume of space under study. If we move closer to the source or the transmitter, we will find that the waves propagate in a direction directly away from the source. Thus, the equal phase surfaces are spheres and the waves are called *spherical waves*. Hence, the power radiated into a solid angle will be distributed over a surface that is inversely proportional to the square of the distance to the source. The *intensity* is therefore given by

$$I(\theta,\phi,r) = \frac{P_t D(\theta,\phi)}{4\pi r^2} \tag{1.14}$$

where I is the intensity (in W/m^2), P_t is the transmitted power, r is the distance to the source, and $D(\theta,\phi)$ is the *directivity* of the source (e.g., transmitter antenna). The directivity is specific to the source and gives the distribution of the radiation with direction. The integral of D over all directions is unity by definition. Because the field is proportional to the square root of the power, we can write for the field in a certain direction from the source:

$$\bar{E}(r) = \bar{E}(r_0) \frac{\exp[-jk(r - r_0)]}{r/r_0} \tag{1.15}$$

where r_0 is an arbitrary reference point in the same direction as r. We can see that far from the source a small change in r has a negligible effect on the field amplitude, and the wave can be locally approximated by a plane wave (Figure 1.5).

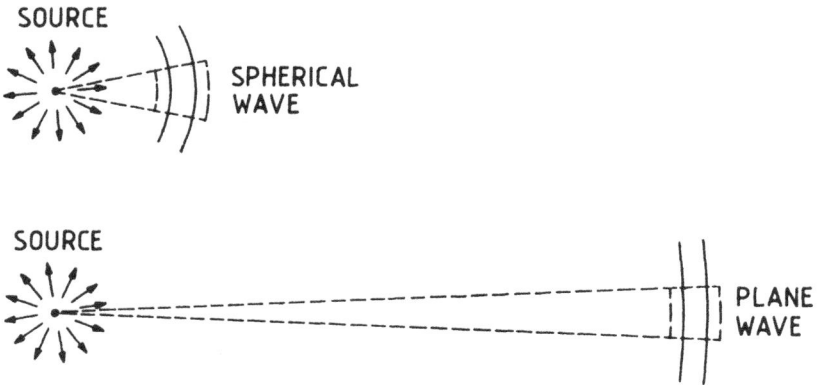

Figure 1.5 Sperhical waves and plane waves. Far from the source, a spherical wave will look like a plane wave.

1.3.4 Guiding of Microwaves and Boundary Conditions for Conducting Surfaces

Microwaves will not penetrate the surface of a perfect conductor. The electric field of the wave induces a current in the conductor that exactly compensates for the tangential electric field at the surface. The induced current radiates another wave observed as a reflection of the incident wave. The boundary condition for the electric field at the surface of a perfect conductor is therefore that *the tangential electric field is zero.* Only a normal component (i.e., one that is perpendicular to the surface) can exist. *The magnetic field is tangential to the surface,* and is related to the induced surface current by

$$\bar{n} \times \bar{H} = \bar{J}_s \qquad (1.16)$$

where $\bar{n}$ is the normal (perpendicular to the surface) unit vector and $\bar{J}_s$ is the surface current density (in A/m). For practical metal surfaces, the conductivity σ is finite, and as a result the electromagnetic field will penetrate a minute distance into the metal. The field decays exponentially from its surface value by $1/e$ at what is called the *skin depth:*

$$\delta_s = \left(\frac{2}{\omega\mu\sigma}\right)^{1/2} \tag{1.17}$$

The finite conductivity also causes a metal surface to be lossy. The loss power can be calculated by using the concept of *surface resistance:*

$$R_m = \frac{1}{\sigma\delta_s} \tag{1.18}$$

$$p_l = \frac{R_m}{2}|J_s|^2 \tag{1.19}$$

where R_m is the surface resistance, p_l is the loss power per unit area, and the bonds denote absolute value.

Because metal reflects microwaves, they can be guided in *transmission lines* made of metal tubes, called *hollow waveguides* (often called simply waveguides). The analysis of the propagation requires that we solve the wave equation in a waveguide. By doing so, we would find that there is an infinite series of solutions. Each solution corresponds to a certain *wave mode,* which has a characteristic standing wave field pattern in the waveguide. There are two separate sets of solutions. One set corresponds to waves with only transverse electric fields (no component parallel to the axis of the waveguide), but having both transverse and longitudinal magnetic field components. These waves are called TE$_{nm}$ modes. The other set of solutions are called TM$_{nm}$ modes. They have a transverse magnetic field and also an electric field with a longitudinal component. The subscripts n and m obtain integer values, and correspond to the number of field maxima in the standing wave pattern along the x and y directions in the transverse plane.

The standard waveguides used in practice are rectangular or circular (Figure 1.6). The rectangular waveguide with a broad wall to narrow wall ratio of 2:1 are the most common. The field equations and other important relations are given in Section 9.1 for rectangular and circular waveguides. The field pattern of any wave mode in the waveguide can be calculated from these equations. Figure 1.7 shows the results for some of the most frequently used modes.

Waveguide modes have some peculiar features. First, each mode has a *cut-off frequency,* $f_{c,nm}$, which depends on the dimensions of the waveguide. Below that frequency, the mode will not propagate. The cut-off frequency can be calculated from the cut-off wavenumber, $k_{c,nm}$, or the cut-off wavelength, $\lambda_{c,nm}$, given in Section 9.1:

$$f_{c,nm} = \frac{k_{c,nm}}{2\pi\sqrt{\mu\varepsilon}} = \frac{1}{\lambda_{c,nm}\sqrt{\mu\varepsilon}} \tag{1.20}$$

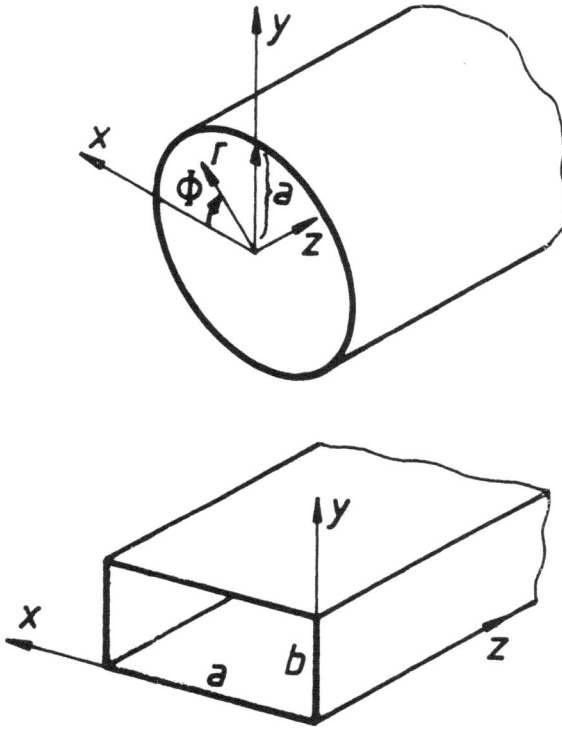

Figure 1.6 Rectangular and circular waveguides.

In rectangular waveguides, the mode TE_{10} (where 1 refers to the broad wall and 0 to the narrow wall) has the lowest cut-off frequency. From (1.20) and Table 9.1, we have for the cut-off wavelength:

$$\lambda_{c,10} = 2a$$

where a is the width of the broad wall. At frequencies below the lowest cut-off, microwave power cannot propagate. Each size of standard waveguide is normally used in the frequency range between the lowest and second lowest cut-off frequencies, where only one mode can propagate. In rectangular waveguides, with the width-to-height ratio of $2:1$, that range is about one octave.

Another characteristic feature of waveguides is that the velocity by which the energy propagates is lower than in free space. This notion is easy to understand if we think of the wave mode as the standing wave pattern between the walls that results

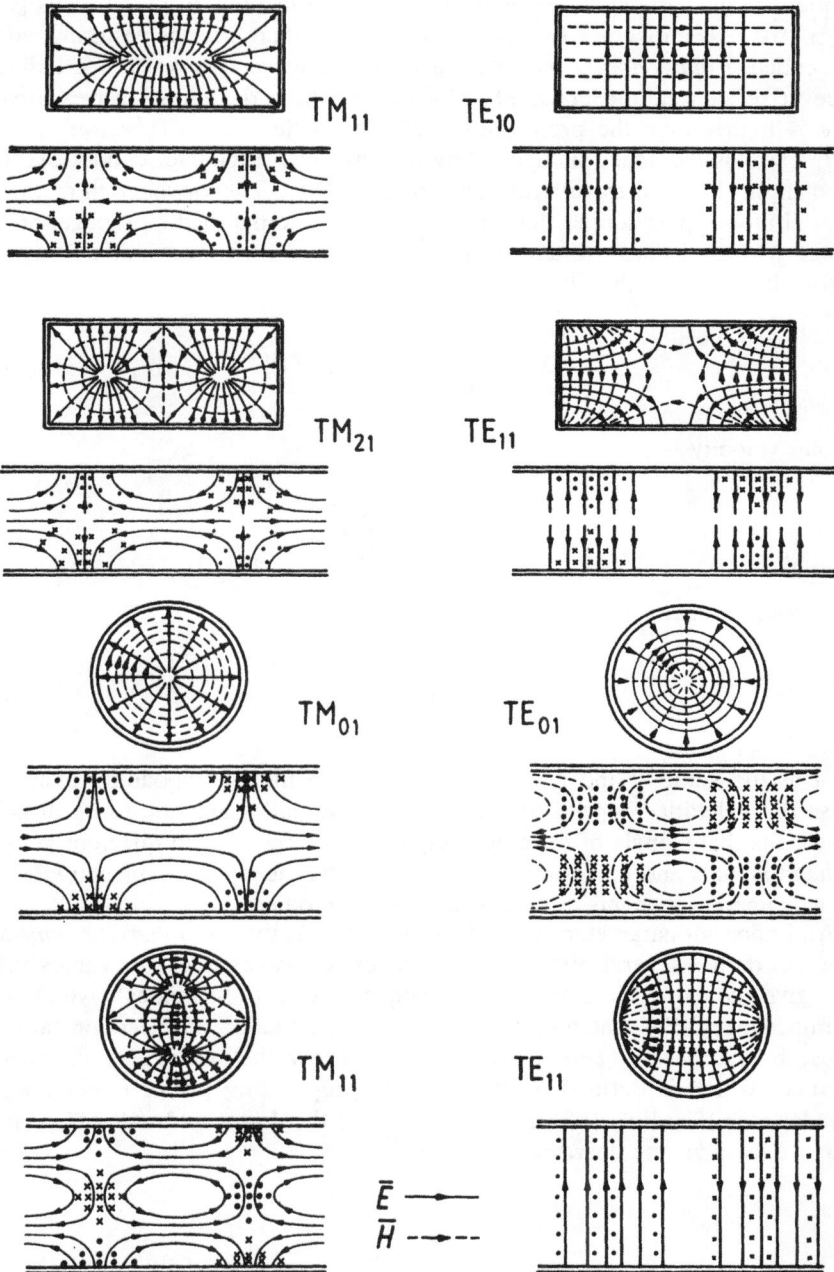

Figure 1.7 The field pattern of some of the wave modes with the lowest cut-off frequencies.

when plane waves propagate by repeated oblique reflections from the walls [Collin, 1966, p. 105]. The distance to travel is larger, and therefore the resulting axial velocity, called *group velocity,* is lower than in free space. However, for oblique incidence at reflection, the apparent velocity by which the phase moves along the surface is higher than the propagation velocity of the wave. Therefore the *phase velocity* exceeds the speed of light. Any information, of course, cannot exceed the speed of light, but will travel with the group velocity. The wavelength of the wave mode will also appear longer than in free space. In terms of the propagation factor $\gamma = \alpha + j\beta$ (see Tables 9.1 and 9.2), which is used for transmission lines instead of jk for plane waves, the *guide wavelength* is

$$\lambda_g = \frac{2\pi}{\beta} = \lambda[1 - (f_c/f)^2]^{-1/2} \qquad (1.21)$$

the group velocity is

$$v_g = \frac{\lambda}{\lambda_g} c \qquad (1.22)$$

and the phase velocity is

$$v_p = \frac{\lambda_g}{\lambda} c \qquad (1.23)$$

where c is the velocity and λ is the wavelength of the corresponding plane wave. Because the velocities depend on frequency, waveguides are said to be dispersive, which means that signals of different frequency will travel with different velocities. Modulated signals and pulses containing several frequency components will be distorted as some components are delayed more than others.

An important parameter of a transmission line is the *characteristic impedance,* but we can define several different impedances for waveguides. In Tables 9.1 and 9.2 we give the expressions for Z_e (TE modes) and Z_h (TM modes), which are the wave impedances. For matching purposes (i.e., elimination of reflections at discontinuities, boundaries, *et cetera*), we need to know the waveguide characteristic impedance, which is defined on the basis of propagating power rather than field amplitudes (see [Gardiol, 1984], pp. 32–33). For the dominant mode (TE$_{10}$) in rectangular waveguide, the characteristic impedance is

$$Z_0 = \frac{2b}{a} Z_e \qquad (1.24)$$

Now, the *field reflection coefficient* at a discontinuity, where the transmission line impedance changes from Z_1 to Z_2, can be calculated from

$$\Gamma = \frac{Z_{02} - Z_{01}}{Z_{02} + Z_{01}} \tag{1.25}$$

and the power reflection coefficient is the square of the absolute value of Γ.

In addition to straight waveguides, typical waveguide components are transitions from coaxial cable to waveguide, filters, circulators, directional couples, isolators, attenuators, phase shifters, and polarizers. For more information on how they work and how they are used, see [Gardiol, 1984 or Collin, 1966], for example. Diagrams and design rules are given in *The Microwave Engineer's Handbook* [Saad, 1971].

Microwaves can also be guided to propagate in structures other than waveguides. Any structure containing at least two separate conductors will support TEM type of modes, which have infinitesimal longitudinal field components or none at all. Some common TEM transmission lines are the coaxial cable and the two-conductor cable shown in Figure 1.8. Other common TEM lines are the microstrip line and the stripline shown in Figure 1.9. In transmission lines like the microstrip, where the wave propagates partly in the dielectric and partly in the air, there will be small longitudinal field components. They are therefore often called quasi-TEM lines. They can, however, with great accuracy be treated as TEM lines. Because they do not have longitudinal field components, the TEM modes propagate with the same propagation factor as plane waves. Therefore, TEM transmission lines are nondispersive.

The impedance of TEM lines is defined as the ratio of voltage and current. The expressions for the impedance and fields of some TEM lines are given in Section 9.1, and diagrams and design rules for microstrip lines and striplines are provided by [Edwards, 1981; Hammerstad *et al.*, 1975; and Saad, 1971]. The reflection coefficient of a discontinuity can then be calculated by using (1.25). When reflections are present, a standing wave pattern will arise in the transmission line. In this case, the ratio of the voltage and current varies along the transmission line, and the apparent input impedance (Z_{in}) depends on the distance (x) to the reflecting discontinuity. The input impedance (i.e., the ratio of the voltage and current at the input terminals) is

$$\frac{Z_{in}}{Z_1} = \frac{Z_2 + Z_1 \tanh\gamma x}{Z_1 + Z_2 \tanh\gamma x} \tag{1.26}$$

There are some types of transmission line that we have not yet mentioned, such as dielectric transmission lines, fin lines, slotlines, and several others, but we will not study them here.

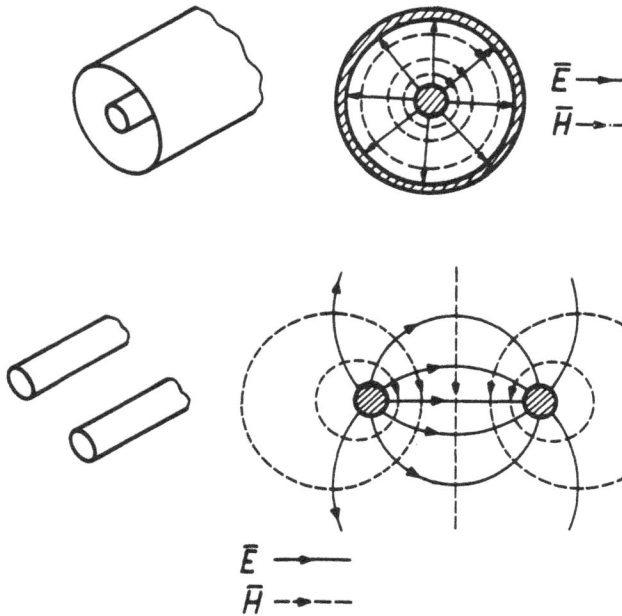

Figure 1.8 The coaxial cable and the two-conductor cable are common TEM transmission lines.

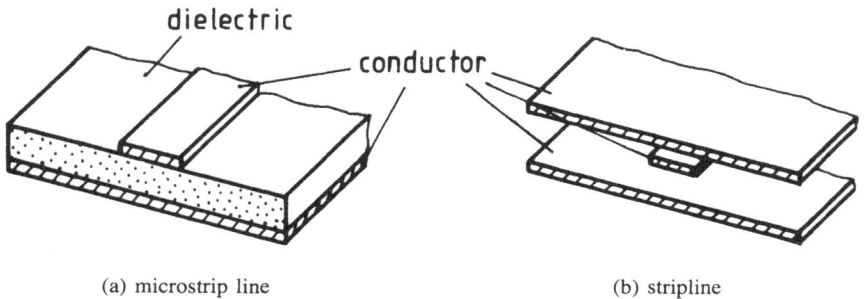

(a) microstrip line

(b) stripline

Figure 1.9 TEM transmission lines

1.3.5 Radiation and Antennas

Electric charges in accelerated movement radiate electromagnetic energy. Thus, a sinusoidal current of frequency *f* in a wire will radiate an electromagnetic wave with the same frequency as the current. This is the way that antennas work.

If we want to calculate the field at a particular location or the directional pattern achieved with a certain current distribution, we can use the formula for the radiation from the current density $\bar{J}$ (Figure 1.10), integrating over the whole source:

$$\bar{E}(r) = -j\omega\mu \frac{\exp(-jkr)}{4\pi r} \bar{P}_r\left[\int_V \bar{J}(\bar{r}') \exp(jk\bar{u}_r \cdot \bar{r}')\, dV'\right] \qquad (1.27)$$

where r is the distance from the source, $\bar{u}_r$ is the unit vector pointing in the direction under study ($\rightarrow \bar{r} = r\bar{u}_r$), $\bar{r}'$ is the position vector in the source area, and $\bar{P}_r$ is the projection operator:

$$\bar{P}_r(\bar{a}) = -\bar{u}_r \times (\bar{u}_r \times \bar{a}) \qquad (1.28)$$

If the source of the radiation is not a current but an electromagnetic field $(\bar{E}, \bar{H})$, for example, in an aperture antenna, we must first calculate the equivalent electric $(\bar{J}_s)$ and magnetic $(\bar{J}_{ms})$ surface currents:

$$\bar{J}_s = \bar{n} \times \bar{H} \qquad (1.29a)$$

$$\bar{J}_{ms} = -\bar{n} \times \bar{E} \qquad (1.29b)$$

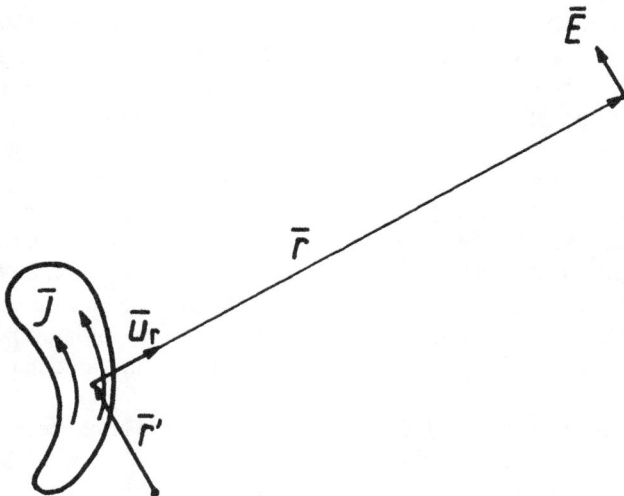

Figure 1.10 The geometry for calculating the radiation from the current density $\bar{J}$ in the direction $\bar{u}_r$.

The electric field radiated by the equivalent currents is given by the surface integrals:

$$\bar{E}(r) = -j\omega\mu \frac{\exp(-jkr)}{4\pi r} \bar{P}_r \left[\int_S \bar{J}_s(\bar{r}')\exp(jk\bar{u}_r \cdot \bar{r}')dS' \right] \qquad (1.30a)$$

$$\bar{E}(r) = jk \frac{\exp(-jkr)}{4\pi r} \bar{u}_r \times \int_S \bar{J}_{ms}(\bar{r}')\exp(jk\bar{u}_r \cdot \bar{r}')dS' \qquad (1.30b)$$

Equations (1.27) and (1.30) are valid in the *far-zone,* where all of the rays from different parts of the antenna can be regarded as parallel. The boundary between the *near-zone* and the far-zone is usually taken as

$$R_{fz} = \frac{2L^2}{\lambda} \qquad (1.31)$$

where L is the largest diameter of the transmitting antenna. Because of the parallax and reactive fields in the vicinity of the antenna, formulas more complicated than (1.27) and (1.30) must be used in the near-zone [Collin *et al.,* 1969]. The reactive fields do not propagate, but are due to the capacitance and the inductance of the antenna.

Microwaves can be transmitted and received by using many different kinds of antennas. To make a rough classification, we can divide them into *line-current sources* (e.g., dipole, Yagi, and wire antennas) and *aperture antennas* (e.g., horn, parabolic dish, and slot antennas), as shown in Figure 1.11. The most important parameters of an antenna are its pattern, gain, directivity, beamwidth, bandwidth, and impedance.

The normalized antenna *power pattern, $P_n(\theta,\phi)$,* describes the directive properties of the antenna. The pattern is defined to be equal to one in the maximum direction (main lobe). The *directivity, $D(\theta,\phi)$,* is the antenna power pattern divided by the integral of the pattern over all directions, and the *gain, $G(\theta,\phi)$,* is the directivity minus the power loss in the antenna. Hence, the gain in the direction of the maximum radiation determines how much stronger the radiated power is in that direction as compared to when all of the power fed to the antenna is evenly radiated in all directions.

The half-power beamwidth, θ_{3dB} or HPBW, is the angle between the half-power directions on either side of the maximum direction (Figure 1.12). The higher the directivity of the antenna, the narrower is the beamwidth. A high directivity and gain with a narrow beamwidth are achieved with a large antenna. Generally, the larger the antenna (as measured in wavelengths), the narrower is the beamwidth. As a rule of thumb, for an aperture antenna or an antenna array with a diameter L, the beamwidth is

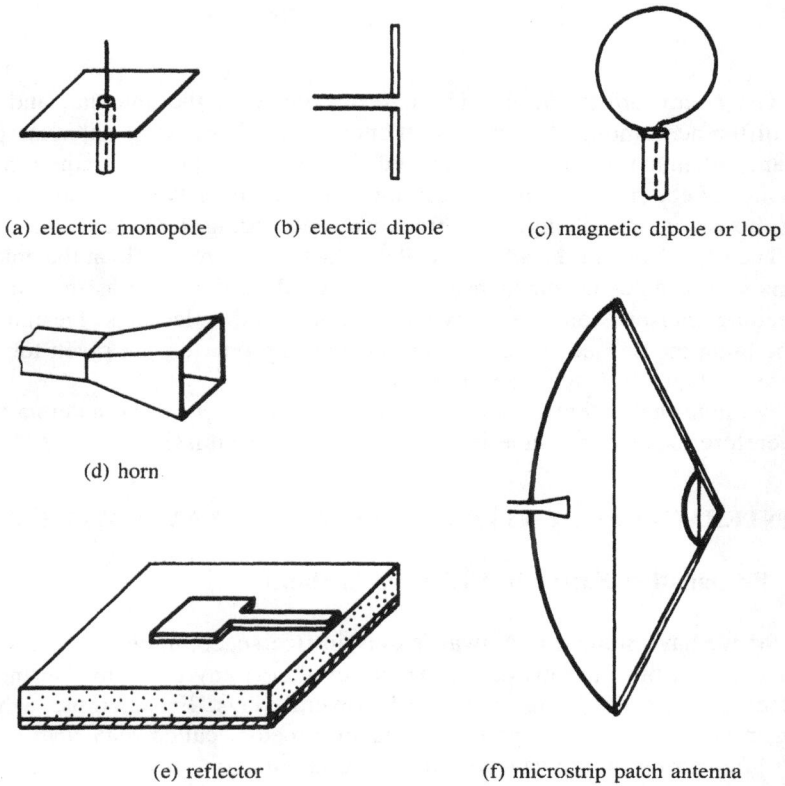

(a) electric monopole   (b) electric dipole   (c) magnetic dipole or loop

(d) horn

(e) reflector   (f) microstrip patch antenna

Figure 1.11 Some examples of antenna constructions

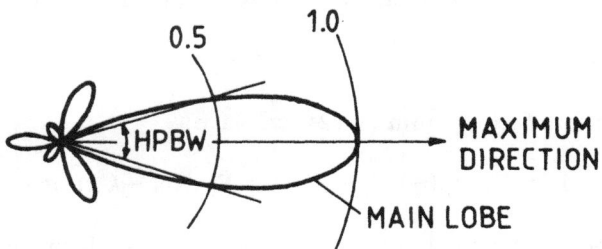

Figure 1.12 Example of an antenna power pattern $P(\theta,\phi)$ in one plane.

$$\theta_{3dB} \geq \frac{\lambda}{L} 70° \qquad (1.32)$$

The *bandwidth* is the usable frequency range of the antenna, and there is a clear difference among the various antenna types. Microstrip antennas (and other resonant antennas) are usually designed for use at a specific frequency, whereas parabolic reflectors, for example, can be employed over two decades. Their beamwidth, however, will change with frequency, according to (1.32).

The *impedance* of the antenna is the voltage-to-current ratio at the antenna input terminals. The antenna impedance ideally should be the same as the impedance of the feeding transmission line to avoid mismatch and reflections. Design principles for obtaining the desired radiation and impedance properties are given for many useful antenna types in [Stutzman *et al.*, 1981].

We note that antennas obey the *reciprocity principle*. The antenna parameters are therefore exactly the same in reception as in transmission.

1.4 INTERACTION OF ELECTROMAGNETIC WAVES WITH MATTER

1.4.1 Propagation Factor in Dielectric Medium

Thus far we have studied microwaves only in free space, at conducting surfaces, or in waveguides, but microwaves propagate in almost any medium. During propagation they are affected by the medium in several ways. For example, while propagating in nonconducting or poorly conducting media, called *dielectric materials* or simply *dielectrics*, microwaves suffer some attenuation. It arises from the friction that results when the electromagnetic field of the wave acts with a force on the electric charges (e.g., ions, electrons, polar molecules) or magnetic dipoles in the medium, causing them to be slightly displaced. We have indicated (Section 1.3.2) that we can take this into account by defining the propagating factor as complex:

$$k = k' - jk'' \qquad (1.33a)$$

where k'' is the loss factor. From (1.12), we have

$$\bar{\mathbf{E}} = \bar{\mathbf{E}}_0 \exp[-j(k' - jk'')x] = \bar{\mathbf{E}}_0 \exp(-jk'x) \cdot \exp(-k''x) \qquad (1.33b)$$

The real exponential function now describes exponential damping with propagated distance. Because k is complex, that ε_r and μ_r also are complex mathematically follows:

$$\mu_r = \mu_r' - j\mu_r'' \qquad (1.34a)$$

$$\varepsilon_r = \varepsilon_r' - j\varepsilon_r'' \tag{1.34b}$$

According to (1.10) and (1.8), the propagation factor is

$$k = \omega\sqrt{\mu\varepsilon} = \omega\sqrt{\mu_0\varepsilon_0}\,\sqrt{\mu_r\varepsilon_r} = k_0\sqrt{\mu_r\varepsilon_r}$$
$$= k_0\,\mathrm{Re}(\sqrt{\mu_r\varepsilon_r}) + jk_0\,\mathrm{Im}(\sqrt{\mu_r\varepsilon_r}) \tag{1.35}$$

where k_0 is the propagation factor in vacuum.

Attenuation and amplification (gain) are often measured in decibels (dB), which is defined as the logarithm of the ratio. If the field amplitude is attenuated from E_1 to E_2 at a certain distance, the attenuation in dB is

$$A_E(\mathrm{dB}) = 20\,\log_{10}\!\left(\frac{E_2}{E_1}\right) \tag{1.36a}$$

Because the power is proportional to the square of the field amplitude, the definition for the power is

$$A_P(\mathrm{dB}) = 10\,\log_{10}\!\left(\frac{P_2}{P_1}\right) \tag{1.36b}$$

If the impedance is unchanged, A_E equals A_P. Attenuation of the power to one-tenth therefore corresponds to -10 dB, or to one-half it is -3 dB. For amplification, the decibels are positive.

Because the permittivity and permeability are complex, we have four practically independent constants $(\varepsilon_r', \varepsilon_r'', \mu_r', \mu_r'')$ describing the electrical properties of the medium. Of course, the four constants depend on the other physical properties (moisture, composition, density, temperature, structure, *et cetera*) of the medium and on the measurement frequency. If we knew the relationships between ε_r and μ_r and the properties of the material, this would allow us to measure, for example, moisture in a material by studying the propagation of microwaves. Such is the foundation for the majority of the microwave sensors described in this book.

Because most materials are nonmagnetic, in the following we will often assume that $\mu_r = 1$ and discuss only measurement of permittivity. The results are easily transferred to magnetic materials because of the symmetry in (1.35). We need only remember that ε_r describes interaction with the electric field, whereas μ_r describes interaction with the magnetic field. This is especially important in resonators and waveguides, where the electric and magnetic field maxima are in different locations.

The concept of a complex propagation factor is powerful when we deal with propagating waves, but in other cases we encounter difficulties. For example, in a resonator, where we have an oscillating electromagnetic field confined in a cavity, or in the case of an oscillating plasma, the energy does not propagate in any direction. The same loss mechanisms (in the dielectric and the metal parts) of course, are present whether the fields propagate or not. If the exciting power is turned off, the field will decay exponentially with time. This is best described by introducing the concept of complex angular frequency:

$$\omega = \omega' + j\omega'' \tag{1.37a}$$

The time dependence of the field is now

$$\bar{E} = \bar{E}(\bar{r}) \, \text{Re}[\exp(j\omega t)] = \bar{E}(\bar{r}) \, \exp(-\omega''t) \, \text{Re}[\exp(j\omega't)] \tag{1.37b}$$

For resonators, a *quality factor* (or Q-factor) is defined, which depends on the losses (see Section 3.2.1):

$$Q = \frac{\omega \times \text{stored energy}}{\text{loss power}}$$

Because the energy is proportional to the square of the field and the loss power is the negative time derivative of the energy, we have

$$Q = \frac{\omega'}{2\omega''}$$

or

$$\omega = \omega'\left(1 + j\frac{1}{2Q}\right) \tag{1.38}$$

In addition to causing attenuation, a dielectric medium affects propagating microwaves in other ways. If we look at the phase component in (1.33b) and the propagation factor in (1.35), we find that the wavelength depends on the material:

$$\text{Phase angle} = -k'x$$

$$\rightarrow k'\lambda = 2\pi \rightarrow \lambda = \frac{2\pi}{k'}$$

From (1.35), we have

$$k' = k_0 \, \text{Re}(\sqrt{\mu_r \varepsilon_r}) \qquad (1.39a)$$

$$\rightarrow \lambda = \frac{2\pi}{k_0 \, \text{Re}(\sqrt{\mu_r \varepsilon_r})} = \frac{\lambda_0}{\text{Re}(\sqrt{\mu_r \varepsilon_r})} \qquad (1.39b)$$

For nonmagnetic ($\mu_r = 1$), low-loss ($\varepsilon_r'' \ll \varepsilon_r'$) materials (which is most often the case), we now have

$$\lambda \approx \frac{\lambda_0}{\sqrt{\varepsilon_r'}} \qquad (1.40)$$

Because $\varepsilon_r' \geq 1$, the wavelength is always shorter in a dielectric medium than in vacuum. For the same reason, the speed of propagation is lower than in vacuum:

$$c = \lambda f \approx \frac{2\pi f}{k_0 \sqrt{\varepsilon_r'}} = \frac{c_0}{\sqrt{\varepsilon_r'}} \qquad (1.41a)$$

or, from (1.6),

$$c = \frac{1}{\sqrt{\mu \varepsilon}} \approx \frac{c_0}{\sqrt{\varepsilon_r'}} \qquad (1.41b)$$

The practical consequences become obvious if we study the propagation of microwaves, first in free space between two antennas, and then after inserting a slab of dielectric in the path (Figure 1.13). We find that the waves are delayed and the phase angle is turned by the slab. In addition, the amplitude may have been attenuated because of ε_r'' and reflections at the surfaces. Reflection is one of the topics of the next section.

1.4.2 Refraction and Reflection

In the previous section, we saw that the wavelength and the velocity of a wave depend on the permittivity and the permeability of the medium. At oblique incidence at boundaries between areas with different ε_r or μ_r, this dependence will cause wavefronts to bend. This is the phenomenon of refraction that is familiar from optics. If we look at the geometry in Figure 1.14, where a wave is incident at the interface from medium 1, we can immediately write the expression for the angles of incidence:

$$\frac{\sin\theta_1}{\sin\theta_2} = \frac{\lambda_1}{\lambda_2} = \frac{k_2'}{k_1'} \qquad (1.42)$$

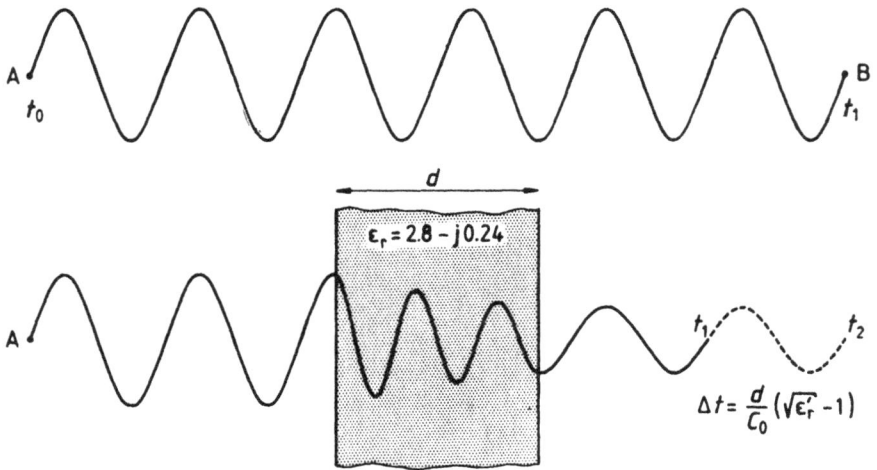

Figure 1.13 If a transmitter in point A is turned on at $t = 0$, the signal will arrive to point B at $t = t_1$ when it travels through free space. If the signal must travel through a dielectric, it will arrive later. (The reflections at the surfaces have been omitted in the figure.)

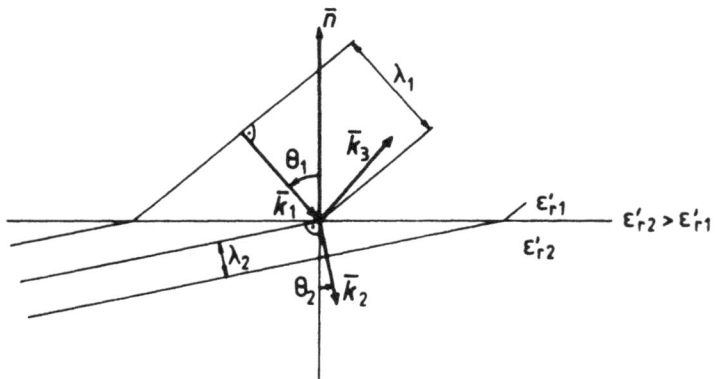

Figure 1.14 Refraction and reflection of wavefronts at a dielectric boundary.

Equation (1.42) is called *Snell's law*. By using (1.40), we have for the lossless nonmagnetic case:

$$\frac{\sin\theta_1}{\sin\theta_2} = \frac{\sqrt{\varepsilon'_{r2}}}{\sqrt{\varepsilon'_{r1}}} = \frac{n_2}{n_1} \tag{1.43}$$

where n is the refractive index used in optics. The general consequence of refraction is that a wavefront will be bent in case of oblique incidence at an interface between two media. The direction of bending is such that the direction of propagation is closer to perpendicular in the medium with higher ε_r' ("denser" medium) than in the other. If the wave is incident from the denser medium, there is a limit for θ, above which no transmission will occur, but the wave suffers *total reflection*. At total reflection, there are no propagating fields on the other side of the interface, but only reactive fields. The $\theta_{\lim}$ for total reflection can be calculated from (1.43) by using the condition $\theta_2 = 90°$:

$$\sin\theta_{\lim} = \sqrt{\frac{\varepsilon_{r2}'}{\varepsilon_{r1}'}} \tag{1.44}$$

Even when total reflection does not occur, the energy of the wave will not totally penetrate the interface, except for special cases. There normally is partial reflection and partial transmission due to the different wave impedances in the media and boundary conditions at the interface (see Section 2.2.1). The *wave impedance* Z_w is the ratio between $\bar{E}$ and $\bar{H}$ and is given by

$$Z_w = \frac{|\bar{E}|}{|\bar{H}|} = \sqrt{\frac{\mu}{\varepsilon}} \approx \sqrt{\frac{\mu_r}{\varepsilon_r}} \cdot 377\Omega \tag{1.45}$$

When ε_r or μ_r changes at an interface, the field strengths of the refracted wave will be different on both sides of the interface. The refraction also changes the normal and tangential components of the field strengths. Fulfillment of the boundary conditions therefore requires a third wave component, a reflected wave. The relative amplitudes of the waves can be calculated from the boundary conditions. The result is the so-called Fresnel field reflection coefficients. The amplitude of the reflected wave is then the reflection coefficient multiplied by the amplitude of the incident wave. For a *vertically polarized* wave (electric field tangential to the plane defined by $\bar{k}_1$ and $\bar{n}$) incident from medium 1, we have

$$\Gamma_v = \frac{-(\mu_{r1}'/\varepsilon_{r1}')^{1/2}\varepsilon_{r2}' \cos\theta_1 + (\mu_{r2}'\varepsilon_{r2}' - \mu_{r1}'\varepsilon_{r1}' \sin^2\theta_1)^{1/2}}{(\mu_{r1}'/\varepsilon_{r1}')^{1/2}\varepsilon_{r2}' \cos\theta_1 + (\mu_{r2}'\varepsilon_{r2}' - \mu_{r1}'\varepsilon_{r1}' \sin^2\theta_1)^{1/2}} \tag{1.46}$$

and for a *horizontally polarized* wave (electric field tangential to the interface):

$$\Gamma_h = \frac{(\varepsilon_{r1}'/\mu_{r1}')^{1/2}\mu_{r2}' \cos\theta_1 - (\mu_{r2}'\varepsilon_{r2}' - \mu_{r1}'\varepsilon_{r1}' \sin^2\theta_1)^{1/2}}{(\varepsilon_{r1}'/\mu_{r1}')^{1/2}\mu_{r2}' \cos\theta_1 + (\mu_{r2}'\varepsilon_{r2}' - \mu_{r1}'\varepsilon_{r1}' \sin^2\theta_1)^{1/2}} \tag{1.47}$$

If the incident wave is neither purely horizontally nor vertically polarized, it can be divided into two such components. Each component is then reflected and refracted independently. The angle of reflection is equal to the angle of incidence and symmetrical with respect to the normal vector.

The transmission coefficients can be calculated from the reflection coefficients:

$$t_v \cos\theta_2 = (\Gamma_v + 1)\cos\theta_1 \tag{1.48a}$$

$$t_h = \Gamma_h + 1 \tag{1.48b}$$

As the power is given by the square of the electric field divided by the wave impedance, we have for the power reflection and transmission coefficients:

$$R = |\Gamma|^2 \tag{1.49}$$

$$T = |t^2| \frac{Z_{w1}}{Z_{w2}} \tag{1.50}$$

Because of the energy conservation principle, we also have

$$T = 1 - R \tag{1.51}$$

The polarization has an interesting effect on the power reflection coefficient. The typical behavior as a function of incidence angle is shown in Figure (1.15). We can see that there is a certain angle for which R_v becomes to zero, called the *Brewster angle*.

Thus far, we have studied only what happens at an interface between two lossless materials. The general case of lossy media is slightly more complicated. The most important consequence of the losses is that the equal amplitude surfaces are no longer exactly parallel to the equal phase surfaces. We therefore must use the concept of wave vectors introduced in (1.13). The real ($\bar{\mathbf{k}}'$) and imaginary ($\bar{\mathbf{k}}''$) parts of the wave vector will be separate vectors pointing in different directions. The wave is called an *inhomogeneous wave*. Such waves cannot be divided into vertically and horizontally polarized components because they contain an electric field component in the direction of propagation. Instead they are divided into *E*- and *H*-waves, which are defined as

$$E\text{-wave:} \quad \bar{\mathbf{n}} \cdot \bar{\mathbf{E}} = 0$$

$$H\text{-wave:} \quad \bar{\mathbf{n}} \cdot \bar{\mathbf{H}} = 0$$

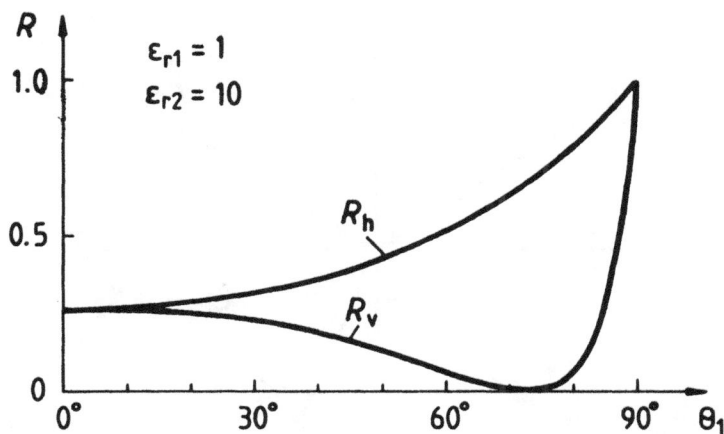

Figure 1.15 The power reflection coefficients for vertically and horizontally polarized waves when ε_{r1} = 1 and ε_{r2} = 10. The Brewster angle is 72.5°.

The E-wave is a generalization of a horizontally polarized homogeneous wave and the H-wave correspondingly is that of a vertically polarized wave. The E- and H-waves are reflected and refracted independently without change of polarization, as are horizontally and vertically polarized waves.

For inhomogeneous waves, the angle of incidence, were we to define it, would inevitably be complex. The angle is not necessary, however, because the wave vectors can be directly calculated from the generalized Snell's law:

$$\bar{\mathbf{n}} \times \bar{\mathbf{k}}_1 = \bar{\mathbf{n}} \times \bar{\mathbf{k}}_2 = \bar{\mathbf{n}} \times \bar{\mathbf{k}}_3 \qquad (1.52)$$

where $\bar{\mathbf{k}}_3$ is the reflected wave vector. For the real parts, (1.52) is the same as (1.42). We can also see that if medium 1 is lossless, both $\bar{\mathbf{k}}_1$ and $\bar{\mathbf{k}}_3$ are real, and $\bar{\mathbf{k}}_2''$ points directly downwards (see Figure 1.16). This is natural because the amplitude of the incident wave is constant over the interface, from which the equal amplitude surfaces in medium 2 hence must be parallel to the interface.

The field reflection coefficients for lossy media are

$$\Gamma_H = \frac{-\varepsilon_2 \bar{\mathbf{n}} \cdot \bar{\mathbf{k}}_1 + \varepsilon_1 [\omega^2 (\mu_2 \varepsilon_2 - \mu_1 \varepsilon_1) + (\bar{\mathbf{n}} \cdot \bar{\mathbf{k}}_1)^2]^{1/2}}{\varepsilon_2 \bar{\mathbf{n}} \cdot \bar{\mathbf{k}}_1 + \varepsilon_1 [\omega^2 (\mu_2 \varepsilon_2 - \mu_1 \varepsilon_1) + (\bar{\mathbf{n}} \cdot \bar{\mathbf{k}}_1)^2]^{1/2}} \qquad (1.53)$$

$$\Gamma_E = \frac{\mu_2 \bar{\mathbf{n}} \cdot \bar{\mathbf{k}}_1 - \mu_1 [\omega^2 (\mu_2 \varepsilon_2 - \mu_1 \varepsilon_1) + (\bar{\mathbf{n}} \cdot \bar{\mathbf{k}}_1)^2]^{1/2}}{\mu_2 \bar{\mathbf{n}} \cdot \bar{\mathbf{k}}_1 + \mu_1 [\omega^2 (\mu_2 \varepsilon_2 - \mu_1 \varepsilon_1) + (\bar{\mathbf{n}} \cdot \bar{\mathbf{k}}_1)^2]^{1/2}} \qquad (1.54)$$

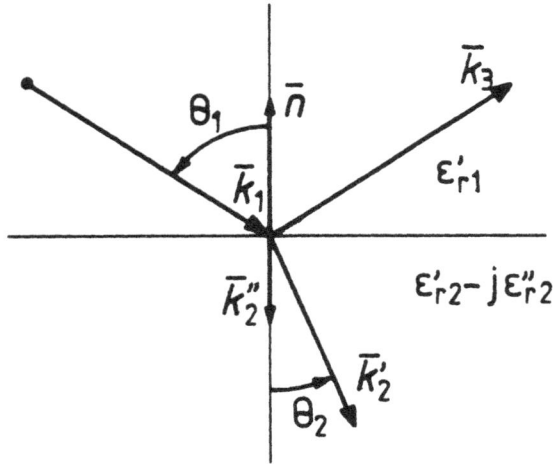

Figure 1.16 The wave vectors at the interface between a lossless and a lossy medium; ñ is the normal unit vector.

where ε_i and μ_i are not the relative values. Equations (1.53) and (1.54) are clearly generalizations of (1.46) and (1.47). The transmission coefficients can now be calculated from (1.48) or (1.51).

Thus far, we have studied only what happens at plane interfaces. If the interface is rough or curved, the wave hits a single object, or it propagates in an inhomogeneous medium, the situation is more complicated. The boundary conditions are still locally valid at the interfaces, but the net effect on the wave is difficult to calculate exactly. Generally, the wave will be *scattered* in many different directions. The scattering from single objects with simple geometry (e.g., spheres or cylinders) is fairly well known. Scattering from rough surfaces and volume scattering is a theoretically difficult problem, partly because of the many parameters needed to describe the medium. An extensive discussion of the scattering problem is found in [Ulaby *et al.*, 1982; Ulaby *et al.*, 1986; and Tsang *et al.*, 1985].

1.5 MICROWAVE SENSORS

1.5.1 How Microwave Sensors Work

In the previous section, we have seen that the medium affects propagating microwaves in many ways. The medium absorbs part of the power, causing the wave to be attenuated; changes the wavelength and the velocity of propagation, causing the wavefront to bend, the phase angle to change, and the signal to be delayed; and changes the wave impedance, causing the wave to be reflected. All of these effects depend on the material constants (permittivity and permeability) of the medium, which, in turn, depend on the other physical properties to be measured. The nature of this dependence between ε_r and μ_r and the other properties is the subject matter of the next chapter. We have also seen that it is possible to transmit, receive, and guide microwaves in a desired way. This offers us many possibilities to measure physical quantities with the help of microwaves. Most of the types of structures used in practice can be classified in five categories, but some are special sensors for specific applications.

Transmission Sensors

The most straightforward sensor construction is the *transmission sensor* (Figure 1.17). It consists of a transmitter, a receiver, and usually a pair of horn antennas. The material to be measured is located between the horns, and causes attenuation of the amplitude of the microwave signal and change of the phase. As we have seen, these changes depend on the propagation factor k and distance of propagation. The propagation factor depends on the permittivity ε_r and permeability μ_r, and therefore contains the desired information. The distance of propagation (i.e., the layer thickness), however, is usually a source of error if it is not kept constant or measured independently. The constant layer of thickness can be achieved with a material stream-forming unit or by forcing the material through a dielectric tube between the antennas. The layer thickness (or the material density) often is measured separately by using, for example, gamma-ray transmission. Another source of error is the reflection at the interfaces between the material stream and the surrounding space. The influence of reflections on the measurements can be eliminated in part by calibration. However, to obtain accurate results generally requires that the change in attenuation by dielectric loss in the material be much higher than the variation of the attenuation by reflections.

The advantages of transmission sensors are their simplicity and generality. The major constraint is the relatively large amount of material required for achieving sufficient sensitivity and for reducing the influence of reflections. This type of sensor is best suited for measurement of high loss materials.

Reflection and Radar Sensors

An obvious approach to microwave measurement is to measure the signal reflected back from an object. We will call such sensors *reflection* or *radar sensors*. The radar principle may be used for a wide range of applications because of its inherent versatility. Measuring material properties is possible by studying the magnitude and

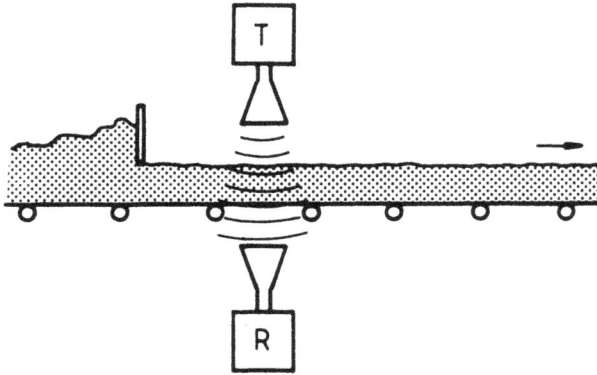

Figure 1.17 The measurement principle of a transmission sensor. The transmitter (T) directs a microwave signal through the material to be measured (on a conveyor). The receiver (R) measures the attenuation and the phase shift, which depend on the permittivity and permeability of the material and the thickness of the material layer.

phase of the reflection coefficient, either with a contacting sensor or from a distance. Measuring surface roughness and orientation is possible by studying the scattering properties (i.e., reflections in different directions), and measuring distance and movement is possible by studying the time of arrival of the reflections or the change of frequency (Doppler effect). We can also measure layer thicknesses of laminated materials by studying the reflection coefficient as a function of frequency or angle of incidence.

Contacting sensors based on the measurement of the reflection coefficient have become important as an application of reflection sensors. The most frequently used sensor structure is a coaxial cable with an open end, which is held against the object. Even miniature-sized sensors have been made, and can be used for intrusive measurements in medicine.

An important application of the use of radar is level sensing (Figure 1.18) in tankers and other places where the liquid is flammable or the surface may be covered by foam.

The greatest advantage with the radar principle is its versatility. The radar sensor may use a continuous wave or pulsed signals, a fixed frequency or swept frequency. The sensor may measure time, amplitude, or change of frequency.

Resonator Sensors

The third major class of sensors is composed of *resonators* (Figure 1.19), where the waves are bouncing back and forth between two reflectors or reflecting discontin-

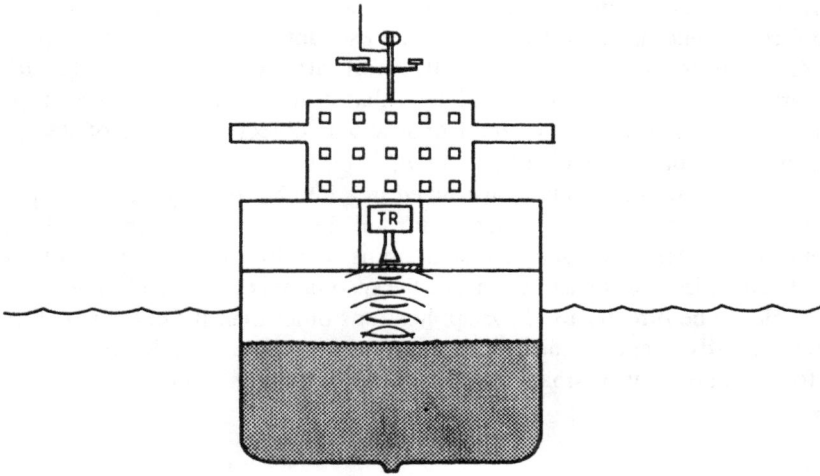

Figure 1.18 A level-sensing radar can be installed behind a hermetic dielectric window to eliminate the danger otherwise caused by electronic equipment in contact with explosive gases.

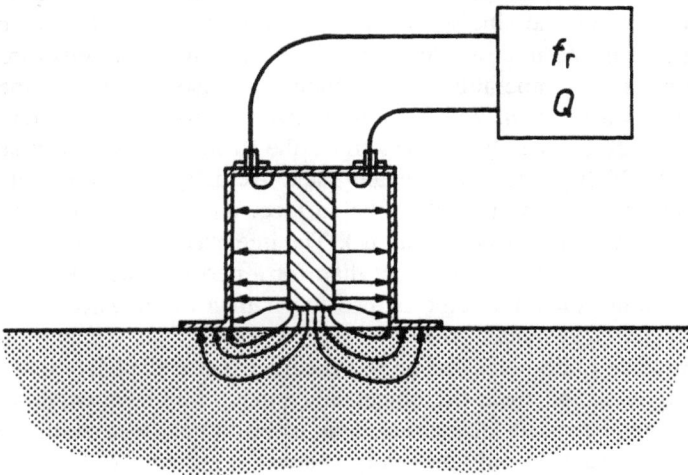

Figure 1.19 A coaxial resonator sensor can be used for measurements at a planar surface. The field penetrates the surface of the object to a certain depth.

uities in a transmission line. At certain excitation frequencies, the waves will combine to form a standing wave pattern. These resonant frequencies depend on the size of the sensor in wavelengths, and are therefore affected by the ε_r' or the μ_r' of the measurement object, which is located such that it affects the wave at each passage, thus amplifying the effect of a small or low-loss object. The rate of decay of the waves, the Q-factor, is affected by ε_r'' or μ_r''.

A resonator sensor can be constructed in a number of ways to fit the particular requirements of the application, depending on the shape, size, and ε_r of the measurement object. An advantage of resonators is therefore the versatility of the measurement principle. For the same reason, the resonator sensors tend to be specialized and can seldom be directly used for applications other than those for which they are intended. An advantage is their high measurement accuracy. Resonators are best suited for measurement of small, thin, or low-loss objects, and for surfaces of large objects.

Radiometer Sensors

As the name implies, a radiometer measures the intensity of radiation. Microwave radiometer sensors measure the thermal black-body radiation emitted over a broad range of frequencies by all matter. The radiation intensity, as given by Planck's law, is thermal noise, the same kind of radiation that hot objects emit in the optical range when they glow. In the microwave range, the radiation intensity is directly proportional to the physical temperature of the object, but also depends on the emissivity. The emissivity is a constant with a value between zero and one, indicating which portion of the black-body radiation penetrates the surface of the object and is emitted in the air, and which portion is reflected back. Actually, the emissivity is equal to the power transmission coefficient of the surface, and depends on both the surface roughness and wave impedance (ε_r and μ_r). The intensity measured with a radiometer is therefore lower for a practical object than for a perfect black body (emissivity = 1). The temperature thus measured is also lower than the physical temperature and is called the *brightness temperature*.

A microwave radiometer sensor is a receiver that is sensitive enough to detect the black-body radiation from the object in a microwave frequency band. The microwave radiometer is used for measurement of brightness temperature and is called a passive sensor because it only passively "listens" to the noise radiation transmitted by the object (Figure 1.20). Radiometers can be used to derive thermographical images of objects, either by turning the antenna in different directions or by using an electronically scanning antenna (i.e., phased array). Another possibility is to use a contacting probe-like antenna, which is moved over the surface of the object.

Radiometers are most commonly used for remote sensing of the earth from satellites and airplanes [Ulaby *et al.*, 1981, 1982, 1986], but that topic is beyond

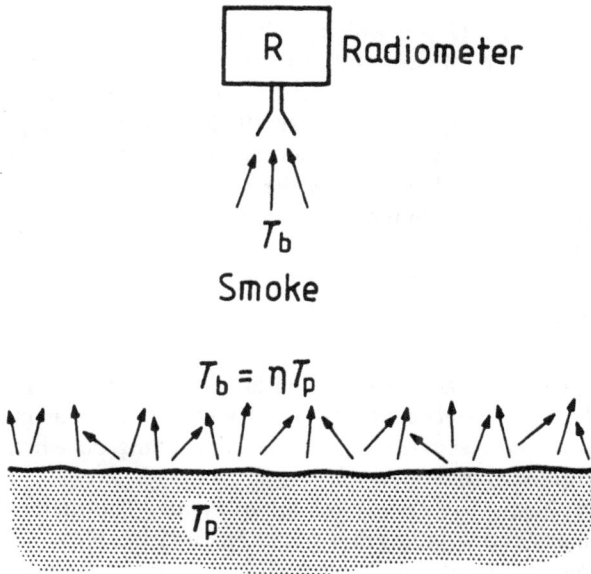

Figure 1.20 The microwave radiometer measures the thermal noise radiation emitted by the object, which gives the brightness temperature T_b. The microwaves are unaffected, for example, by smoke between the object and the antenna. T_p is the physical temperature of the object and η is the emissivity.

the scope of this book. In the industry, radiometers can be used for "remote" measurement of temperature in ovens, converters, kilns, and other places where use of conventional contacting temperature sensors or IR radiometers is impossible because of high temperature, smoke, or water vapour. In other places, microwave radiometers cannot compete because conventional sensors are cheaper and simpler, and IR radiometers are less sensitive to emissivity variations than are microwave radiometers. In the factories, microwave radiometers have not yet become common, but the potential applications are numerous. The advantages are the possibility to measure through optically thick smoke and vapor, and the fact that thermal microwave noise radiation comes from a thicker surface layer than does IR radiation.

Medicine is another field of potential application for microwave radiometers. Research activity since the mid-1970s has emphasized the development of thermographical means for detecting "hot spots" in the human body by using radiometers. The hot spots correspond to cancerous tumors or other abnormalities. Microwave thermography has also been used to monitor the changes in temperature during microwave hyperthermia (i.e., treatment of tumors by raising the temperature with microwave power). The advantage of microwave thermography as compared to IR thermography is that microwave radiation comes from a thicker layer (about 1 cm in muscle or 5 cm in fat at 5 GHz), which enables the microwave radiometer to "see" below the skin. However, the emissivity varies from place to place, depending on the thickness and moisture of the skin, and the differences in temperature between the hot spots and surrounding tissues are small. Monitoring changes in temperature during hyperthermia treatment is easier, but the absolute temperature is still difficult to measure as a function of depth below the skin. Correlation radiometery (two or more antennas focused to a certain depth) or multifrequency radiometry may provide a solution in the future.

Active Imaging Sensors

A special group of sensors are those which are used for imaging of the interior of objects, or for detecting concealed objects. These active imagers are divided into *holographic sensors* and *tomographic sensors,* depending on whether they measure reflected or transmitted radiation.

Holographic sensors are closely related to radars. The major difference is that the reflection is measured from several directions and the phase also is detected. A computer-generated image of the target is then produced. Holographic techniques, for example, can be used for detecting concealed objects in walls or underground.

Microwave tomography is based on the transmission principle. In the simplest case, we move the object or the antennas and look for differences. The other possibility is to illuminate the whole object with one antenna (Figure 1.21). The wave is then diffracted, refracted, reflected, attenuated, and scattered by the inhomogeneous object. If we measure the amplitude and phase of the wavefront in all directions on the other side, we can reconstruct the distribution of ε_r or μ_r in the object by using some rather complicated mathematics. This is the methodology of microwave tomography. Its applications are thought to be found primarily in medicine, but at present the technology is in the laboratory stage.

Special Sensors

In addition to the five major groups of microwave sensors, constructing special sensors is possible for specific applications. By special sensors we mean those which may have attributes in common with some of the major groups, but also utilize a specific feature of the object. A good example is the knot detector for timber, which is described in Chapter 5. The sensor is based on the fact that a knot acts like a dielectric waveguide supporting a specific wave mode, which can be detected.

Very little can be said in general about special sensors. Each application is based on an essentially original idea. In Chapter 5, we will give more examples in hope of inspiring the creativity of the reader, because only the imagination limits what can be invented.

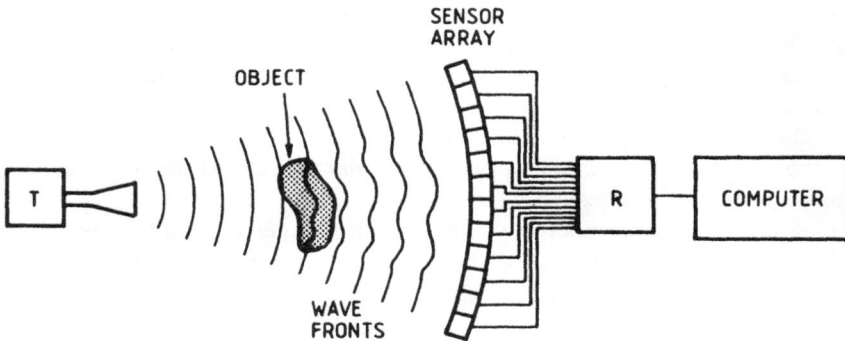

Figure 1.21 Microwave tomography.

1.5.2 The Measurement System

The system needed for performing measurements with a microwave sensor consists of several parts (Figure 1.22). First, there is often a *material stream-forming* or *sample preparation unit,* which physically prepares the material to be measured. The purpose is to eliminate or decrease the variations in density, layer thickness, temperature, or location.

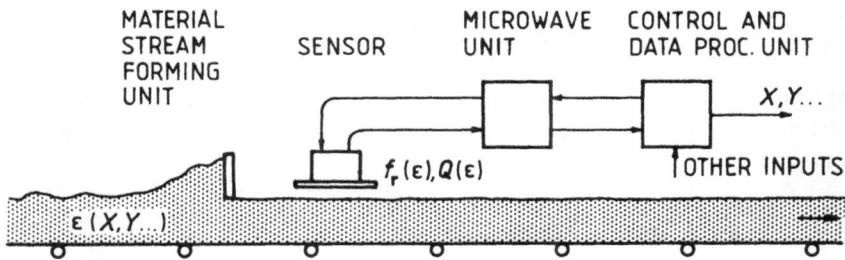

Figure 1.22 The microwave meter.

The measurement device comprises three parts: *sensor, microwave unit,* and *control and data processing unit.* In the sensor, the interaction occurs between the material to be measured and the microwaves. The sensor consists of a resonator, one or several antennas, or some specialized structure. The microwave unit produces the microwave signal fed to the sensor (except for radiometers), and receives and detects the signal that contains the desired information. Many identical sensors can be connected to the same microwave unit for imaging or profile determination purposes. The control and data processing unit directs the whole measurement procedure, calculates the results, and transforms the output message into the desired form. The unit may also receive inputs from other than microwave sensors, for example, to measure temperature or density. The output may be connected to an optical display, recording device, or control system.

In the following, we will refer to the microwave unit and the control and data processing unit by the common name of *electronic unit,* and we will call the measurement system the *microwave meter.*

The requirements on microwave meters vary, depending on the kind of use for which they are intended. We can distinguish four categories of meters:

- Portable field meters: lightweight; robust and simple construction; easy to use; cheap; limited measurement range; specialized; often of limited accuracy.
- Industrial, automatic, on-line meters: reliability and stability are more important than absolute accuracy or price.
- Industrial laboratory meters: used for measurement of samples taken from the production line; easy to use and fast; moderate to high accuracy.
- Laboratory scientific instruments: accuracy is most important; often complicated to use; large in size; expensive.

Microwave sensors are probably best suited for constructing industrial, automatic, on-line meters because of their robustness and relatively high price of microwave components. However, construction of microwave meters has been shown to be possible in all of the categories above.

1.5.3 How to Choose the Type of Sensor

We cannot give any absolute rules for when a resonator is the best choice or when, for example, a transmission sensor is better, but we have tried to list some general hints in Table 1.1.

1.5.4 Interference (RFI or EMI) and Safety

Most microwave sensors are open structures, which means that there is a risk of interference. The sensor may radiate microwave energy, or may collect disturbing signals from the environment if this possibility is not taken into account in the design. Sensors (except radars) are usually classified as electrical devices (not radio equipment) because they are not intended to radiate into the environment. In most countries there are regulations limiting the radiated emission from such devices (for example, United States: FCC-15J; Federal Republic of Germany: VDE 0871 and 0875; Canada: CSA C108.8-M1983; CISPR member countries: CISPR/PUB 22). The radiation limits will usually not cause serious problems if they are taken into account in the sensor design because the microwave power needed by the sensor is usually

Table 1.1
How to Choose the Microwave Sensor

Unknown	Circumstances	Reflection Sensor	Resonator Sensor	Transmission Sensor	Radar	Radiometer
Temperature						x
Distance, long					x	
Distance, short		x	x			
Level					x	
Movement			(x)		x	
Speed					x	
Size			x			
Material properties (via ε_r, μ_r)	Single object (<λ)		x	(x)		
	Single object (>λ)	(x)	(x)	x		
	Sheet-like material	(x)	x	(x)		
	Thick layer (>λ) (measurement through)			x		
	Thick layer or filled "half-space" (surface measurement)	x	x	x		
	Thread (<λ)		x			
	Liquid or granular, solid material in pipe	(x)	x	x		
	Gas		x	(x)		
	Solid particles in pneumatic transportation		x		x	
	Sensor inserted in soft material	x	x			

very small. By using resonators with high Q-factors or by suitably directing the antennas the radiation can be decreased considerably. This also protects against reception of disturbing radiation from the environment. The protection against signals at frequencies other than those used in the sensor can be further improved by filters.

If the interference radiated by the sensor is a problem, it is also possible to use an ISM (industrial, scientific, and medical) frequency. ISM comprises frequency bands allocated by the International Telecommunication Union (ITU) in the *Radio Regulations* (Geneva, 1982) primarily for power applications. In those bands there are no exact limits (except for those of safety) for the stray radiation, which, of course, also means that there is a greater risk that the environment will disturb the sensor. The international ISM frequencies are given below:

13.560 MHz ± 7   kHz
27.120 MHz ± 163 kHz
40.680 MHz ± 20   kHz
433.92 MHz ± 870 kHz (some European countries)
915 MHz ± 13   MHz (North and South America)
2450 MHz ± 50   MHz
5.8 GHz ± 75   MHz
24.125 GHz ± 125 MHz

Another aspect to consider with microwave sensors is safety. Nobody has been able to prove the existence of any safety risks from microwave radiation, other than those caused by thermal effects. The possible existence of nonthermal effects, however, has been taken into account by most countries in formulating the exposure limits. Because of the uncertainty, there is a discrepancy among the imposed limits in different countries. Limits, however, normally will not cause any problems with microwave sensors. The International Non-Ionizing Radiation Committee of the International Radiation Protection Association has published a proposal [IRPA, 1988] for international exposure limits. The proposal classifies all small devices (i.e., radiating less than 7 W) as safe under all circumstances. The power used in microwave sensors is much lower, usually in the milliwatt range.

1.6 ADVANTAGES AND DISADVANTAGES OF MICROWAVE SENSORS

The benefits and drawbacks of different sensors depend to a large degree on the specific application, but here we list some general remarks:

* Microwaves do not need mechanical contact with the object. Therefore, performing on-line measurements from a distance is usually possible, without interference to the industrial process.
* Microwaves penetrate all materials except for metals. The measured result therefore represents a volume of the material, not only the surface.
* Microwave sensors are insensitive to environmental conditions, such as water vapor and dust (contrary to infrared methods), and high temperatures (contrary to semiconductor sensors).
* At low frequencies (capacitive and resistive sensors), the dc conductivity often dominates the electrical properties of a material. The dc conductivity depends strongly on temperature and ion content. At microwave frequencies, the influence of the dc conductivity disappears.
* At the power levels used for measurements with microwave sensors, micro-

waves (nonionizing radiation) are safe (contrary to radioactive (ionizing) radiation).
* Microwave sensors are fast (contrary to radio active sensors).
* The microwaves do not affect the material under test in any way.

However, we note the following disadvantages:

* The higher is the frequency, the more expensive are the electronic components.
* Microwave meters must be calibrated separately for different materials.
* The sensors are often adapted to a specific application, resulting in low universal applicability.
* The sensors are sensitive to more than one variable. Other measurements are therefore in some cases necessary for compensation.
* Because of the relatively long wavelengths, the achievable spatial resolution is limited.

Clearly, the microwave sensors have many attractive features and will often provide the best option, but not always. The possibilities of the different sensor types are treated in more detail in Chapters 3 through 8.

REFERENCES

Collin, R.E., *Foundations for Microwave Engineering,* New York: McGraw-Hill, 1966, 589 p.

Collin, R.E., and F.J. Zucker, *Antenna Theory,* Parts I and II, New York: McGraw-Hill, 1969, 1249 p.

Edwards, T.C., *Foundations for Microstrip Circuit Design,* New York: John Wiley and Sons, 1981, 265 p.

Gardiol, F.E., *Introduction to Microwaves,* Norwood, MA: Artech House, 1984, 495 p.

Hammerstad, E.O., and F. Bekkadal, *Microstrip Handbook,* ELAB Report STF44 A74169, University of Trondheim, Norwegian Institute of Technology, 1975, 118 p.

Harrington, R.F., *Time-Harmonic Electromagnetic Fields,* New York: McGraw-Hill, 1961, 480 p.

IRPA (International Non-Ionizing Radiation Committee of the International Radiation Protection Association), "Guidelines on Limits of Exposure to Radiofrequency Electromagnetic Fields in the Frequency Range from 100 kHz to 300 GHz," *Health Physics,* Vol. 54, No. 1, January 1988, pp. 115–123.

Saad, T.S. (editor), *Microwave Engineer's Handbook,* Vols. I and II, Norwood, MA: Artech House, 1971, 401 p.

Stutzman, W.L., and G.A. Thiele, *Antenna Theory and Design,* New York: John Wiley and Sons, 1981, 598 p.

Tsang, L., J.A. Kong, and R.T. Shin, *Theory of Microwave Remote Sensing,* New York: John Wiley and Sons, 1985, 613 p.

Ulaby, F.T., R.K. Moore, and A.K. Fung, *Microwave Remote Sensing, Active and Passive, Vol. I, Microwave Remote Sensing Fundamentals and Radiometry,* Norwood, MA: Artech House, 1981, 456 p.

_____, *Microwave Remote Sensing, Active and Passive, Vol. II, Radar Remote Sensing and Surface Scattering and Emission Theory,* Norwood, MA: Artech House, 1982, pp. 457–1064.

_____, *Microwave Remote Sensing, Active and Passive, Vol. III, From Theory to Applications,* Norwood, MA: Artech House, 1986, pp. 1065–2162.

Chapter 2
Dielectric Properties of Materials

2.1 INTRODUCTION

Chapter 1 showed that, in the case of nonionizing radiation and nonmagnetic materials, the permittivity (ε_r) determines the interaction of electromagnetic waves with matter. Propagating microwaves do not "see" any matter, but only a space filled with ε_r. By studying the propagation of microwaves using some sort of sensing device, we are able to derive information in the variation of the permittivity with time and place in the medium. The permittivity, nonetheless, depends on the physical properties of the material, such as moisture, density, and temperature. If the relation between permittivity and, for example, density is known, we are able to measure density with a microwave sensor. Understanding permittivity is therefore of primary importance for using microwave sensors.

In Chapter 1, we introduced ε_r without an exact physical definition or description. In this chapter, we define the permittivity and study the physical mechanisms influencing it. The discussion is limited to that which is useful for understanding and predicting the behavior of permittivity as a function of the variables of interest. Special attention will be given to the measurement frequency, basic mechanisms of polarization, bound water, heterogeneous materials, and internal structure, as well as usage of both parts (real and imaginary) of the permittivity in the case of two simultaneously occurring variables. Water, air, and wood will be studied in a detailed manner. Water and air are ubiquitous and therefore important. Wood is an important material for the industry and a good example of a complicated biological material exhibiting many different polarization phenomena.

2.2 POLARIZATION PHENOMENA IN MATTER

2.2.1 Relative Permittivity ε_r and Boundary Conditions

Physical materials are normally composed of electrically charged particles arranged in such a way that any macroscopic region of the material is electrically neutral.

When an external electric field $\bar{E}_e$ is applied to the region, the field will act with a force on each individual electric charge, displacing it slightly from its previous equilibrium location. The positively and negatively charged particles will move in opposite directions, polarizing the whole region. The polarization partly compensates for the external field, making the resultant internal electric field strength $\bar{E}_i$ in the polarized region lower than the external field strength. The relative permittivity ε_r is a measure of the polarizing effect on the field strength; that is, how easily the medium is polarized.

To derive the proper relations, we may conveniently think of a plate capacitor (Figure 2.1). The plates of area A are charged with the charges $+Q$ and $-Q$. The charges create an electric flux density $\bar{D}$ between the plates. $\bar{D}$ is a vector directed from the positive charge toward the negative charge. Neglecting the curvature of the field lines near the margin, the magnitude of $\bar{D}$ is

$$D = \frac{Q}{A} \tag{2.1}$$

between the plates. For the sake of clarity, we can think of field lines originating at the positive charges and ending at the negative charges. Then, the magnitude of D is related to the local field line density and the direction is parallel to the field lines.

The electric flux density is related to the amount of charge, but the forces between charges, given by Coulomb's law, are proportional to the electric field strength $\bar{E}$. By definition, the relation between $\bar{E}$ and $\bar{D}$ in vacuum is

$$\bar{E} = \frac{1}{\varepsilon_0} \bar{D} \tag{2.2}$$

The constant ε_0 is the permittivity of vacuum. Its value follows from the choice of electrical units. The value in the International System of Units, *Système Internationale* (SI) was given in (1.7a): $\varepsilon_0 = 8.854 \times 10^{-12}$ F/m.

If a slab of dielectric material is inserted between the capacitor plates (Figure 2.2), the electric field will cause the charged particles in the slab to move slightly in the direction of the plates, the negative charges toward the positively charged plate, and *vice versa*. This displacement of charges creates an electric dipole moment per unit volume $\bar{P}$ in the slab. The electric dipole moment, or *polarization*, is a vector pointing in the opposite direction (in isotropic materials) than the electric flux density, and therefore partly counteracts it. This results in a lower mean electric field strength $\bar{E}_i$ inside the dielectric:

$$\bar{E}_i = \frac{1}{\varepsilon_0} (\bar{D} + \bar{P}) \rightarrow E_i = \frac{1}{\varepsilon_0} (D - P) \tag{2.3}$$

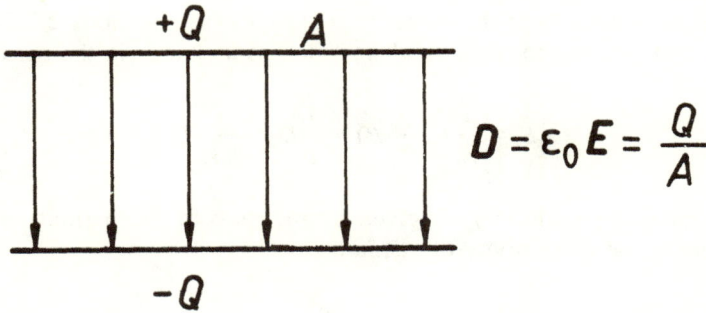

Figure 2.1 A plate capacitor of area A, charged with the charge Q.

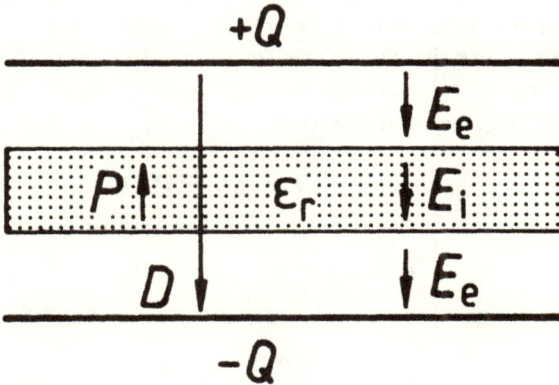

Figure 2.2 The plate capacitor of Figure 2.1, partly loaded with a dielectric slab of permittivity ε_r. The electric field strength is lower in the slab than in the air gap.

In the following we will use the scalar form of (2.3). If there are N charges q of each kind per volume, and the positive and negative charges are displaced by the distance δ from each other by the field, the magnitude of the polarization will be

$$P = Nq\delta \tag{2.4}$$

The distance δ is proportional to the local internal electric field E_l, which is not necessarily the same as E_i, but depends on the internal structure of the material. Therefore, the polarization is usually written

$$P = N\alpha E_l \tag{2.5}$$

where α is the *polarizability* of the atoms or molecules (see Sections 2.2.3–2.2.7). The effect of polarization is normally written as a change in the permittivity:

$$E_i = \frac{1}{\varepsilon_0}(D - P) = \frac{1}{\varepsilon}D = \frac{1}{\varepsilon_r \varepsilon_0}D \qquad (2.6)$$

where ε_r, the relative permittivity, is given by the ratio of the permittivity of the medium relative to the permittivity of vacuum:

$$\varepsilon_r = \frac{\varepsilon}{\varepsilon_0} \qquad (2.7)$$

Throughout the rest of this book, the relative permittivity ε_r will mainly be used. We will continue to use the subscript r for the relative value, but we will call it simply permittivity for the sake of brevity.

From (2.5) and (2.6), the permittivity is

$$\varepsilon_r \varepsilon_0 E_i = D = \varepsilon_0 E_i + P \qquad (2.8)$$

$$\varepsilon_r = 1 + \frac{1}{\varepsilon_0} \cdot \frac{P}{E_i} \qquad (2.9)$$

$$\varepsilon_r = 1 + \frac{N\alpha}{\varepsilon_0} \cdot \frac{E_l}{E_i} \qquad (2.10)$$

From (2.10) we clearly see that ε_r is really a measure of the polarizability of the medium and that the permittivity depends on the internal structure of the material.

Between the slab and plates, the electric field strength is still given by (2.2), resulting in the following ratio of the mean electric field strength in the slab E_i and the external electric field strength E_e:

$$E_i = \frac{E_e}{\varepsilon_r} \qquad (2.11)$$

This relation is always valid locally for the normal components (those perpendicular to the surface) of the electric field strengths at the surface of a dielectric material, regardless of how small the granule or curved the surface may be. If the surface is a boundary between two different dielectric media, the relation is transformed to

$$\frac{E_{n1}}{E_{n2}} = \frac{\varepsilon_{r2}}{\varepsilon_{r1}} \qquad (2.12)$$

Here, the subscript n stands for the normal component. Relation (2.12) is called the boundary condition for the normal component of the electric field strength at a dielectric boundary.

To derive the boundary condition for the tangential component of the electric field strength, we must turn the slab 90°, as shown in Figure 2.3. In this case, the voltage V between the plates measured inside the slab must be equal to that measured outside it due to the energy conservation principle and because $\nabla \times \bar{E} = 0$ for static fields (see (1.3)). Because the voltage is defined as the electrical potential energy difference per unit charge and the electric field strength is defined as the force per unit charge, the electric field strength can be thought of as voltage per distance. Therefore, the electric field strength E_i inside the dielectric is clearly equal to the field strength E_e outside the dielectric. This might seem contradictory to the previous case, as if there were no polarization at all. The situation is caused by a rearrangement of the charge in the plates, causing charge to accumulate at the location of the slab to such an extent that the higher electric flux density compensates the polarization. The result is that the boundary condition for the tangential component of the electric field strength is

$$E_{t1} = E_{t2} \tag{2.13}$$

So far, our analysis has used static fields. There is, however, no difference between the static case and the time-harmonic (sinusoidal time-dependent) one, as long as the physical dimensions of the region under study are small compared to the wavelength. The boundary conditions are always locally valid. The only difference when using the complex number notation for the time dependence (see Section 1.3.2) is that the permittivity will also be a complex quantity, written as

$$\varepsilon_r = \varepsilon_r' - j\varepsilon_r'' = \varepsilon_r'(1 - j\tan\delta) \tag{2.14}$$

where $j = \sqrt{-1}$ is the imaginary unit and $\tan\delta$ ($= \varepsilon_r''/\varepsilon_r'$) is the *loss tangent*. The imaginary part ε_r'' (or, more traditionally, the loss tangent) is a measure of how dissipative a medium is, and gives the rate of attenuation to a propagating wave (see Section 1.4). In a lossy medium, the electromagnetic energy is gradually turned into heat by the friction due to displacing the internal charges when the material is polarized in pace with the alternating electric field of the propagating microwave. The loss factor ε_r'' is always positive and usually much smaller than ε_r'. The minimum value is $\varepsilon_r'' = 0$ (lossless medium). The minus sign in (2.14) is a direct consequence of the choice of signs in (1.11) and the fact that physical media attenuate waves rather than amplify them.

The real part of the permittivity, ε_r', affects the electric field of a propagating wave, and so changes the ratio between the electric and magnetic field strengths,

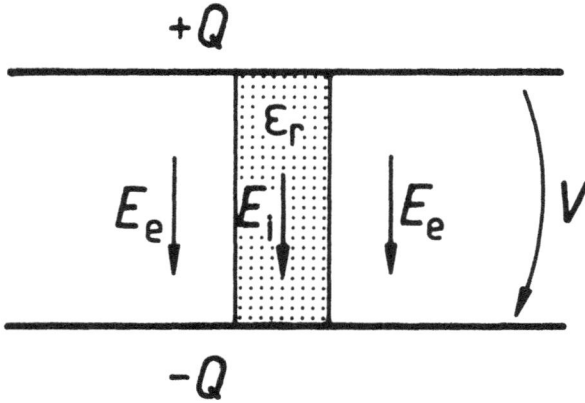

Figure 2.3 The plate capacitor of Figure 2.1, partly loaded with a vertical dielectric slab of permittivity ε_r. The electric field strength in the slab is equal to the field in the air.

said ratio being the wave impedance. At the same time, ε_r' decreases the speed of propagation. This is not evident from the discussion above, but can readily be seen from the wave equation (1.5) or the propagation factor (1.35). In a nonmagnetic, lossless medium, the speed of propagation is given by (1.41):

$$c = \frac{c_0}{\sqrt{\varepsilon_r'}} \tag{2.15}$$

where c_0 is the speed of light in vacuum, $c_0 = 2.998 \times 10^8$ m/s.

Because the product of frequency and wavelength for microwaves is equal to the velocity, low velocity means that the wavelength in the dielectric medium (λ_ε) is shorter than that in free space (λ_0):

$$\lambda_\varepsilon = \frac{\lambda_0}{\sqrt{\varepsilon_r'}} \tag{2.16}$$

The change in wavelength leads to refraction of wavefronts at the interface between two media with different ε_r'. In optics the bending of light rays is described by the refractive index (n), which in lossless media equals the square root of ε_r'.

For vacuum $\varepsilon_r' = 1$, and for air it is on the order of $\varepsilon_r' = 1.0006$. In all solids and liquids $\varepsilon_r' > 1$. Its value depends on many different factors, including moisture, density, composition, microstructure, and temperature, and is always frequency dependent. These factors will be examined in the sections to follow.

2.2.2 Polarization in Dense Materials

So far, only macroscopic effects have been studied with the exception of the polarizability α introduced in (2.5). We only mentioned that the locally acting electric field strength E_l might deviate from the mean electric field strength E_i. In fact, $E_l \approx E_i$ only for materials with low density, such as gases for which $E_l \approx E_i \approx E_e$. For dense materials, $E_i < E_l < E_e$.

To approximate E_l in dense materials, we may assume that the particle (e.g., atom) to be polarized is located in a spherical hole in the material. We can easily show [von Hippel, 1954, p. 20] that the field in such a hole (regardless of the size) in a dielectric is

$$E_l = E_i + \frac{P}{3\varepsilon_0} \tag{2.17}$$

Combining (2.8), (2.10), and (2.17) leads to

$$\frac{N\alpha}{3\varepsilon_0} = \frac{\varepsilon_r - 1}{\varepsilon_r + 2} \tag{2.18}$$

which is called the *Clausius-Mossotti equation*. In this form, it contains the number of particles per volume N, which is inconvenient. In many cases, a more practical approach is to calculate the polarizability per mole Π by substituting N for N_0, which is Avogadro's number or the number of particles per mole.

$$N_0 = \frac{NM}{\rho} = 6.023 \times 10^{23} \tag{2.19}$$

where M is the molecular weight and ρ is the density. This leads to

$$\Pi = \frac{N_0\alpha}{3\varepsilon_0} = \frac{\varepsilon_r - 1}{\varepsilon_r + 2} \cdot \frac{M}{\rho} \tag{2.20}$$

or

$$\varepsilon_r = 1 + \frac{3\Pi\rho}{M - \Pi\rho} \tag{2.21}$$

Equations (2.20) and (2.21) give the relation between the polarizability of a mole of particles and the permittivity of the solid or liquid material of known density. For many materials, we may measure α when the material is in the gas phase and then calculate ε_r for the corresponding solid or liquid. Some discrepancy between calculated and measured ε_r values may occur, depending on how good the spherical hole model is for that material.

The Clausius-Mossotti equation is perhaps most useful in predicting the dependence of ε_r on the density and *vice versa*. Equation (2.20) gives

$$\varepsilon_r = 1 + \frac{3\rho/\rho_0}{1 - \rho/\rho_0} \qquad (2.22a)$$

or

$$\rho = \rho_0 \frac{\varepsilon_r - 1}{\varepsilon_r + 2} \qquad (2.22b)$$

where $\rho_0 = M/\Pi$ can be treated as an empirical constant. Equation (2.22) is one of the simplest possible mixing formulas, or one for calculation the permittivity of a mixture of some material and air. In Section 2.4, this matter will be studied in more detail.

2.2.3 Typical Frequency Dependence of ε_r in Solids and Liquids

Dense materials contain several different kinds of charge carriers that can be displaced by an external electric field. The electrons are displaced with respect to the nucleus in the atoms, the atoms are moved in crystals or molecules, polar molecules are turned around to form a net dipole moment, and free ions travel through the material causing loss and sometimes polarization of conducting regions. All of the mechanisms mentioned above are frequency dependent.

Figure 2.4 qualitatively shows a possible behavior of ε_r' and ε_r'' as a function of frequency. In the low frequency range, ε_r'' is dominated by the influence of ion conductivity. It is caused by the free ions, which can only exist if some solvent is present. Usually, the solvent is water. If the material contained conducting regions that were not in contact with each other, the ion conductivity would have an influence on ε_r' as well, called the *Maxwell-Wagner effect*.

The variation of ε_r in the microwave band is caused by turning polar molecules. The kind of frequency dependence shown in Figure 2.4 is described by the *Debye relation*. A good example of a substance exhibiting strong *orientation polarization* is water. Both ion conductivity and the Debye relation are strongly temperature dependent.

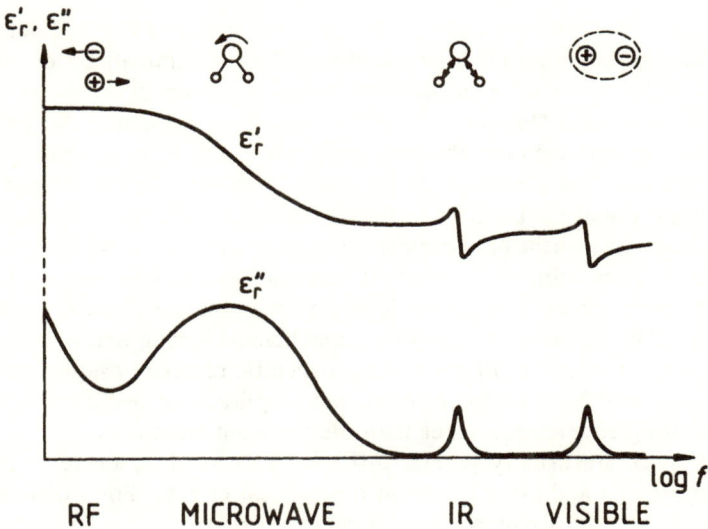

Figure 2.4 A qualitative representation of the real and imaginary part of ε_r as a function of frequency. The effect of ion, orientation, atomic and electronic polarization are shown.

The resonant phenomena occurring in the infrared region and above are caused by electronic and atomic polarization. The resonant frequencies are normally so high that they are of little practical importance in the microwave range. The polarization mechanisms are still important, however, because they produce a small, frequency and temperature independent, and almost lossless polarization for all frequencies below resonances. For many dry solids, these are, in fact, the dominant polarization mechanisms at microwave frequencies. Electronic polarization occurs in every medium containing chemical elements as ions, molecules, or crystals.

2.2.4 Electronic Polarization

In a neutral atom the positive charge is confined to the nucleus and an equal negative charge is carried by the electrons surrounding the nucleus. According to modern quantum physics, the electron population is best described by electron orbitals, a kind of probability density functions, which makes the exact calculation of the polarization complicated. One will, however, gain quite a good understanding of the phenomenon by using the classical model of electron particles orbiting the nucleus.

Taking the simplest possible atom, the hydrogen atom, as an example (Figure 2.5a), there is only one electron orbiting the nucleus. The distance between the nucleus and electron is constant so that the mean point of gravity for the negative charge coincides with that of the positive charge. An external electric field will act on the electron with a force trying to pull it from the atom. The Coulomb attraction between the nucleus and electron will act as a restoring force, resulting in only a small distortion of the electron orbit (Figure 2.5b). The distortion separates the points of gravity for the positive and negative charges, giving the atom a dipole moment. This form of polarization is called *electronic polarization*. For normal field strengths, the polarization is proportional to the electric field strength so that the polarizability α is constant. Because the inertia of the electron is very small, the above described situation holds in an alternating field, even at reasonably high frequencies. Only at frequencies in the optical band and above does the inertia of the electron need to be taken into account. The situation is then like a mechanical spring and mass system. Such a system is a mechanical oscillator having a specific resonant frequency. When the oscillator is activated by a periodic force, the amplitude of oscillations will be small for all activating frequencies other than the resonant frequency.

In practice, there are usually several different resonant frequencies related to different electron orbitals and other quantum mechanical effects. For s different oscillators, the permittivity of a material can be written as

$$\varepsilon_r = 1 + \sum_s \frac{N_s e^2 / \varepsilon_0 m_s}{\omega_s^2 - \omega^2 + j\omega^2 \alpha_s} \tag{2.23}$$

where N_s is the number of electrons per volume, with the resonant angular frequency ω_s, e is the charge of the electron, m_s is the mass of the electrons, ω is the measurement frequency ($\times 2\pi$), and α_s is a damping factor. The usefulness of (2.23) for quantitative calculations is limited, but it clearly shows the nature of the phenomenon. At frequencies far below the lowest resonant frequency (at microwaves), each oscillator type contributes with a constant value, resulting in a constant permittivity:

$$\varepsilon_r = 1 + \sum_s \frac{N_s e^2}{\varepsilon_0 m_s \omega_s^2} \tag{2.24}$$

Notice that ε_r in (2.24) is a real number, and it is independent of temperature and frequency. In reality, however, small and equally constant losses are normally associated with this type of polarization in the microwave range.

Far above each resonant frequency, the contribution from this type of oscillator vanishes. In Figure 2.6, the behavior of ε_r in the vicinity of one resonant frequency ω_n is depicted.

Electronic polarization occurs in all atoms in all materials and it is therefore important. In combination with the atomic polarization, which is of the same nature, the polarizations give most dry solids a permittivity on the order of $1 < \varepsilon_r' < 10$. When only these mechanisms are present, the material is almost lossless at microwave frequencies.

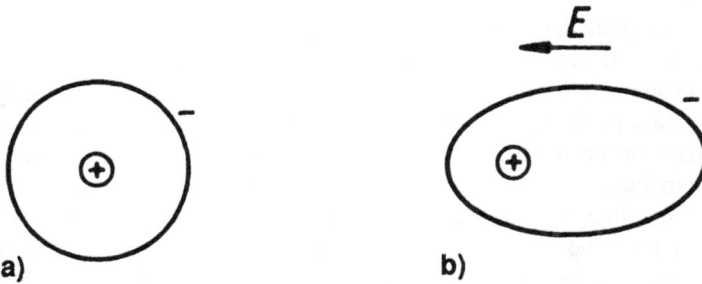

Figure 2.5 A simplified model of the distortion of the electron orbit of the hydrogen atom in the presence of an electric field.

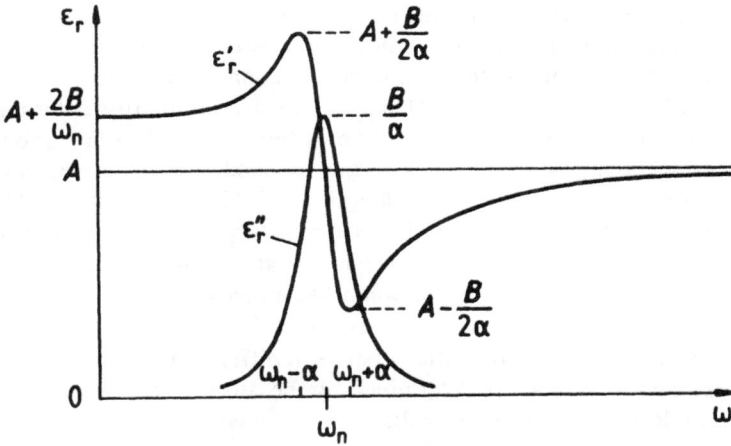

Figure 2.6 The behavior of the permittivity in the vicinity of the resonant frequency ω_n due to electronic polarization. A is the contribution of the higher resonances and $2B/\omega_n$ is the contribution at low frequencies of the resonance shown in the picture.

2.2.5 Atomic Polarization

Free atoms are electrically neutral, except for ions, which are atoms that have lost or captured one or several electrons and therefore carry a net positive or negative charge. The case of free ions in solids and liquids will be treated separately in Section 2.2.8. Because of the neutrality of free atoms, electronic polarization is the only type of polarization. When atoms combine to form molecules or crystals, the situation may change. If the molecule is composed of two identical atoms, it still is electrically neutral. This is also true for many other combinations of atoms, be they identical or different. Despite the neutrality, such molecules exhibit new electronic polarization resonances because the combination of atoms distorts and fuses the electron orbitals, and many electrons may be shared by several atoms.

In many cases, the distribution of electrons in the molecule is uneven. Chemists distinguish between several different types of chemical bonding, depending on how the outer electrons (valence electrons) are shared or rearranged, but that concept is outside the scope of this book. Of importance is the fact that, because of the rearrangement of electrons in the formation of molecules, many molecules have an imbalance in charge distribution, and therefore carry a permanent dipole moment. The case of such polar molecules will be studied in the next two sections. Other molecules have no permanent dipole moment because of symmetry. However, because some parts of the molecule are positively charged and some negatively, they will feel different attraction in an electric field. The molecule will therefore be bent and a dipole moment induced. This also applies to essentially ionized atoms in crystals. Such atomic polarization is closely related to electronic polarization, and is the source of similar resonances. Because of the much greater mass to be moved, the resonant frequencies of atomic (or vibration) polarization are lower and can be found in the infrared band. As mentioned in Section 2.2.4, the resonant frequencies of electronic polarization typically occur in the optical band, and they primarily determine the optical color of materials by selective absorption of light. Atomic polarization, conversely, produces specific absorption lines in the IR band. For example, the water molecule has such absorption lines at wavelengths of 1.43 μm (209 600 GHz) and 1.94 μm (154 500 GHz), which are used in IR moisture sensors based on reflectometry. Such resonances are quite sharp for gases, but are broadened in solids and liquids because of the close interaction between the molecules.

According to quantum physics, the absorption or radiation of energy at the resonances must be interpreted as transitions between different quantized energy levels. The exact resonant frequency is determined by the difference in energy between the levels. Free molecules in gases, in addition to the electronic and atomic (sometimes called *vibration*) polarization, exhibit another form of quantized energy, which produces resonances in the ε_r curve as far down as the microwave band. It is the rotation of the molecule around an axis. These rotational spectral lines, for example, to the radio astronomer, form the fingerprint of a specific molecule in interstellar space. Such rotational resonances can also be observed on earth as strong attenuation of microwaves in the atmosphere in certain frequency bands: 22 GHz (water vapor); $\approx$60 GHz (oxygen). The lines are much broader than those found in space because of the high temperature and pressure, causing collisions and fluctuations in the thermal motion of the molecules in the atmosphere. Under low pressure conditions, the rotational spectral lines are quite sharp and they have been used for gas analysis with microwave spectrometers and as frequency standards in atomic clocks. In solids and liquids, the rotation of molecules is severely limited and the rotational spectral lines disappear.

2.2.6 Orientation Polarization in Gases

In the previous section, we indicated that, due to the rearrangement of the valence electrons of the atoms in forming molecules, many molecules carry a permanent dipole moment. In an electric field, such a molecule experiences a torque that tries to align the dipole axis with the field (Figure 2.7). The thermal agitation, however, tries to restore disorder. The result is that the molecules statistically tend to point slightly more in the direction of the force produced by the field than in the opposite direction, resulting in orientation polarization. From the nature of the mechanism, clearly, the polarization will be saturated when all of the dipoles are totally aligned with the field lines. Because the polarizability is defined as the polarization per electric field strength, α must be nonlinear for the case of orientation polarization. However, for practical measurement or power applications of microwaves, the nonlinearity can be ignored as it usually requires millions of volts per centimeter to be significant [Hasted, 1973, p. 16].

Assuming that the density is so low that the dipolar interaction between molecules is small in comparison to the thermal energy (kT) and the permanent dipole moment is unaffected by the field, the total polarization can easily be calculated. Based on the energy of a dipole in an applied electric field and Boltzmann's statistics describing the orientation distribution, the following result has been derived [von Hippel, 1954, pp. 31–32].

If the permanent dipole moment of the molecule is μ, the electric field strength is E, the temperature is T, and k is Boltzmann's constant ($k = 1.3805 \times 10^{-23}$ J/K), the induced apparent mean dipole moment per molecule μ_d is

$$\mu_d = \mu\left(\coth x - \frac{1}{x}\right) = \mu L(x) \tag{2.25}$$

where $x = \mu E/kT$.

$L(x)$ is called the *Langevin function* and contains the information of the nonlinear feature of the polarizability. As mentioned above, the function can be linearized by using its first derivative at zero field strength:

$$\mu_d \approx \frac{\mu^2}{3kT} E \tag{2.26}$$

The same molecule also exhibits electronic (α_e) and atomic (α_a) polarization, resulting in a total permittivity of

$$\varepsilon_r = 1 + \frac{N}{\varepsilon_0}\left(\alpha_e + \alpha_a + \frac{\mu^2}{3kT}\right) \tag{2.27}$$

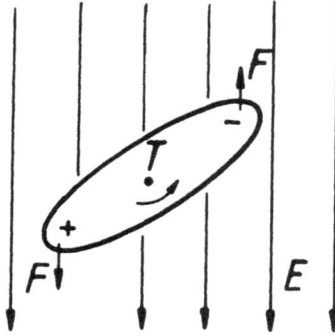

Figure 2.7 A molecular dipole experiences a torque T, caused by the force F, in an external electric field.

If the gas contains several different components, the corresponding terms can simply be added:

$$\varepsilon_r = 1 + \sum \frac{N_i}{\varepsilon_0} \left(\alpha_{ei} + \alpha_{ai} + \frac{\mu_i^2}{3kT} \right) \tag{2.28}$$

The fact that the orientation polarization is strongly temperature dependent, whereas α_e and α_a are not, offers a convenient way to measure the dipole moment of molecules by varying the temperature of the gas (provided that the gas contains only one kind of polar molecules):

$$|\mu| = \left(\frac{3\varepsilon_0 kT^2}{N} \cdot \left| \frac{d\varepsilon_r}{dT} \right| \right)^{1/2} \tag{2.29}$$

Equation (2.29) assumes that N is kept constant during the measurement of the derivative of the permittivity with respect to the temperature.

If the pressure is kept constant instead, the dipole moment will be given by

$$|\mu| = \left\{ \frac{3\varepsilon_0 kT^3}{N_0 T_0} \left[-\frac{d\varepsilon_r}{dT} - \frac{1}{T}(\varepsilon_r - 1) \right] \right\}^{1/2} \tag{2.30}$$

where N_0 is the value of N at a temperature T_0.

Each molecule has a finite mass, and therefore inertial forces will cause the orientation polarizability to fall off at high frequencies. For free molecules in a gas, this is of little importance in the microwave band. In solids and liquids, however, where internal friction may be significant, the dipoles turn more slowly than in a gas. This is the topic of the next section.

2.2.7 Orientation Polarization in Liquids and Solids, the Debye Relation

The molecules of materials in the condensed state are so close to their neighbors that freedom to move is severely limited by constant collisions, causing "friction." Orientation polarization of polar molecules still occurs, but is slowed down by the friction. To obtain a qualitative understanding of the polarization, let us think of an electric field being switched on in an infinitely short time. The electronic and atomic polarizations develop instantly, but the polar molecules turn slowly and exponentially approach the final state of polarization with a time constant τ, called the *relaxation time*. This kind of time function is called the *step response*. When the field is switched off, the sequence is reversed and disorder is restored with the same time constant. Deriving the permittivity as a function of frequency is possible by applying the Fourier transform to the step response. We will then get the Debye relation for polar substances:

$$\varepsilon_r = \varepsilon'_{r\infty} + \frac{\varepsilon'_{rs} - \varepsilon'_{r\infty}}{1 + j\omega\tau} \tag{2.31a}$$

or, as separated into real and imaginary parts:

$$\varepsilon'_r = \varepsilon'_{r\infty} + \frac{\varepsilon'_{rs} - \varepsilon'_{r\infty}}{1 + \omega^2\tau^2} \tag{2.31b}$$

$$\varepsilon''_r = \frac{(\varepsilon'_{rs} - \varepsilon'_{r\infty})\omega\tau}{1 + \omega^2\tau^2} \tag{2.31c}$$

In (2.31), $\varepsilon'_{r\infty}$ is the permittivity at "infinite" frequency, where orientation polarization has no time to develop. The static permittivity, denoted ε'_{rs}, corresponds to the low frequencies, where the orientation polarization has enough time to develop fully. The values of the constants in (2.31) are difficult to calculate theoretically, and for practical purposes are best regarded as empirical. The frequency dependence of the permittivity of a polar material, as given by the Debye relation, is shown in Figure 2.8. We can see that the real part ε'_r is constant at low and high frequencies, and the transition occurs slowly in the vicinity of the *relaxation frequency*, which is $f_{rel} = 1/(2\pi\tau)$. The imaginary part ε''_r is small for both low and high frequencies and large only in the transition region. The behavior of ε''_r can be understood in the following way.

In the low frequency range, where the polarization always develops fully, the loss per cycle is directly proportional to the frequency because the friction in a viscous fluid is directly proportional to the speed of movement. In the high frequency

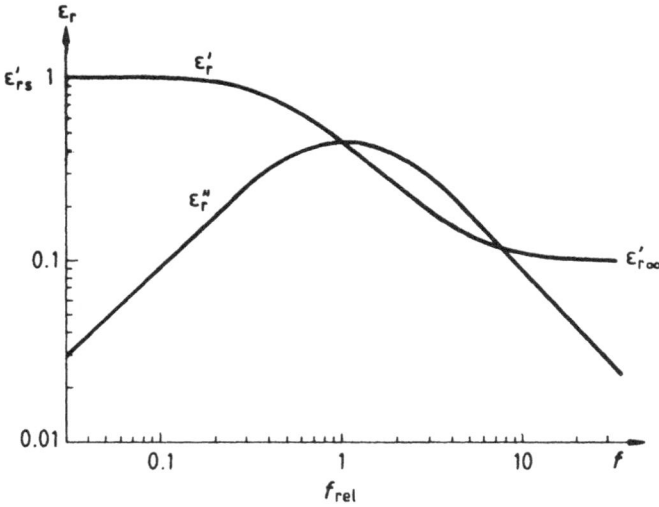

Figure 2.8 The frequency dependence of the complex permittivity according to the Debye relation (2.31) ($f_{rel} = \omega_{rel}/2\pi = 1/2\pi\tau$).

range, the loss disappears with frequency because the polarization does so, too. In fact, the two charges of the dipole can be regarded as separate ones acting like free conducting charges (see Section 2.2.8). In reality, only polar losses disappear and total losses settle in the millimeter-wave range on a low and constant level, normally associated with the atomic and electron polarization mechanisms. At the relaxation frequency, ε_r'' reaches a maximum value:

$$\varepsilon_{r \, max}'' = \frac{\varepsilon_{rs}' - \varepsilon_{r\infty}'}{2} \tag{2.32}$$

The Debye relation contains three different constants, ε_{rs}', $\varepsilon_{r\infty}'$, and τ. The permittivity at infinite frequency $\varepsilon_{r\infty}'$ is due to electronic and atomic polarization, and is therefore independent of the temperature. The other two constants are affected by the thermal agitation. The static permittivity ε_{rs}' decreases with rising temperature because of the increasing disorder. The relaxation time, τ, is inversely proportional to temperature, due to the fact that all movement is faster at higher temperatures.

The permittivity as a function of frequency is often presented as a two-dimensional diagram, a so-called *Cole-Cole diagram*. For the Debye relation, this leads to a semicircle as shown in Figure 2.9.

In many cases, the relaxation phenomenon cannot be described by using only a single relaxation frequency. This is the case, for example, when a material contains

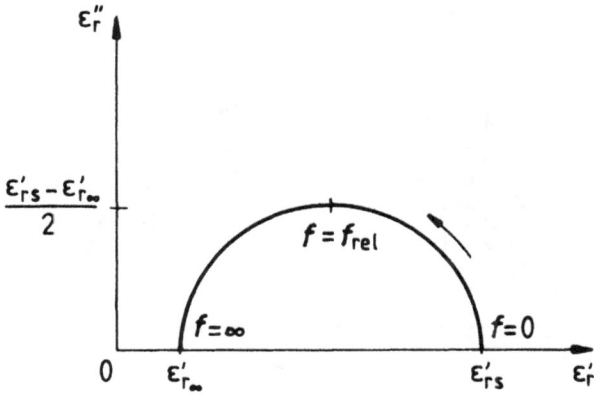

Figure 2.9 The Cole-Cole representation of the Debye relation (2.31).

bound water (water absorbed on the surface of fibers, crystals, molecules, *et cetera* through hydrogen bonds). A moist material contains water molecules bound with different strength and also free water, perhaps, depending on the moisture. Because the relaxation frequency depends on the mobility of the molecules, bound molecules will have a lower relaxation frequency than will free water molecules. Depending on the ability to bind water and on the moisture, the material will exhibit a distribution of relaxation frequencies. This situation can be approximated by introducing the empirical constant, α. Two different models are widely used. The *Cole-Cole equation:*

$$\varepsilon_r = \varepsilon'_{r\infty} + \frac{\varepsilon'_{rs} - \varepsilon'_{r\infty}}{1 + (j\omega\tau)^{1-\alpha}} \tag{2.33}$$

corresponds to a symmetrical distribution of relaxation times. In the Cole-Cole diagram, the curve remains a semicircle, but the center is below the ε'_r axis (Figure 2.10). The *Cole-Davidson equation:*

$$\varepsilon_r = \varepsilon'_{r\infty} + \frac{\varepsilon'_{rs} - \varepsilon'_{r\infty}}{(1 + j\omega\tau)^{\alpha}} \tag{2.34}$$

corresponds to an unsymmetrical distribution of relaxation times, which results in a skewed semicircle in the two-dimensional diagram (Figure 2.10). The subject of relaxation times and bound water are studied in more detail in Sections 2.3 and 2.6.

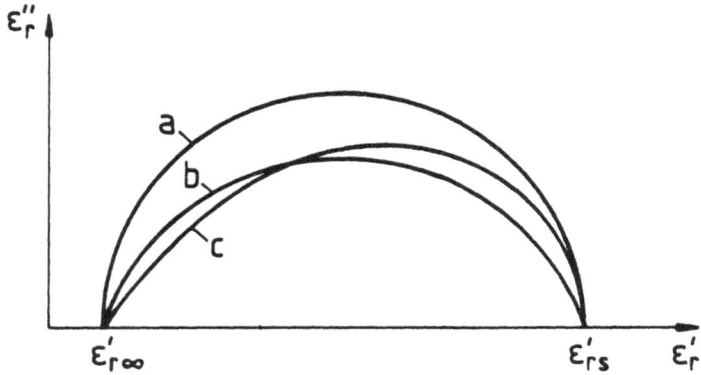

Figure 2.10 Cole-Cole diagrams for (a) the Debye relation (2.31); (b) the Cole-Cole equation (2.33) for $\alpha = 0.2$; (c) the Cole-Davidson equation (2.34) for $\alpha = 0.6$.

2.2.8 Ion Conductivity and the Maxwell-Wagner Effect

Ion conductivity may occur in both solids and liquids. If a solvent (e.g., water) is involved, the conductivity is due to electrolytic conduction, but ions, electrons, and lattice defects may also propagate in crystals (see, for example, [Hasted, 1973, Ch. 4]). Several mechanisms of conduction often are involved in the same material, resulting in a net conductivity, σ, but in moist materials the ion conductivity is by far the most important.

In a medium possessing a conductivity σ, an applied electric field $\bar{E}$ will induce a current density $\bar{J}$ given by

$$\bar{J} = \sigma\bar{E} \tag{2.35}$$

The effect of the current on a time-harmonic field can be incorporated in the complex permittivity, as we can see from Maxwell's equations (1.3). For time-harmonic fields, the time derivative in (1.3b) can be substituted through multiplication by $j\omega$, resulting in

$$\nabla \times \bar{H} = j\omega\varepsilon\bar{E} + \sigma\bar{E}$$

From the right-hand side, we get

$$j\omega\varepsilon_r\varepsilon_0\bar{E} + \sigma\bar{E} = j\omega\varepsilon_0\bar{E}\left(\varepsilon_r - j\frac{\sigma}{\omega\varepsilon_0}\right) \tag{2.36}$$

Because the term introduced by the conductivity is purely imaginary, it only contributes to the imaginary part of the permittivity. Ion conductivity therefore introduces losses into a material, but, to the first-order approximation, does not affect the wavelength of a propagating microwave. If the dc conductivity is known, we can write

$$\varepsilon_r = \varepsilon'_r - j\left(\varepsilon''_r + \frac{\sigma}{\omega\varepsilon_0}\right) \tag{2.37}$$

but ε''_r is most often defined to contain all of these losses:

$$\varepsilon_r = \varepsilon'_r - j\varepsilon''_r = \varepsilon'_r - j\left(\varepsilon''_{rd} + \frac{\sigma}{\omega\varepsilon_0}\right) \tag{2.38}$$

where ε''_{rd} represents the dielectric losses. Especially if ε_r is measured on a single frequency, the part of ε''_r due to conductivity is impossible to know. In earlier literature, common practice was to convert the whole measured loss into a conductivity term. It was justified when only low frequencies were used, where the loss by conductivity dominated, but otherwise this approach would yield a nonphysical interpretation.

The effect of the conductivity is inversely proportional to frequency because of the ω in the denominator in (2.37). Conductivity therefore greatly influences measurements made with sensors employing low frequencies, but not measurements made at GHz frequencies.

The frequency where the other losses begin to dominate conductive loss depends on many factors (ion concentration and mobility, temperature, moisture, and other losses). For example, for moist wood, the frequency is in the 100–400 MHz range and for sea water the frequency is about 10 GHz. In dry materials, it is much lower. Figure 2.11 shows ε''_r of margarine. The $1/f$ behavior caused by the salt is clearly visible, as is the Debye relaxation of the water. The effect of the fat is negligible.

The conductivity is usually fairly independent of frequency in the range where σ can be observed. Hence, its presence can be detected from the $1/f$ slope in ε''_r on low frequencies. Because σ depends on temperature and several other factors, the dc conductivity is difficult to exploit in measurement sensors.

If a heterogeneous material contains electrically conducting isolated regions embedded in nonconducting material, the material as a whole has zero dc conductivity. However, at high frequencies, an applied electric field generates currents in the conducting regions. The length of travel of an ion or other charge carrier during one cycle is generally short compared to the dimensions of the conducting region.

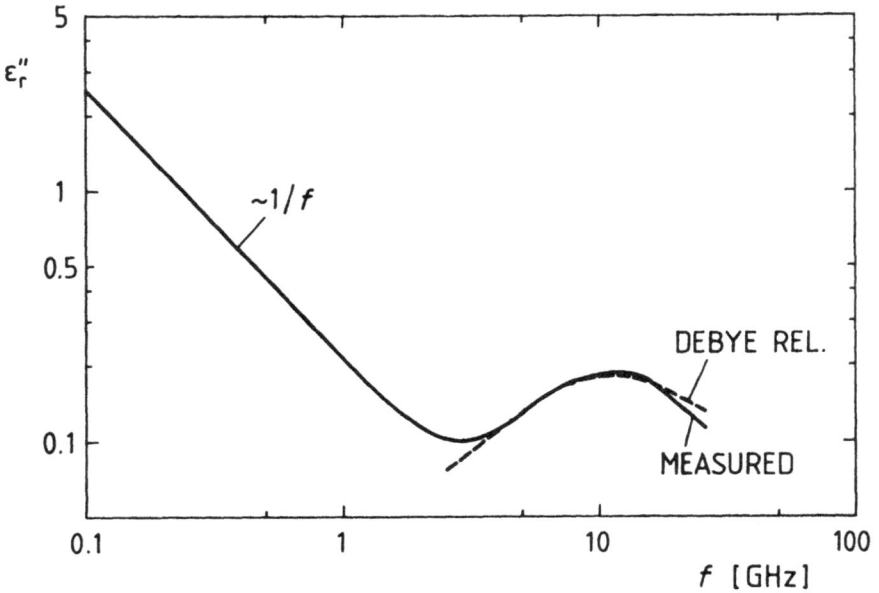

Figure 2.11 The loss factor of margarine. The effect of the ion conductivity, caused by the salt, dominates below 4 GHz and the Debye relaxation of the water dominates above that frequency. The moisture is unknown. (After [Meyer *et al.*, 1982].)

Therefore, the fact that the current only exists in the conducting regions has no macroscopic effect. Polarization does not occur and the losses exhibit the same $1/f$ behavior as with normal conductivity. At low frequencies, the charge carriers have time to accumulate at the borders of the conducting regions, thus causing polarization. Therefore, ε_r' increases at low frequencies due to the isolation of the conductive regions. This is the Maxwell-Wagner effect. It is encountered in many organic materials such as moist wood, where the cellulose crystals contain both conducting and nonconducting regions. Also, the cell structure in biological materials may, in some cases, be the origin of the variations in conductivity.

To derive the frequency dependence of ε_r for the Maxwell-Wagner effect, we can use the simple model of a plate capacitor filled with a two-layered dielectric (Figure 2.12), of which one layer is conducting and the other is not. For the capacitance, we have

$$C = \frac{\varepsilon_0 \varepsilon_{\text{eff}} A}{d}$$

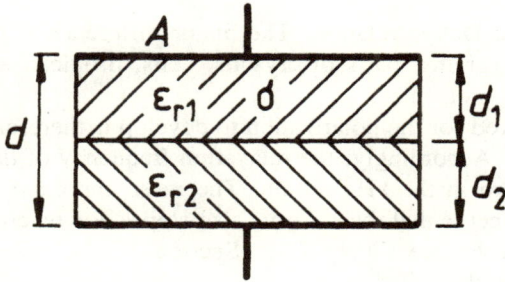

Figure 2.12 The Maxwell-Wagner effect can be demonstrated by using a plate capacitor filled with two dielectric slabs. Slab 1 has conductivity σ, slab 2 is nonconducting.

$$\varepsilon_{\text{eff}} = \frac{\varepsilon_{r2}\left(\varepsilon_{r1}' - j\,\dfrac{\sigma}{\omega\varepsilon_0}\right)}{\left(\varepsilon_{r1}' - j\,\dfrac{\sigma}{\omega\varepsilon_0}\right)\dfrac{d_2}{d} + \varepsilon_{r2}\dfrac{d_1}{d}} \tag{2.39}$$

At the low and high frequency limits, ε_{eff} approaches

$$\varepsilon_{rs}' = \lim_{\omega \to 0} \varepsilon_{\text{eff}} = \varepsilon_{r2}\left(\frac{d}{d_2}\right) \tag{2.40a}$$

$$\varepsilon_{r\infty}' = \lim_{\omega \to \infty} \varepsilon_{\text{eff}} = \frac{\varepsilon_{r2}\varepsilon_{r1}'d}{\varepsilon_{r1}'d_2 + \varepsilon_{r2}d_1} \tag{2.40b}$$

If we further note,

$$\tau = \frac{\varepsilon_0(\varepsilon_{r1}'d_2 + \varepsilon_{r2}d_1)}{\sigma d_2} \tag{2.41}$$

and substitute (2.40) and (2.41) into (2.39), the effective permittivity transforms to

$$\varepsilon_{\text{eff}} = \varepsilon_{r\infty}' + \frac{\varepsilon_{rs}' - \varepsilon_{r\infty}'}{1 + j\omega\tau} \tag{2.42}$$

which is equivalent to the Debye relation (2.31) for polar substances. Hence, we can draw the conclusion that if a material contains electrically conducting regions separated by nonconducting material, the frequency behavior of ε_r will be similar to that described by the Debye relation. The major difference is that in this case we have whole regions (grains, parts of crystals, cells, droplets, *et cetera*) becoming polarized, instead of molecules.

The time required for the polarization to develop is therefore much longer than for polar substances. Accordingly, the relaxation frequency of the Maxwell-Wagner effect is usually down in the kHz region. The most important consequence of the Maxwell-Wagner effect is that the ε_r' curve also slopes downward in the MHz region as well as the ε_r'' curve (see Figure 2.15, Section 2.3.4). For normal conduction, only the ε_r'' curve has the $1/f$ slope.

From (2.41), we can see that the relaxation time is inversely proportional to the conductivity. A better conductivity gives a shorter relaxation time. It is also dependent on the size (d_1) of the conducting regions. The smaller are the regions, the shorter is the relaxation time. This is utilized in the well known method of producing materials with a high ε_r' by mixing metal filings with some conventional dielectric. Such mixtures have a very high and constant ε_r', even at millimeter-wave frequencies.

In heterogeneous materials, there will normally be a distribution of sizes of the conducting regions. This leads to a distribution of relaxation times, and ε_r can be described by using (2.33) or (2.34), instead of the Debye relation (2.39).

We have used the model of conducting regions embedded in nonconducting material. We would, of course, have obtained similar results with a mixture of two components with finite but different conductivities. When there are local deviations in the conductivity, the induced currents will always cause a build-up of charge at the interfaces leading to polarization of the medium. Only in the extreme case of paths with constant conductivity stretching along the field lines of the applied electric field will there be no polarization.

2.3 WATER

2.3.1 The Water Molecule

Water is present everywhere, in the atmosphere, in and on the ground, absorbed into substances, adsorbed onto most surfaces; water is used in most industrial processes, and it even makes up the major part of our brain. Water is present in almost every measurement situation and one of the most important applications of microwave sensors is measurement of moisture. We will therefore give a short review of the electrical properties of water and behavior of the water molecule. In Sections 2.4, 2.5, and 2.6, we will deal with mixtures of water and other substances. For those readers who want more details on the physics of water, there is a good book, *Aqueous Dielectrics*, written by J.B. Hasted [Hasted, 1973].

The water molecule consists of one oxygen (O) and two hydrogen (H) atoms. Water molecules are nonlinear so that the mean HOH angle is 104.5° (Figure 2.13),

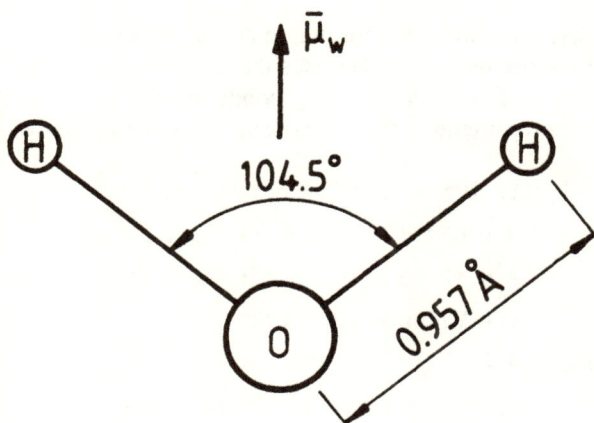

Figure 2.13 The water molecule (1 Å = 10^{-10} m).

and the mean OH distance is 0.95718×10^{-10} m. The chemical bonds are such that the electrons are closer to the oxygen atom than the hydrogen atoms, leaving the oxygen atom more negatively charged than the hydrogen atoms. Because of the bent structure of the molecule, this produces a permanent dipole moment $\mu_w = 6.14 \times 10^{-30}$ Asm. It also results in small quadrupole and octopole moments, but these are of no interest here. The total polarizability (atomic + electronic) of the molecule is $\alpha_w = \alpha_a + \alpha_e = 1.89 \times 10^{-34}$ Asm^2/V. These are values for the free molecule, and are not exactly maintained for bound molecules (liquid water, ice, or bound water).

2.3.2 Liquid Water

Because the water molecule possesses a permanent dipole moment, liquid water is a polar substance. As shown in Section 2.2.7, the permittivity of a polar substance obeys the Debye relation (2.31). If the liquid also contains ions in electrolytic solution, we must add the conductivity term from (2.38):

$$\varepsilon_r = \varepsilon'_{r\infty} + \frac{\varepsilon'_{rs} - \varepsilon'_{r\infty}}{1 + j\omega\tau} - j\frac{\sigma}{\omega\varepsilon_0} \qquad (2.43)$$

The constants in (2.43) could, in principle, be calculated approximately from the properties of the water molecule. The difficulty of estimating the inner field sensed by a molecule in the liquid, the restricted mobility of the molecules, and the inner viscocity would reduce the accuracy. There is, however, a substantial amount of measurement data available. Thrane has made a survey of both the theoretical calculations and measurement data. He uses the composition of sea water for the evaluation of the effect of free ions causing conductivity. He gives the following best-fit polynomial approximations for the constants usable up to 40° C [Thrane, 1976]:

$$\varepsilon'_{rs} = 88.195 - 4.3917 \times S + 0.16738 \times S^2 - 0.40349 \times t$$
$$+ 0.65924 \times 10^{-3} \times t^2 + 0.43269 \times 10^{-1} \times St$$
$$- 0.42856 \times 10^{-2} \times S^2 t + 0.4441 \times 10^{-5} \times S^2 t^2$$
$$- 0.92286 \times 10^{-4} \times St^2 \tag{2.44a}$$

$$\varepsilon'_{r\infty} = 4.9 \tag{2.44b}$$

$$\tau \times 10^{12} = 19.39 - 1.137 \times S + 0.11417 \times S^2 - 0.6802 \times t$$
$$+ 0.95865 \times 10^{-2} \times t^2 + 0.58629 \times 10^{-1} \times St$$
$$- 0.54577 \times 10^{-2} \times S^2 t + 0.82521 \times 10^{-4} \times S^2 t^2$$
$$- 0.87596 \times 10^{-3} \times St^2 - 0.65303 \times 10^{-17} \times \exp(t) \tag{2.44c}$$

$$\sigma = 0.87483 \times S + 0.25662 \times 10^{-2} \times S^2$$
$$+ 0.45802 \times 10^{-1} \times St - 0.37158 \times 10^{-2} \times S^2 t$$
$$+ 0.39288 10^{-4} \times S^2 t^2 - 0.16914 \times 10^{-3} \times St^2 \tag{2.44d}$$

where S is the salinity in percent by weight and t is the temperature in degrees centigrade (°C). Figure 2.14 shows some curves calculated from (2.44). We can see that ε'_r reaches values around 80 at frequencies below 5 GHz, whereas the permittivity of most dry solids, originating primarily from atomic and electronic polarization, is usually <10. This great difference suggests that the moisture in a material will have a substantial effect on the permittivity. This fact is usually the reason that has made sensing of moisture one of the most important applications of microwave sensors.

In the submillimeter range, there is some experimental evidence [Hasted, 1973, pp. 50–61] for a second relaxation process, the origin of which is not clear. The best-fit relaxation frequency is about 3.8 THz. The accompanying broad absorption band stretches down to about 300 GHz, where the principal relaxation phenomenon dominates.

At infrared wavelengths, several atomic polarization resonances occur (see Figure 2.6). They correspond to intramolecular vibration in the OH bonds (both liquid and vapor) and to intermolecular vibration in the O...H hydrogen bonds between molecules (only liquid). The only effect of these vibrations at microwave frequencies is the high frequency permittivity $\varepsilon'_{r\infty}$, which is independent of both frequency and temperature.

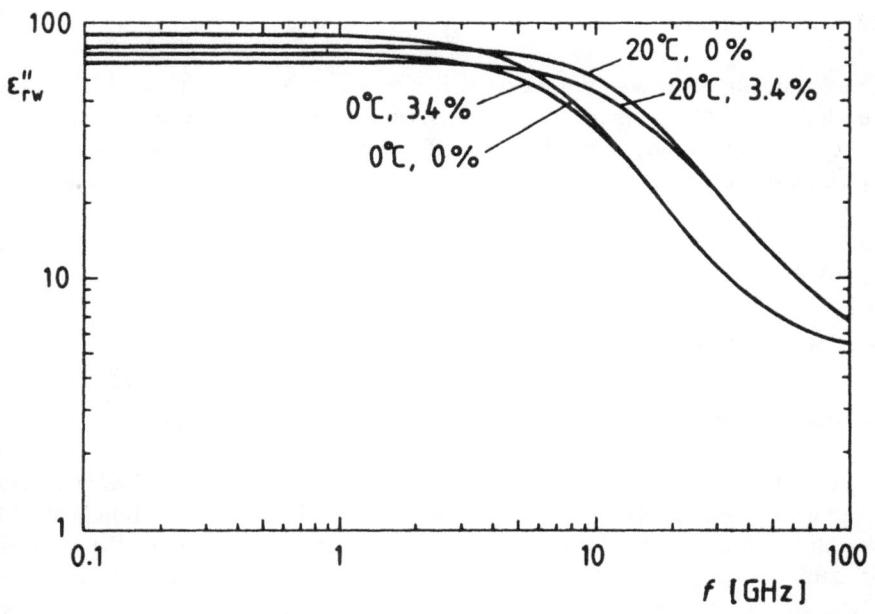

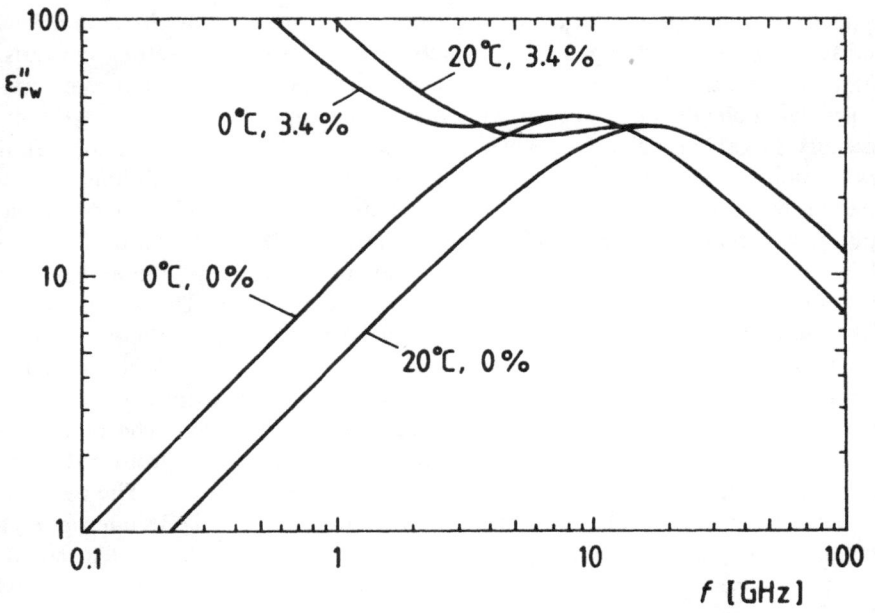

Figure 2.14 The permittivity of water as calculated from (2.44).

2.3.3 Water Vapor

The molecules in water vapor are essentially free and the interaction with other molecules has little effect on the permittivity of the gas. At microwave frequencies, (2.27) is therefore valid, using the values for α_w and μ_w given in Section 2.3.1. For practical purposes, combining (2.27) and (2.19) is convenient, resulting in

$$\varepsilon_r = 1 + \frac{\rho N_0}{M\varepsilon_0}\left(\alpha_w + \frac{\mu_w^2}{3kT}\right) \tag{2.45}$$

In Section 2.6, this is applied to the case of humidity in air.

2.3.4 Bound Water

Because of the hydrogen atoms, the water molecules tend to form *hydrogen bonds* between each other and on the surface of other molecules. Hydrogen bonds are rather weak compared to chemical bonds, but restrict the free movement of the molecules. Hydrogen bonds therefore affect the permittivity of the water. Water molecules usually tend to stick to all surfaces of solids, both external and internal ones in porous materials. This is called *adsorption*. Because water is always available in the atmosphere, a balance normally exists between the moisture in an object and relative humidity in the surrounding air. At which level the moisture settles depends, in addition to the humidity, on the total internal area per volume of the material to which water molecules can adsorb. Wood is an example of a material with an exceptionally large internal area. Water is adsorbed on the surface of the cellulose fibers to such an extent that it causes expansion. Under normal conditions, the moisture settles to between 5% and 10% of the dry mass of the wood, depending on the humidity. The permittivity of wood is treated in more detail in Section 2.6.

When bound to another molecule, a water molecule is no longer free to turn around and to participate in the polarization of the medium. There are, however, different degrees of binding. A molecule may be tied to its neighbors with up to three bonds. Both hydrogen atoms may bind to any other atom, and the oxygen atom may bind to a hydrogen atom of a neighboring molecule. The mean number of bonds active per molecule is, of course, a matter of quantum statistics. The molecules in liquid water are statistically tied with one bond and able to turn around. In ice, all three bonds are active and the adsorbed water is between these two. The mean number of active bonds depends on both the temperature and available internal area of the material compared to the amount of water molecules (i.e., the structure and the moisture). A measure of the strength of binding is the energy released per mole when the bonds are formed, the activation energy Q. For ice it is 54.4 kJ/mol, and for liquid water 18.8 kJ/mol. When the molecules are bound, they move slower and the polarization involves breaking of bonds. The relaxation frequency is therefore lower than for free water. The dependence of the relaxation frequency on the activation energy and temperature has been shown to be

$$f_{\text{rel}} = \frac{\exp(-Q/RT)}{2\pi\tau_0} \tag{2.46}$$

where R is the gas constant ($R = 8.3143 \text{ J/mol} \cdot \text{K}$) and τ_0 is a proportionality factor. Figure 2.15 shows the permittivity of tobacco. The Debye relaxation is clearly distinguished at 390 MHz. For free water at the same temperature, the relaxation frequency would be 17 GHz.

In a moist material, some of the water molecules will have different activation energies from the others. Those molecules added first are tightly bound, and according to (2.46) have a low relaxation frequency. The more water that is added, the higher its relaxation frequency will be, until no more water can be adsorbed. If more water still is added, it will remain as liquid water, so-called *capillary water*, in the cavitites of the material. If bound water is present in a material, there will be a distribution of relaxation frequencies. The mean relaxation frequency consequently depends on the moisture. The drier is the material, the lower is the mean relaxation frequency of the water. From (2.46), we can also see that the effect of the temperature is such that an increase in temperature causes an increase in the mean relaxation frequency. Kent and Meyer have studied microcrystalline cellulose in the moisture range 1.6% to 13.75% (dry basis) [Kent *et al.*, 1983]. They found a distribution of relaxation times dependent on temperature and moisture. The permittivity could best be described by the Cole-Davidson equation (2.34), and the mean activation energy was estimated between 31 kJ/mol (13.75%) and 42 kJ/mol (1.6%).

The real part of the permittivity, ε'_r, of course, is also affected by the binding of the water molecules. The static permittivity, ε'_{rs}, for the most tightly bound molecules obtains a value of the same order as that for ice (3.15). This means that measurement of moisture with microwaves is more difficult if the moisture is low than if the material contains capillary water. The moisture range with reduced sensitivity due to bound water depends on the inner area (i.e., capability to bind water) of the material to be measured. Usually, the range stretches from 0% to 2–10% (30% for wood and related biological substances).

The adsorption of water as a function of temperature and moisture is an essentially continuous process, and therefore no discontinuity in the ε_r for bound water is observed at the freezing point (0° C). This offers a means for roughly measuring the ratio of capillary water to bound water in a material sample.

Water molecules also bind to other molecules or ions added to liquid water. The permittivity of aqueous solutions or mixtures is therefore decreased by two mechanisms: the displacement of water by a substance with a lower permittivity and the binding of water molecules.

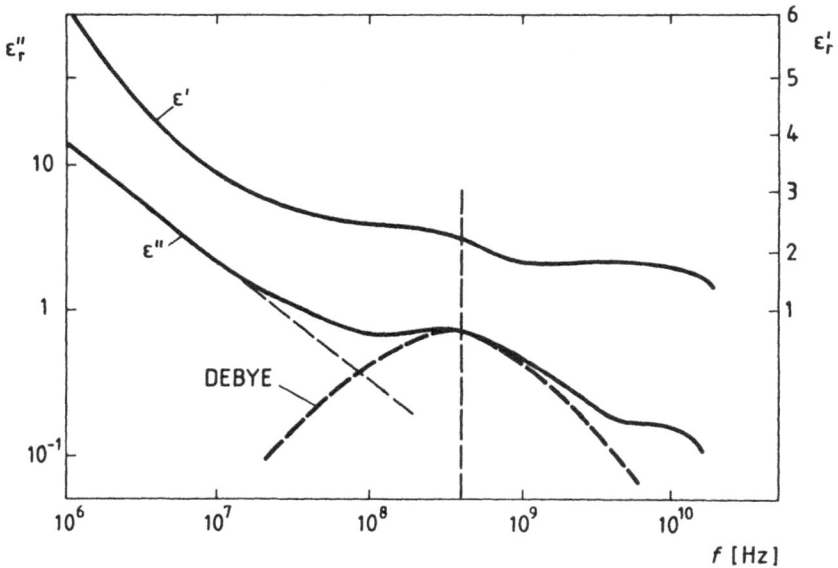

Figure 2.15 The permittivity of tobacco. The Debye relaxation is clearly discernible but because of the bound water the relaxation frequency is lower and the peak is broader than for free water (compare Figure 2.11). The Maxwell-Wagner effect has raised the ε_r'-curve at the lower frequencies. Moisture = 17%, density = 260 kg/m^3. (After [Meyer *et al.*, 1979].)

2.3.5 Ice

The water molecules in ice are strongly bound with three hydrogen bonds and arranged in the form of a crystal lattice. The activation energy is so high (54.4 kJ/mol) that the relaxation frequency is far down in the kHz region at all temperatures. At microwave frequencies, $\varepsilon_{r,\text{ice}}' = 3.15$. The losses are low because of the negligible polarization movement of the molecules in ice, as shown in Figure 2.16. At frequencies below 1 GHz, the Debye relaxation losses dominate, giving rise to a $1/f$ behavior, and above 10 GHz other loss mechanisms occur. Between 1 GHz and 10 GHz, $\varepsilon_{r,\text{ice}}''$ is on the order of 10^{-3}, depending on the temperature. The losses are so low, in fact, that measuring the ground profile of Greenland through the 3 km thick ice cover has been possible using radar echo sounding [Gudmandsen *et al.*, 1976]. A comprehensive review of the dielectric properties of ice appears in [Mätzler *et al.*, 1987].

Snow is a mixture of ice and air as well as water if the snow is wet. The dielectric properties of snow can be calculated from the properties of the constituents by using mixing formulas [Sihvola *et al.*, 1985], this being the topic of the next section.

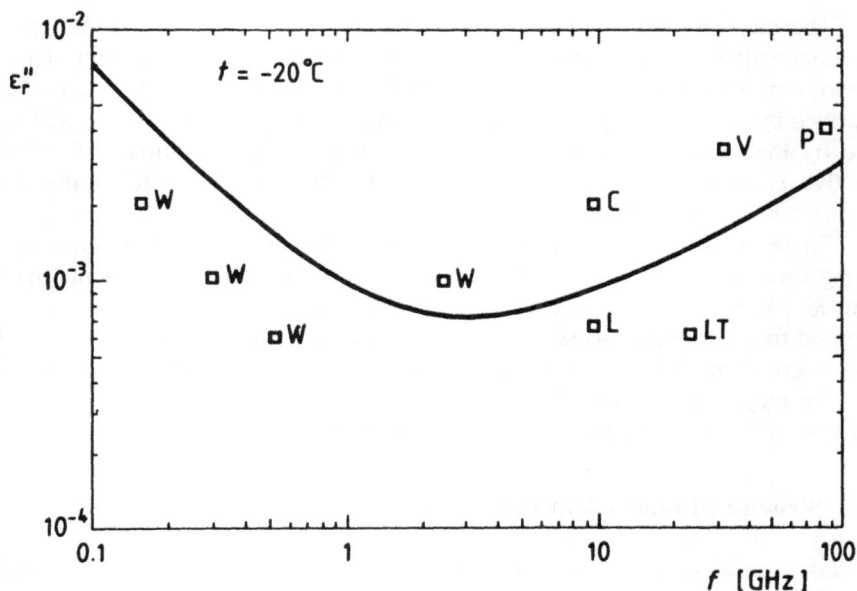

Figure 2.16 The loss factor of ice at $-20°$ C. The line is calculated on the basis of measurements of snow [Nyfors, 1982]. The dots are measurement points reported by Westphal (W), Cumming (C), Lamb (L), Lamb and Turney (LT), Perry and Straiton (P), and Vant *et al.* (V), as reported in [Stiles *et al.*, 1981].

2.4 MIXING FORMULAS AND STRUCTURE DEPENDENCE OF PERMITTIVITY

2.4.1 General

So far, we have studied the polarization phenomena and permittivity of homogeneous materials, but in reality the majority of the materials we want to measure are mixtures of several materials. The effective permittivity of such heterogeneous mixtures can be approximated from the permittivity of constituents by using *mixing formulas*. Such formulas help us to predict and understand the dependence of the effective permittivity on mixing ratio, moisture, density, and structure. They also provide good choices for a mathematical model when we wish to fit our measurement data.

Two types of mixing formulas have been derived, those which depend on the structure of the material and those which do not. The latter are merely suitable mathematical functions, whereas the former are based on theoretical calculations of the mean polarization in a material with a certain microstructure.

The main condition for the mixing formulas to be valid is that the size of the inhomogeneities (grains, particles) is much smaller than the wavelength. This is not a serious limitation if the condition is not fulfilled because the medium cannot be described by an effective permittivity only, as a propagating microwave will be scattered by the inhomogeneities. However, as long as the condition is fulfilled, the effective permittivity of the mixture will not depend on the size of the inhomogeneities, but only on their shape.

There are two main difficulties that may arise in some cases when applying mixing formulas. One is the fact that the permittivity of the constituents may have changed when they were mixed together. This is especially true for water if it is adsorbed to form bound water. The other difficulty is that the structure is not necessarily constant. When, for example, the moisture or the density is changed, the structure may also change. However, despite the difficulties, the formulas usually give reasonably good approximations for the permittivity of mixtures.

2.4.2 Structure-Independent Formulas

The most useful structure-independent mixing formula is the *exponential model:*

$$(\varepsilon_{rm})^a = \sum_i f_i (\varepsilon_{ri})^a \qquad (2.47)$$

where ε_{rm} is the effective permittivity of the mixture; ε_{ri} is the permittivity and f_i the *volume fraction* of the ith constituent, $\Sigma_i f_i = 1$. The exponent a is an empirical constant called the *degree of the model*, and $0 < a \leq 1$. For $a = 1$, the permittivity is linearly dependent on the volume fractions. Such a formula has been presented, for example, by Brown [Wang *et al.*, 1980]. Birchak and others [Birchak *et al.*, 1974] have used $a = 1/2$ to describe soil-water mixtures and $a = 0.4$ seems to be the optimum for wet snow [Sihvola *et al.*, 1985]. For $a = 1/3$, it reduces to the mixing formula derived by Looyenga [Looyenga, 1965].

In Figure 2.17 we show some curves for a two-component mixture as calculated from (2.47) for different values of a. The constituents are air, ice, and some measurement points of dry snow [Nyfors, 1982] are included for comparison.

2.4.3 Structure-Dependent Formulas

General

The structure-dependent formulas are derived from the mean polarization in a medium. One constituent is regarded as the *host material* in which the other constituents

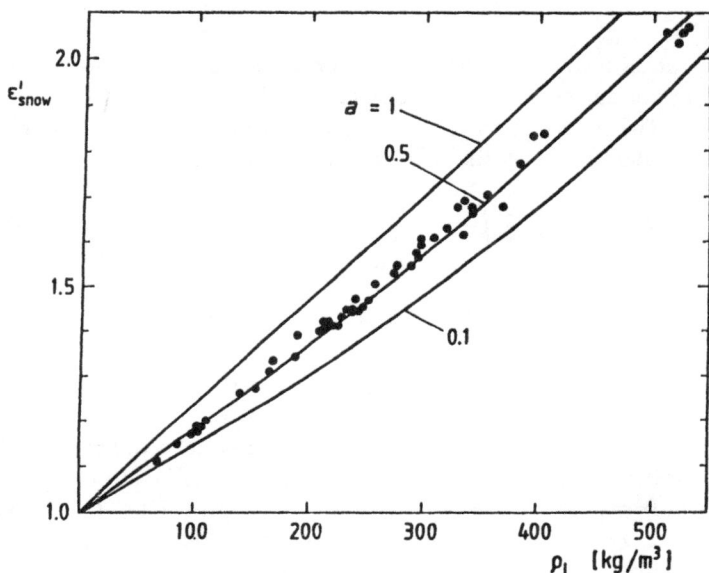

Figure 2.17 The effective permittivity of dry snow as calculated from the structure-independent exponential mixing formula (2.47) for different values of a. Measurement points are shown for comparison.

are expected to form *inclusions*. The only shape of the inclusion for which the internal field can be exactly calculated from the applied field is the ellipsoid. All formulas therefore treat the inclusions as ellipsoids of different eccentricity. Inclusions of any other shape can be described by a sum of ellipsoids with different eccentricities if necessary. The improvement in accuracy thus gained, however, is small compared to the total inaccuracy. In most cases, to use only one shape of inclusions for each constituent is therefore practical.

The internal field of an ellipsoid in an applied field is uniform and exactly known, provided that the wavelength of the applied field is much larger than the dimensions of the ellipsoid. This is the so-called *quasistatical approximation*. The difficulty of deriving mixing formulas is how to account for the disturbances in the applied field caused by adjacent inclusions. This is usually done by ascribing the environment of an ellipsoid an *apparent permittivity* ε_{ra}. The correct value of ε_{ra} is somewhere between the permittivity of the host material (ε_{rh}) and the resulting effective permittivity of the mixture. We will derive some simple formulas for mixtures with small volume fractions of inclusions in air ($\varepsilon_{ra} = \varepsilon_{rh}$) and list some of the reported formulas for arbitrary mixture ratios.

Ellipsoids

The shape of an ellipsoid is described by the axes a_1, a_2, and a_3. Only the ratios $a_1 : a_2 : a_3$ are important because the absolute size does not affect the permittivity as long as $a_i \ll \lambda$. The polarization properties of the ellipsoid in a field applied parallel to an axis a_i is determined by the *depolarization factor* N_i:

$$N_i = \int_0^\infty \frac{a_i a_j a_k \, ds}{2(s + a_i^2)^{3/2}[(s + a_j^2)(s + a_k^2)]^{1/2}} \tag{2.48}$$

where

$$N_1 + N_2 + N_3 = 1$$

Important special cases are those with two or three equal axial ratios. Three equal ratios give a sphere (depolarization factors $1/3$, $1/3$, $1/3$). Two equal ratios (rotational ellipsoid) result in an oblate (pill-shaped, Figure 2.18) or a prolate (cigar-shaped) spheroid. For closed-form expressions for the depolarization factors of the rotational ellipsoids, see, for example, [Sihvola *et al.*, 1985]. The extreme cases of these are the *disk* (depolarization factors 1, 0, 0) and the *needle* (depolarization factors, 0, $1/2$, $1/2$).

Formulas for Small Volume Fractions of Inclusions

From (2.9) we have the relation between the field E_i and the polarizations P_i inside an inclusion:

$$P_i = (\varepsilon_{ri} - 1)\varepsilon_0 E_i \tag{2.49}$$

Because the field (and polarization) inside an ellipsoid is uniform, the mean polarization in the mixture is

$$P_m = P_i f_i \tag{2.50}$$

where f_i is the volume fraction of the inclusions. The effective permittivity of the mixture is now given by (2.8):

$$\varepsilon_{rm} = 1 + \frac{P_m}{\varepsilon_0 E_a} = 1 + (\varepsilon_{ri} - 1)f_i \frac{E_i}{E_a} \tag{2.51}$$

where E_a is the field applied to an individual inclusion. If the volume fraction of the inclusions is small and the host material is air, the field applied to an inclusion is

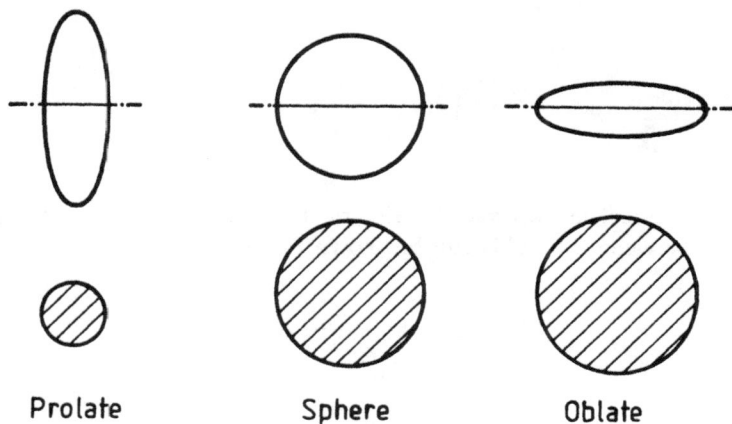

Figure 2.18 Types of ellipsoids with two equal axes.

equal to the external field applied to the mixture. The ratio E_i/E_a is well known for all kinds of ellipsoids. For spheres, we have

$$E_i = \frac{3}{\varepsilon_{ri} + 2} E_a \qquad (2.52)$$

resulting in the mixing formula:

$$\varepsilon_{rm} = 1 + 3f_i \frac{\varepsilon_{ri} - 1}{\varepsilon_{ri} + 2} \qquad (2.53)$$

Equation (2.53) is valid for example in the case of water drops in the atmosphere. If the inclusions are disks or needles, all equally aligned in such a way that the applied electric field is parallel to the surface of the inclusions, according to (2.13), we have

$$E_i = E_a$$

resulting in

$$\varepsilon_{rm} = 1 + f_i(\varepsilon_{ri} - 1) \qquad (2.54)$$

If the field is perpendicular to the surface of disks, we have, from (2.11):

$$E_i = \frac{E_a}{\varepsilon_{ri}}$$

resulting in

$$\varepsilon_{rm} = 1 + f_i \frac{\varepsilon_{ri} - 1}{\varepsilon_{ri}} \tag{2.55}$$

In the case of aligned needles and the electric field perpendicular to the principal axes, the ratio between the fields can be shown to be

$$E_i = \frac{2}{\varepsilon_{ri} + 1} E_a$$

giving

$$\varepsilon_{rm} = 1 + 2f_i \frac{\varepsilon_{ri} - 1}{\varepsilon_{ri} + 1} \tag{2.56}$$

If the host material is not air, ε_{rm} and ε_{ri} in the formulas above should be regarded as normalized with respect to ε_{rh}.

The reader may note that the effective permittivity of a mixture of aligned discs or needles depends on the direction of the applied field with respect to the inclusions. This is also true for denser materials, as we will see from the formulas below. Such media with aligned inclusions are said to be *anisotropic*. In these cases, a single permittivity is not enough to describe the medium, but the *permittivity is a tensor* $\|\varepsilon_{rm}\|$. If the coordinate axes are chosen so as to coincide with the three axes of the ellipsoids, only the elements on the diagonal of the tensor differ from zero. The elements can be calculated from the formulas given here. Practical examples of anisotropic media are wood and other fibrous materials (needle-like inclusions) and laminated materials (disk-like inclusions).

Formulas for All Kinds of Mixtures

The formulas derived above for small volume fractions fail at high volume fractions because the apparent permittivity in the vicinity of an inclusion is affected by the other inclusions. A good mixing formula should also give $\varepsilon_{rm} = \varepsilon_{ri}$ for $f_i = 1$. Several authors have reported different mixing formulas that account for the environment and fulfill the requirement for $f_i = 1$. Sihvola and Kong have made a comparative review of them [Sihvola *et al.*, 1988a]. In the same publication, they derived a general mixing formula, from which most other formulas can be derived as special cases. For the case of randomly oriented ellipsoids in a mixture of n constituents, the formula is

$$\varepsilon_{rm} = \varepsilon_{rh} + \cfrac{\dfrac{1}{3} \sum_{j=1}^{n} f_j(\varepsilon_{rj} - \varepsilon_{rh}) \sum_{i=1}^{3} \dfrac{\varepsilon_{ra}}{\varepsilon_{ra} + N_{ji}(\varepsilon_{rj} - \varepsilon_{rh})}}{1 - \dfrac{1}{3} \sum_{j=1}^{n} f_j(\varepsilon_{rj} - \varepsilon_{rh}) \sum_{i=1}^{3} \dfrac{N_{ji}}{\varepsilon_{ra} + N_{ji}(\varepsilon_{rj} - \varepsilon_{rh})}} \qquad (2.57)$$

The shape of the inclusions may be different for the different constituents. In the case of aligned ellipsoids, the divisions by 3 and summations over i should be dropped. In that case, the different N_{ji} terms give the different elements on the diagonal in the $\|\varepsilon_{rm}\|$ tensor, and the other terms are zero if the coordinate axes are chosen to coincide with the axes of the ellipsoids. The apparent permittivity ε_{ra} may obtain any value between ε_{rh} and ε_{rm}. We can write it as a function of a parameter a:

$$\varepsilon_{ra} = \varepsilon_{rh} + a(\varepsilon_{rm} - \varepsilon_{rh}) \qquad (2.58)$$

For small volume fractions of inclusions, (2.57) is independent of a. With the choice $a = 0$, leads for spheres (2.57) to the *Maxwell-Garnett* (MG) formula. In a form containing the polarizability α the same formula is called the *Lorentz-Lorentz* or the Clausius-Mossotti formula, and in the symmetrical form it is known as the *Rayleigh formula:*

$$\frac{\varepsilon_{rm} - \varepsilon_{rh}}{\varepsilon_{rm} + 2\varepsilon_{rh}} = \sum_{i=1}^{n} f_i \frac{\varepsilon_{ri} - \varepsilon_{rh}}{\varepsilon_{ri} + 2\varepsilon_{rh}} \qquad (2.59)$$

For $a = 1$, (2.57) reduces to the so-called *quasicrystalline approximation with coherent potential* (QCA-CP) [Kohler et al., 1981], also known as GKM after Gyorffy, Korringa, and Mills. The QCA-CP formula always predicts higher values for the permittivity than does the MG formula. Values between those two are obtained with the choice $a = 1 - N_i$, in which case for two-component mixtures (2.57) leads to the well known *Polder–van Santen* formula [Polder et al., 1946]. It has also been derived in other ways by different authors (see [de Loor, 1956], [Taylor, 1965], [Tsang et al., 1985, pp. 378–379]). The formula has been used with success in explaining and predicting the dielectric behavior of many natural media, such as snow and soil. For n types of inclusions, the formula is

$$\varepsilon_{rm} = \varepsilon_{rh} + \cfrac{\dfrac{1}{3} \sum_{j=1}^{n} f_j(\varepsilon_{rj} - \varepsilon_{rh}) \sum_{i=1}^{3} \dfrac{\varepsilon_{rm} + N_{ji}(\varepsilon_{rh} - \varepsilon_{rm})}{\varepsilon_{rm} + N_{ji}(\varepsilon_{rj} - \varepsilon_{rm})}}{1 - \dfrac{1}{3} \sum_{j=1}^{n} f_j(\varepsilon_{rj} - \varepsilon_{rh}) \sum_{i=1}^{3} \dfrac{N_{ji}}{\varepsilon_{rm} + N_{ji}(\varepsilon_{rj} - \varepsilon_{rm})}} \qquad (2.60)$$

Equation (2.60) is implicit and therefore best solved by computer iteration. For special cases of two-component mixtures, the formula simplifies considerably. For spheres, we have

$$\varepsilon_{rm} = -\frac{b}{4} + \left[\left(\frac{b}{4}\right)^2 + \frac{\varepsilon_{ri}}{2} \right]^{1/2}$$

$$b = \varepsilon_{ri} - 2 + 3f_i(1 - \varepsilon_{ri})$$

(2.61)

for randomly oriented needles,

$$\varepsilon_{rm} = -\frac{b}{2} + \left[\left(\frac{b}{2}\right)^2 + \varepsilon_{ri} + \frac{f_i}{3}\varepsilon_{ri}(\varepsilon_{ri} - 1) \right]^{1/2}$$

$$b = (\varepsilon_{ri} - 1)\left(1 - \frac{5}{3}f_i\right)$$

(2.62)

and for randomly oriented disks,

$$\varepsilon_{rm} = \frac{1 - \dfrac{2}{3}f_i(1 - \varepsilon_{ri})}{1 + \dfrac{f_i}{3\varepsilon_{ri}}(1 - \varepsilon_{ri})}$$

(2.63)

Here, ε_{ri} and ε_{rm} are normalized with respect to ε_{rh}. The formulas presented above are for mixtures in which the different constituents form separate inclusions. Tinga, Voss, and Blossey have derived a formula for mixtures where the inclusions are made of confocal layers of the constituents [Tinga et al., 1973b]. For two-component mixtures, the formula reduces to the MG formula ($a = 0$) and therefore gives lower values for ε_{rm} than do the Polder–van Santen formulas, and at least in the case of snow the values are too low. Tinga has, however, applied the three-component formula with success to mixtures of cellulose and water.

Sihvola and Lindell have derived a general formula for n-component mixtures with layered spherical inclusions [Sihvola et al., 1988b], also corresponding to $a = 0$. For three- or two-component mixtures with spherical inclusions, theirs reduces to the Tinga formula. They have also derived it in a form allowing the permittivity to change continuously with the distance from the center of the inclusions [Sihvola et al., 1989].

In Figure 2.19, the Polder–van Santen formulas and the Tinga formula for spheres give lower values than those for disks and needles. This is a general result. We can also see that the difference between the curves is so small that to use any other models than those for needles, disks, or spheres will usually be unnecessary.

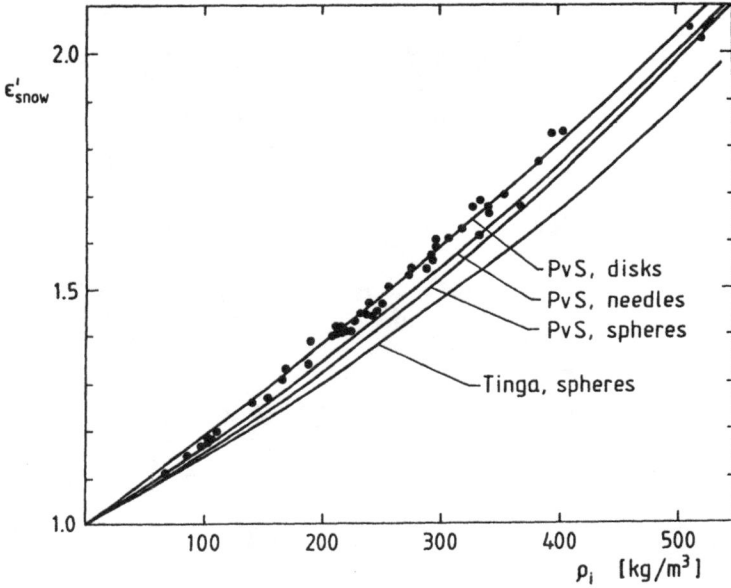

Figure 2.19 The permittivity of dry snow as calculated with the Polder–van Santen (PvS) and Tinga mixing formulas. The points were measured [Nyfors, 1982], using fine-grained snow, aged snow, and coarse old snow compressed to different densities.

However, the greater is the difference between ε_{ri} and ε_{rh}, the greater is the effect of the shape of the inclusions.

In both Figures 2.17 and 2.19, we can see that the permittivity increases slower than linearly as a function of the volume fraction of inclusions. This is also generally true when the host material has a lower permittivity than the inclusions. If the permittivities are reversed, the behavior of the curve is also reversed in the sense that the permittivity decreases faster than linearly with increasing volume fraction of inclusions. As a result, the permittivity of a mixture is always lower than the linear combination of the permittivities of the constituents, regardless of which of them is the host material. The calculated result, however, is not always the same for different choices of host, only the trend. Figure 2.20 shows the permittivity of a mixture of two components as calculated with the Maxwell-Garnett formula (2.59) for the two alternative cases. The Polder–van Santen formula for spheres (2.61), however, predicts equal permittivity for the two cases, whereas those for disks or needles do not.

We have not made any assumptions regarding losses, which means that if the constituents are lossy, the formulas can be treated as complex equations.

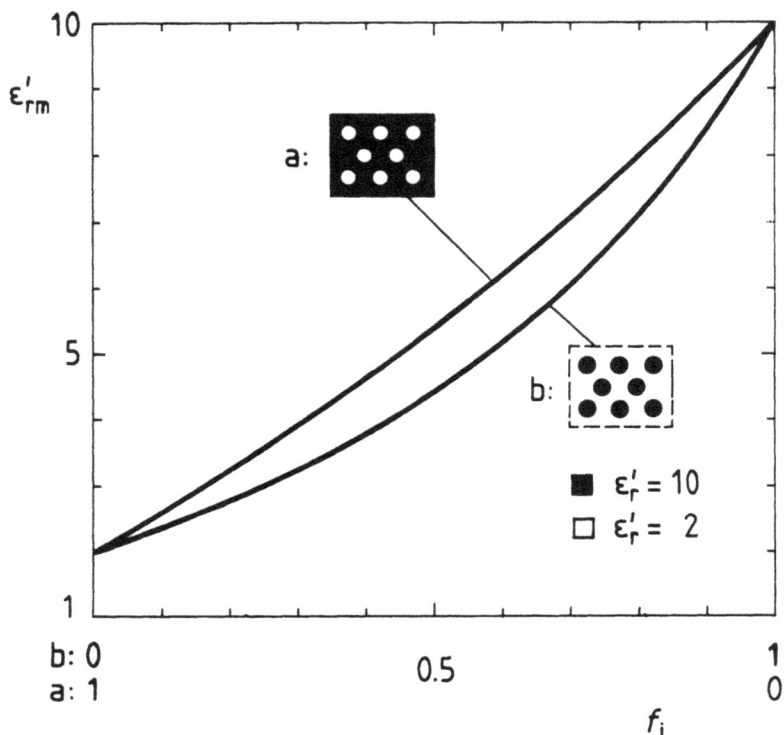

Figure 2.20 The permittivity of a mixture with two constituents ($\varepsilon'_{r1} = 2$, $\varepsilon'_{r2} = 10$) as calculated by the Maxwell-Garnett formula with different choices of host.

Some Consequences of the Mixing Formulas

The most important consequence of the mixing formulas is the dependence of ε_r on the structure of a material. The higher is the permittivity of the inclusions, the stronger is the dependence. Therefore, the way in which the water is absorbed (apart from the binding effect) by a material has an especially great effect on the permittivity. A change in surface tension, for example, as a result of adding a small amount of detergent, may dramatically increase the permittivity. The structure may also change as a result of a change in the moisture or density, but normally the moisture is predictably related to the permittivity, which is the foundation for microwave moisture sensors. Even in some cases when the structure does change, we can make structure-independent measurements by using the multiparameter technique (see Section 2.5).

Another consequence of the mixing formulas is a shift in the apparent relaxation frequency. If we substitute ε_{ri} for (2.31) in one of the mixing formulas, the resultant frequency dependence is altered by some mixing formulas. In most practical cases, this effect is not significant, but for spherical inclusions the effect is considerable. If we, for example, combine (2.53) and (2.31), we get

$$\varepsilon_{rm} = \varepsilon_{rm\infty} + \frac{\varepsilon_{rms} - \varepsilon_{rm\infty}}{1 + j\omega\tau_m}$$

$$\varepsilon_{rms} = 1 + 3f\frac{\varepsilon_{rs} - 1}{\varepsilon_{rs} + 2}$$

$$\varepsilon_{rm\infty} = 1 + 3f\frac{\varepsilon_{r\infty} - 1}{\varepsilon_{r\infty} + 2}$$

$$\tau_m = \tau\frac{\varepsilon_\infty + 2}{\varepsilon_s + 2}$$

This result is applicable for the case of rain in the atmosphere, where we have (0° C):

$$f_{rel} \approx 8.2 \text{ GHz} \frac{88.2 + 2}{4.9 + 2} = 107 \text{ GHz}$$

For example, (2.54), for disks or needles parallel to the field, does not lead to any such effects, and in wet snow only a negligible shift has been observed in relaxation frequency, as compared to that of water.

2.5 MULTIPARAMETER MEASUREMENT

2.5.1 The Measurement Situation

The permittivity of the materials to be measured depends on many factors, as we have seen. The most important factors are the molecular physics of the constituents, composition (density, moisture, mixing ratio), binding of molecules, microstructure, temperature, and measurement frequency. Depending on the type of sensor, additional factors influencing the measurement may be the sample shape (thickness, roughness), homogeneity, background or nearby objects, and location of the sample. The large number of variables makes possible the use of microwave sensors for a great variety of applications, but, if several of the factors appear simultaneously as variables, the measurement result will depend on all of them. If this is not taken into account, the measurement accuracy of the desired variable will be degraded by the unwanted variables. Therefore, knowing the dependence among the measurement quantities and all the occurring variables is important. Only then is development of a measurement system possible with the necessary stabilizing mechanisms and auxiliary measurements.

The development of a specialized sensor application should normally start with a theoretical and practical investigation of the permittivity of the material to be measured. On the basis of this information combined with knowledge of the measurement situation, we can choose the measurement frequency and type of sensor. Here, a remark on caution is necessary. We must be careful not to overlook any of the questions regarding the environment, unknowns, and measurement accuracy. The communication is often difficult between those who need the meter and those who design it. The real purpose of the meter should be clear, as well as which measurement accuracy is desirable and sufficient for the intended purpose. The final price of a meter largely depends on the demanded accuracy.

When we know the material, the measurement situation, and the designed sensor, we can make rough calculations of the influence of the different variables. The calculations can be based on laboratory tests and mathematics presented in the chapters dealing with the sensors. The questions related to the environment and outer appearance of the material in relation to the sensor can usually be handled separately. If necessary, a material stream-forming or sample preparation unit will to a large degree eliminate the disturbing variations in the sample thickness, location, density, and sometimes even temperature. When all that is possible has been done to stabilize the measurement situation, we are usually left with more than one unknown. The unknowns are usually related to the permittivity of the material to be measured, but sometimes also, for example, related to sample thickness or location. In the following sections, we will study some of the techniques that can be used to collect additional information and to process the measurement data in a convenient way.

2.5.2 The Number of Measurements

The mathematical truth associated with the multivariable case is that *if there are n unknowns affecting our measurements, we need at least n independent measurements to be able to solve even one of them.* This can be characterized by a set of equations. If there are three variables x, y, and z (for example, moisture, density, and temperature), we need three independent measurements A, B, and C (for example, resonant frequency, quality factor, and temperaure):

$$A = f_1(x,y,z)$$
$$B = f_2(x,y,z) \qquad (2.64)$$
$$C = f_3(x,y,z)$$

where the functions f_1, f_2, and f_3 can be either derived from the characteristics of the sensor and the permittivity data of the material or determined by regression analysis. In many cases, to choose the former approach and treating the functions related to the permittivity of the material separately from those related to the sensor is wise because they will stay unchanged, even when the sensor is changed. Calibration is then used to improve the final accuracy and to eliminate variations between individual sensors.

The latter way of handling the multiparameter situation is the *multivariable regression technique*. For calibration of the technique, a set of measurements is performed. A polynomial model for the wanted quantity is then fitted to the data set by using the multivariable regression technique. Hence, the desired quantity can directly be calculated from new measurements by using the model.

This technique has been used, for example, by Albert Klein for moisture measurements of coal with transmission sensors [Klein, 1984]. The difficulty lies in the choice of the regression model. It must not be too complicated (depending on the capacity of the computer program performing the fitting), but it must contain all of the necessary powers of each variable and the mixed terms of several variables to varying powers.

2.5.3 Auxiliary Measurements

If the microwave measurements alone are not enough, we must add to the measurement system other meters to measure quantities that depend on the same variables as do the microwave measurements. It is important that they do not introduce new variables to the system. The most frequently used auxiliary measurements are of temperature or of density or mass per area.

The measurement of temperature can be performed either by contacting sensors, such as thermocouples, resistive sensors, semiconductor devices, or by noncontacting IR thermometers. Contacting sensors are usually considerably cheaper.

The mass-related quantities can be measured with either some kind of belt weigher or with γ-ray attenuation. The achievable accuracy with belt weighers is often relatively poor because the influence of the variations in belt tautness is difficult to avoid, but scales for weighing individual samples are usually very accurate. Scales are conventional technology and no safety risks are involved. Certain kinds of γ-rays, for example, the rays emitted by the cesium 137 isotope, are attenuated at a rate proportional to the density. The total attenuation is therefore proportional to the mean density (or the total mass per area) between the radioactive source and the detector. If the geometry of the sample or material stream is constant, the reading gives both the density and total mass per volume or area. If the meter is used on a conveyor, the reading gives only the mass per area. The advantages of the radioactive meters are their ability to measure through the conveyor belt or walls of a container and the simple construction without moving parts. Disadvantages are the safety risks (shielding and safety zones are necessary) and, in some cases, the slowness.

2.5.4 Microwave-Multiparameter Measurement

The permittivity is a complex quantity; that is, it has a real part (ε_r') and an imaginary part (ε_r''). Theoretically, they are connected through the *Kramers-Kronig relations* (see, for example, [Jackson, 1975, pp. 310–312]), which means that if the other part is completely known from zero to infinite frequency, the other is also known. In practice, when measured on a single or a few frequencies, however, they can be regarded as independent. This makes a two-parameter measurement possible by using microwaves on a single frequency only. In terms of interaction between microwaves and medium, we thus measure both the rate of attenuation and speed of propagation.

Based on laboratory measurements of permittivity as a function of the variables, a mathematical model, such as (2.64) with $A = \varepsilon_r'$ and $B = \varepsilon_r''$, can be constructed from which the unknowns can be solved as functions of ε_r' and ε_r''. A graphical representation of the model easily reveals possible ambiguities and values for which the measurement accuracy will be poor.

Figure 2.21 shows such a diagram for wet snow at 1 GHz, which is based on the empirical snow model [Sihvola *et al.*, 1986]:

$$\varepsilon_{rs}' = 1 + 1.7\rho_d + 0.7\rho_d^2 + 8.7f_w + 70f_w^2$$

$$\varepsilon_{rs}'' = 0.9f_w + 7.5f_w^2$$

$$\varepsilon_{rs}''(1 \text{ GHz}) = \frac{\varepsilon_{rs}''(f_r)}{f_r/10^9}$$

$$\rho_s = \rho_d + f_w$$

The unknowns are wetness f_w (volume fraction of water) and dry density ρ_d (liquid water excluded). Curves are shown for constant values of ε_r' and ε_r''. Any pair of ε_r', ε_r'' values unambiguously corresponds to only one pair of f_w, ρ_d values. If the net of curves had been folded, each measurement resulted in two possible solutions. The crossing angle between the curves is always relatively large in Figure 2.21, which gives good measurement accuracy. For small crossing angles, a small measurement error in the other quantity moves the obtained crossing point far from the correct point. In this case, however, the accuracy is somewhat degraded by the fact that the permittivity depends on the structure of the snow. Nonetheless, microwaves provide one of the most accurate, and certainly the fastest, techniques available for measuring the wetness and density of snow. For this purpose, a portable meter employing a

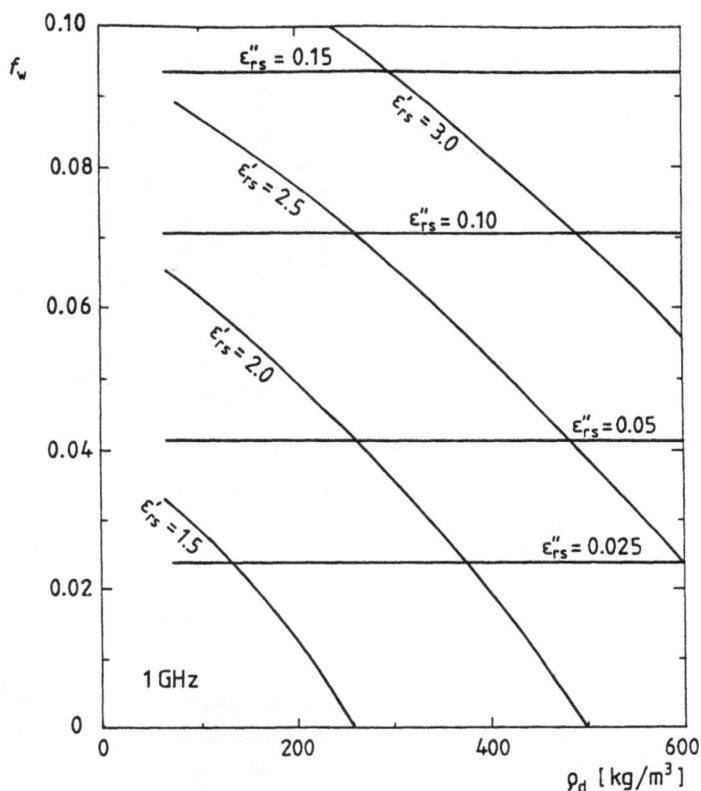

Figure 2.21 A mathematical model for the complex permittivity of snow as a function of the dry density (ρ_d) and volume fraction of water (f_w). The diagram is for $f_r = 1$ GHz. (After [Sihvola *et al.*, 1986].)

resonator sensor has been developed at the Radio Laboratory of the Helsinki University of Technology [Sihvola *et al.*, 1986]. The sensor is a fork resonator (see Section 3.3.5, Figure 3.22), which can be characterized with another diagram shown in Figure 2.22. The diagram is based on calibration with materials of known permittivity because the open construction causes radiation losses, which are difficult to calculate exactly. If we, for example, were to measure a resonant frequency f_r of 595 MHz and a peak width B_{hp} of 12.9 MHz, ($Q_l = 595/12.9 = 46$) we could see from Figure 2.22 that the permittivity of the snow would be $\varepsilon_r = 2.0 - j0.0149$ (at 595 MHz). From the snow model, we can calculate the permittivity at 1 GHz: $\varepsilon_r = 2.0 - j0.025$, and from the diagram we then obtain $\rho_d = 376$ kg/m^3, $f_w = 0.0237$, leading to the total density $\rho_s = 400$ kg/m^3, moisture $m_s = 6\%$ (by weight, wet basis). The same sensor could, of course, be used to measure materials other than

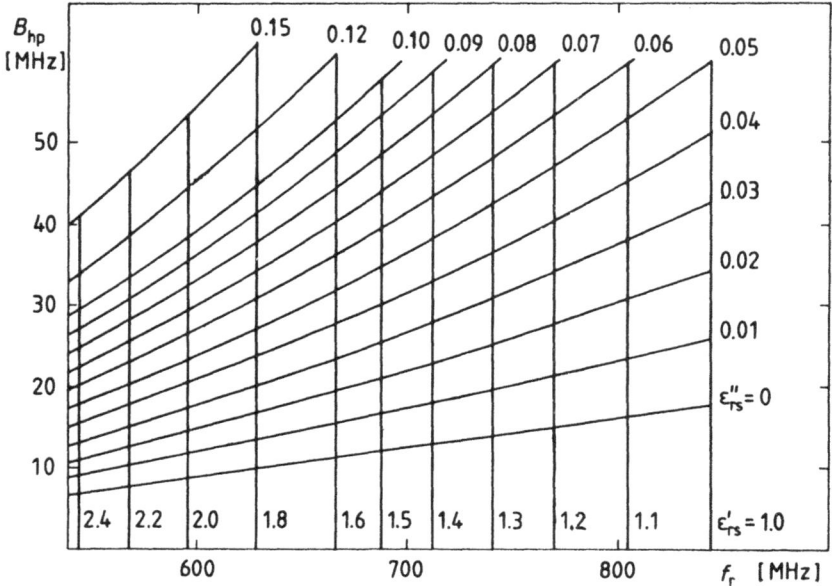

Figure 2.22 Diagram for the snow sensor based on calibration with materials of known permittivity. (After [Sihvola *et al.*, 1986].)

snow, but then we would need other diagrams, rather than Figure 2.21, corresponding to the materials in question. For the convenience of the user, the diagrams in Figures 2.21 and 2.22 can be combined into one linking B_{hp} and f_r with f_w and ρ_d, but in practice all diagrams are clumsy. A meter should therefore be designed for directly calculating the material properties.

The example above was for a resonator sensor, but the principle is the same for a transmission sensor. If we use a single diagram, it gives the dry mass per area and mass of the water per area from the attenuation and phase shift.

Other ways of acquiring more information about the material by using only microwaves are the *multifrequency technique* and the *orthogonality technique*. The multifrequency technique means that the same kind of measurement is performed at two or more different frequencies. The frequencies must be spread far from each other so that the measurements will contain at least partly different information. Three measurement points of ε_r'' may be enough, for example, to determine the relaxation frequency.

The orthogonality technique is especially well suited for thin slab or sheet-like materials. The sensor is a resonator (for example, a two-conductor stripline resonator; see Section 3.3) with two resonant modes and the electric fields perpendicular to each other in such a way that one is tangential to the material surface and the other is perpendicular to it (Figure 2.23). Because of the boundary conditions (2.12)

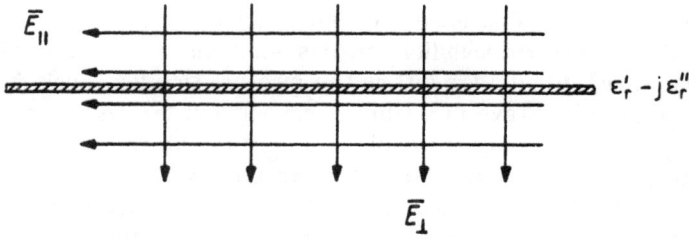

Figure 2.23 Two-parameter measurement with the orthogonality method. Two different resonant modes are used, one with the electric field perpendicular $(E_\perp)$ to the surface of the unknown material and one with the field tangential $(E_\parallel)$ to the surface.

and (2.13), the material sample affects the two modes differently, enabling measurement of the complex permittivity independent of the thickness of the sheet material. The formulas for this case are derived in Section 3.2.2. A thin sheet causes only a small change in the resonant frequency. If we indicate the change by Δf_r, the ratio R_1 between the relative frequency shifts of the tangential ($\parallel$) and the perpendicular ($\perp$) resonant modes is given by (3.50):

$$R_1 = \frac{\Delta f_\parallel / f_\parallel}{\Delta f_\perp / f_\perp} \approx \varepsilon_r' \frac{S_\parallel}{S_\perp}$$

where $S_\parallel$ and $S_\perp$ are the *filling factors* of the modes. The ratio $S_\parallel / S_\perp$ is constant in practice, and can be either calculated (see Section 3.2.2) or calibrated. If we also measure the quality factor (Q) of, for example, the tangential mode, we can calculate the ratio R_2 from (3.44):

$$R_2 = \frac{\Delta f_\parallel / f_\parallel}{\Delta(1/Q_\parallel)} \approx -\frac{\varepsilon_r' - 1}{2\varepsilon_r''} \tag{2.65}$$

Now, we have for the permittivity:

$$\varepsilon_r = \varepsilon_r' - j\varepsilon_r'' = \frac{S_\perp}{S_\parallel} R_1 - j\frac{1 - R_1(S_\perp / S_\parallel)}{2R_2} \tag{2.66}$$

From the permittivity, for example, the moisture and density can be found by using a model or diagram like the one for snow in Figure 2.21. The advantage of the orthogonality technique is that it is independent of small variations in the thickness of the sheet or slab. A slight disadvantage is that it works well only with thin slabs for which the perturbation formulas in Section 3.2.2 are valid. This requires high sensitivity of the measurement electronics.

So far, we have discussed only utilizing the permittivity, but in magnetic materials ($\mu_r \neq 1$) the permeability contains additional information. All four material constants (ε_r', ε_r'', μ_r', μ_r'') are difficult to measure simultaneously because ε_r and μ_r affect a propagating wave in a similar manner, but the resonators provide a possibility. If we put the material sample at a location where the electric field is strong but the magnetic field is zero, only the permittivity affects the resonator characteristics; however, if we put material at a location where the magnetic field is strong and the electric field is zero, we can measure the permeability. The use of two different resonant modes is also possible, one with a strong electric field and the other with a strong magnetic field at the location of the sample.

2.5.5 Short Cuts

The theory of the multiparameter techniques presented in the previous section describes the normal route to the desired results, but, in some cases, reaching similar results in a less laborious way is possible. Such means have been used, for example, for density-independent and structure-independent moisture measurements.

If the density of a material changes, both ε_r' and ε_r'' also change. Because $\varepsilon_r' \geq 1$, and $\varepsilon_r'' \geq 0$, and both primarily depend on the amount of polarizing material in a volume, a fair assumption is that within normal limits of density variations the ratio:

$$A = \frac{\varepsilon_r' - 1}{\varepsilon_r''} \qquad (2.67)$$

is independent of the density. However, if the moisture changes, A will also change. This can readily be seen if we add a small amount of water to a dry and almost lossless material. The real part of the permittivity (and $\varepsilon_r' - 1$) increases slightly, as does ε_r'' (both ε_r' and ε_r'' usually increase monotonically as a function of increasing moisture). The relative increase in ε_r'' is large because water is a lossy substance and the initial value is close to zero. Therefore, the addition of water results in a decrease in A. Because of the binding of water and other effects, A usually varies as a function of the moisture at all levels of interest. The function, however, is not always monotonic as it changes with the frequency. On the basis of laboratory tests, to choose the measurement frequency in such a way that the function is monotonic in the moisture range of interest is therefore usually possible. At least for veneer sheets, the authors have also found possible the use of other functions of ε_r, which are related to A, but with larger unambiguous range for low moisture than can be had by using A (Figure 2.24a).

The usefulness of A for density-independent microwave moisture measurement of foodstuffs and other materials has been demonstrated, for example, by M. Kent, W. Meyer, and W. Schilz in [Kent, 1977; Meyer et al., 1981; and Kent et al., 1982]. The authors of this book have used the method for density-independent and thickness-independent moisture measurement of veneer sheets by stripline resonators [Vainikainen *et al.*, 1987]. This is possible because the sheets can be considered to have only two variables, the dry mass per area and the mass per area of the water. The measurement is performed by using the tangential field of the "even" resonant mode (see Sections 3.3.4 and 3.5.3) and A is calculated from (2.65). The same sensor can be used for all thicknesses of veneer with only one calibration, but, because of the broad dynamic range, the sensor must be tuned differently for measurement before (green veneer) and after the dryer. Figure 2.24 shows A^{-1} for dry and A for green veneer. In the range of 15–20% moisture, the function is ambiguous. The six measurement points far above the others in Figure 2.24(b) are measured with veneer once dried and then remoistened. The difference is due to the fact that the pores in wood become closed when it is dried, considerably reducing its ability to adsorb water.

Similar methods for density-independent microwave moisture measurement can also be used with transmission sensors. Kent and Kress-Rogers have shown that the ratio $R = k''/k'$ (see (1.33)) is approximately equal to A^{-1} [Kent *et al.*, 1986]. It is obtained from the total attenuation divided by the total phase shift after transmission. Figure 2.25 shows measurement points for fish meal.

Chaloupka, Ostwald, and Schiek have shown that considerably reducing the structure dependence under certain conditions is also possible using simple expressions for the permittivity [Chaloupka *et al.*, 1980]. If the dry material is almost lossless and most of the water is free, a calibration factor Δ can be constructed as a linear combination of the measured permittivity (ε_r, possibly on several frequencies) and the permittivity of the dry material (ε_d) in such a way that Δ is independent of structure for a broad range of depolarization factors (of the water ellipsoids). For example, the following calibration factor has proved to be useful with quartz-sand, gypsum, coal-dust, and polyfoam:

$$\Delta = \frac{\varepsilon_r' - \varepsilon_d' - b\varepsilon_r''}{f_w}$$

where $b = 0.94$ and f_w is the volume fraction of water.

2.6 EXAMPLES AND REFERENCES

2.6.1 Humid Air

Air is a mixture of nonpolar gases and water vapor. The mixing ratio of the nonpolar gases (mainly nitrogen and oxygen) is fairly constant, but the humidity varies. The dependence of the permittivity of air as a function of the humidity and temperature can be calculated from (2.27). For the nonpolar gases, we have

$$\varepsilon_{rd}' = 1 + \rho_d A$$

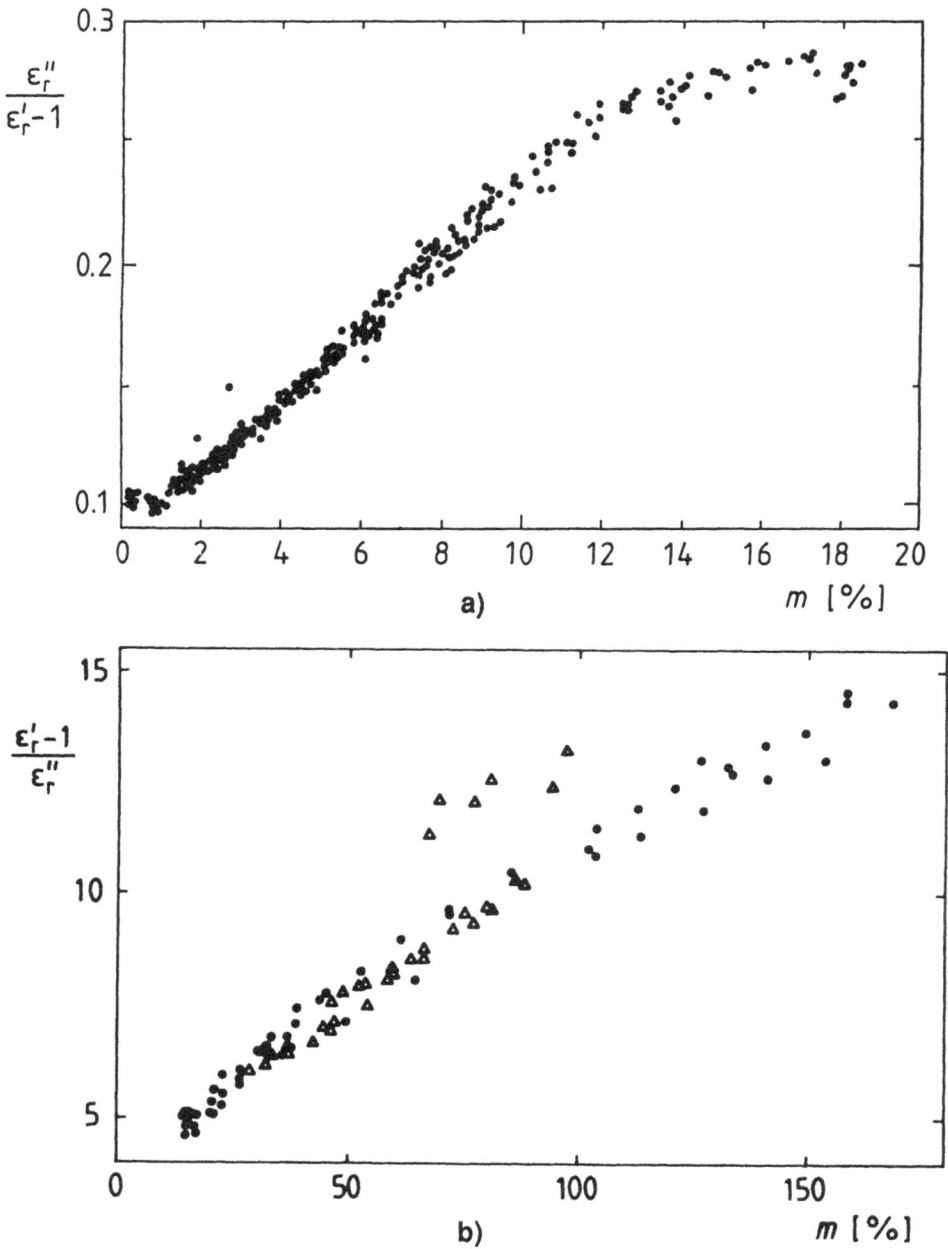

Figure 2.24 Two-parameter measurement results of using A for veneer sheets at 350 MHz: (a) dried 1.5 mm and 2.6 mm thick fir veneer (birch veneer gives a slightly shifted curve); (b) green 1.5 mm thick fir ($\bigcirc$) and birch ($\triangle$) veneer. (After [Vainikainen *et al.*, 1987].)

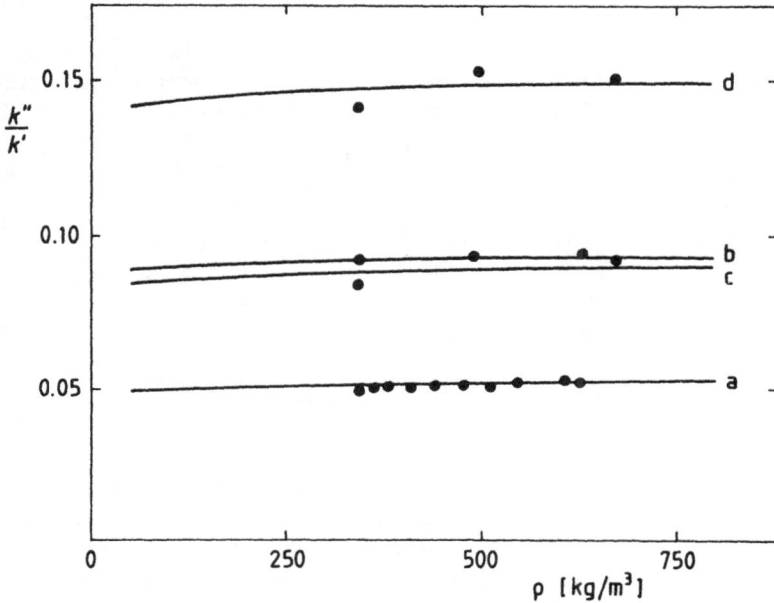

Figure 2.25 Two-parameter measurement results of using the attenuation to phase ratio for fish meal at 9.78 GHz: (a) $t = 24°$ C, $m = 7.2\%$; (b) $86°$ C, 7.2%; (c) $24°$ C, 10.6%, (d) $86°$ C, 10.6%. (After [Kent et al., 1986].)

where ρ_d is the density of the dry component of the air and A is a constant. For the water vapor, we have

$$\varepsilon'_{rw} = 1 + \rho_w \left(B + \frac{C}{T} \right)$$

where ρ_w is the density of water vapor, and B and C are constants. For air, we now have

$$\varepsilon'_{ra} = 1 + \rho_d A + \rho_w \left(B + \frac{C}{T} \right)$$

Normally, we do not know ρ_d, but rather the total pressure p_t, whereas ρ_w is the water content that we either want to measure or to use as an input quantity for the calculation of ε'_{ra}. Because the total density is $\rho_t = \rho_d + \rho_w$ and the partial pressures $p_i \propto \rho_i T$, we have

$$\varepsilon'_{ra} = 1 + A' \frac{p_t}{T} + B' \frac{\rho_w}{T} + C' \rho_w \qquad (2.68)$$

Measured values for the constants A', B', and C' have been reported in the literature by many authors, and seem to be independent of frequency up to at least 24 GHz. The mean values [Hasegawa et al., 1975] are $A' = 1.552 \times 10^{-6}$ Km^2/N, $B' = 3.456$ Km^3/kg, and $C' = -76.57 \times 10^{-6}$ m^3/kg. Figure 2.26 shows the graph of (2.68) at different temperatures. We can see that for 50 g/m^3 and 20° C, $\varepsilon'_{ra} = 1.00112$, which causes a relative frequency change (Equation (3.6)):

$$\frac{f_{ro}}{f_r} = \frac{f_{ro}}{f_{ro} - \Delta f_r} \approx 1 + \frac{\Delta f_r}{f_{ro}} \approx \sqrt{\varepsilon'_{ra}}$$

$$\frac{\Delta f_r}{f_{ro}} \approx \sqrt{\varepsilon'_{ra}} - 1 = 5.6 \times 10^{-4}$$

This change is so great that we must take it into account if the temperature or the humidity changes when we make measurements with resonators using the perturbation technique and the filling factor is very small. In such cases, however, we must also take into account the thermal expansion in the metal parts. We can also utilize the phenomenon for making microwave humidity sensors that measure the permittivity of the air. These sensors can be built to be robust and to withstand high temperatures and a dirty environment. Such a prototype meter has been built at the Radio Laboratory of the Helsinki University of Technology (see Section 3.5.1) [Toropainen et al., 1987].

2.6.2 Moist Wood

Wood consists mainly of cellulose fibers, which consist of cellulose molecules. They are polyalcohols and therefore have plenty of OH groups, which effectively bind water molecules. Because of the fibrous structure, the internal surface to which water molecules can bind is very large. Wood normally binds water up to about 30% moisture (dry basis). For higher moistures, some of the water stays as capillary water between the fibers. The relaxation phenomena related to bound water therefore dominate the permittivity in the frequency range between 100 MHz and 10 GHz. Below 100 MHz, the Maxwell-Wagner effect is encountered because the cellulose fibers contain both electrically nonconducting crystalline parts and conducting amorphous parts.

Wolfgang Trapp has measured the permittivity of different species of wood at a broad range of frequencies by using coaxial sample holders [Trapp, 1954]. Figures 2.27 and 2.28 show some of his results for fir wood. The relaxation of bound water and the ion conductivity effect are very clearly seen. The upward movement in frequency of the relaxation-loss maximum with rising temperature and moisture are well

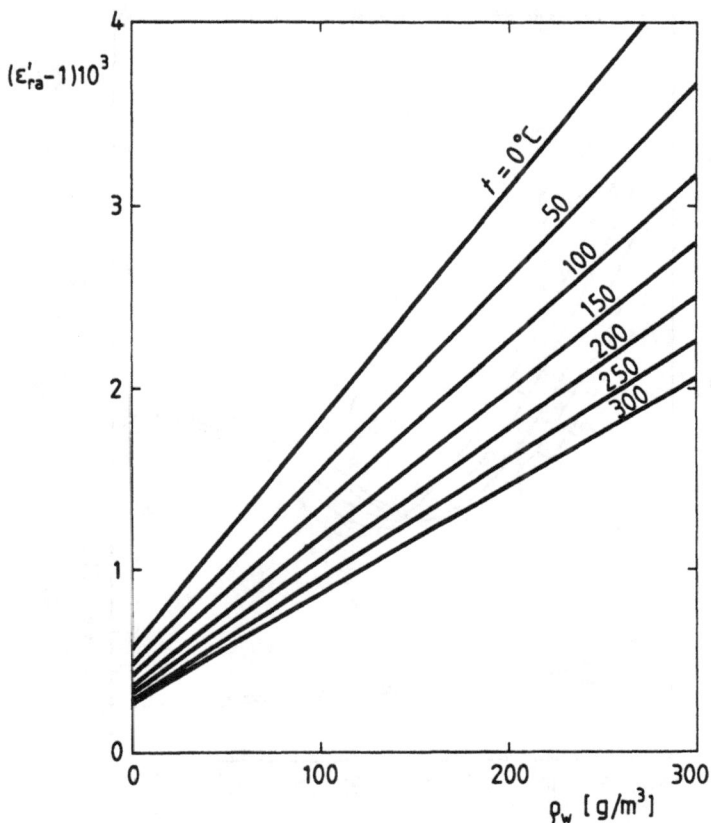

Figure 2.26 The permittivity of air as a function of temperature (t) and density of the water vapor (ρ_w).

in accordance with (2.46). We can see that the losses and ε'_r are fairly independent of temperature in the range of 100 MHz to 1 GHz, which makes this frequency range especially suitable for microwave moisture measurement of dry wood. This range has been utilized in the moisture meter for veneer sheets described in Section 3.5 (see Figure 2.24). The Maxwell-Wagner effect is not detectable in Figure 2.27 because of the low moisture (6.5%, dry basis), but for higher moisture values Trapp found a clear effect. For example, for $m = 65\%$, $t = 20°$ C, ε'_r increased from 7.5 to 72 in the range from 80 MHz to 10 kHz. He also found that the differences between various species of wood are small, but clearly detectable. The main features of the permittivity curves, however, were common for all species of wood that he measured. Trapp's measurements also clearly showed an anisotropy of a few tens of percent with the direction of the field with respect to fibers.

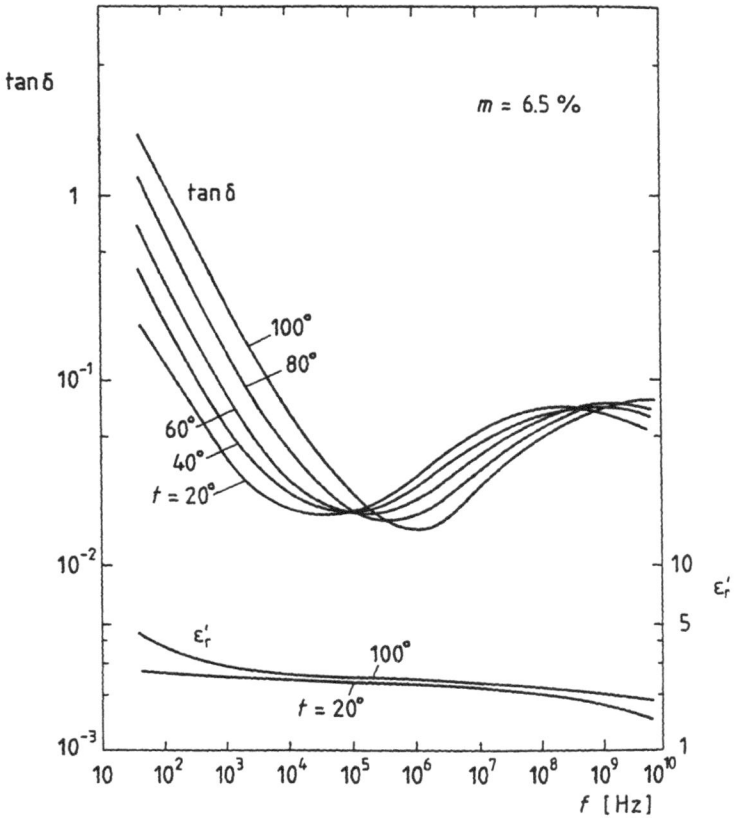

Figure 2.27 The permittivity and loss tangent ($\tan\delta = \varepsilon_r''/\varepsilon_r'$) of fir as a function of temperature and frequency. (After [Trapp, 1954].)

2.6.3 Short Summary and References

- Dry solids usually have a low ε_r' (<10), independent of frequency.
- Free water has high permittivity, bound water has low. Degree of binding depends on temperature and moisture.
- The permittivity (ε_r' and ε_r'') usually increases with increasing moisture.
- Free ions cause high temperature-dependent losses at low frequencies.
- Polar molecules cause high frequency and temperature dependent ε_r'.
- Laminated or fibrous structure gives higher ε_r' than does granular. Anisotropic structure gives anisotropic permittivity.

Tabulated values for the permittivity of different materials can be found in, for example, [von Hippel, 1954; Meyer *et al.*, 1982; Tinga *et al.*, 1973a; Nelson, 1973; To *et al.*, 1974; and Stuchly *et al.*, 1980a].

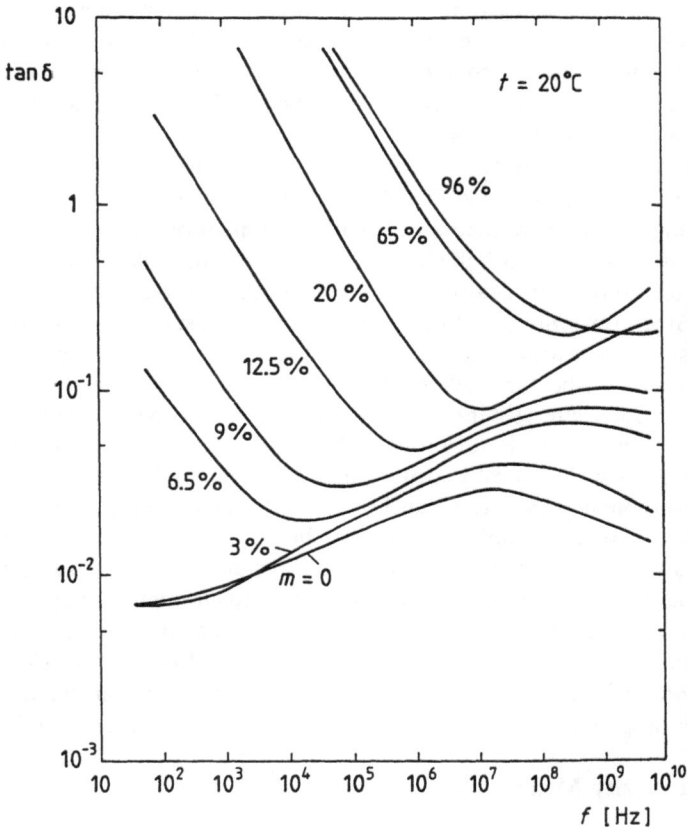

Figure 2.28 The permittivity and the loss tangent of fir as a function of moisture (dry basis) and frequency. (After [Trapp, 1954].)

2.7 PERMITTIVITY MEASUREMENTS IN THE LABORATORY

2.7.1 Introduction

Earlier in this chapter we noticed that the dielectric properties of materials often depended on many parameters in a quite complicated way. Therefore, the dielectric behavior of a material was usually difficult to estimate, although the composition, chemical properties, and structure were known. Also, the data found from the literature, if any, were usually only adequate for deciding if the properties of the material could be measured with microwaves. However, more information on the material is usually needed for choosing the best measurement principle (e.g., resonance method, transmission), frequency of operation, or possible auxiliary measurements (e.g., temperature, density).

To obtain this information, a set of laboratory measurements is needed. Their main purpose is to give an overview of the dielectric behavior of the material in the industrial measurement environment. However, to simulate the practical situation so well that the laboratory measurements can be used for the final calibration of the measurement equipment is usually not possible.

Therefore, a moderate accuracy is sufficient for the laboratory permittivity measurement system, provided that the system is fast, easy to use, and permits convenient change of the parameters of the measurement environment. The demand for short measurement time originates from the fact that because of several affecting parameters the number of measurement points is great, in many cases several hundreds. The most time-consuming part of the laboratory measurements is usually the preparation of samples. Further, if measurement at several frequencies is necessary, we ought to minimize the time taken to change the measurement frequency by using wideband measurement systems or the same sample at several frequencies (although the measurement system is different). The ease of use of the system reduces measurement time as well as cost because qualified personnel are not necessary. Control of the measurement parameters can be difficult (for example, the temperature) and time-consuming if the sample holder must be changed to change frequency.

In this section, we present some basic laboratory methods for permittivity measurements and examine their use in different measurement situations. The presentation is not a comprehensive review, but is limited to those methods and problems that seem important in solving industrial measurement problems. More detailed and complete information on the measurement of permittivity can be found in the literature [von Hippel, 1954, pp. 47–147; Sucher *et al.*, 1963, Ch. IX].

2.7.2 Permittivity Measurement Techniques

Numerous permittivity determination techniques can be found in the literature. The techniques can be divided into groups according to their working principles. The main groups of measurement methods are given below:

- *Lumped Circuits*—In this low-frequency method, the sample is a part of the insulator of a lumped capacitor.
- *Resonance Methods*—The sample is a part of a resonating transmission line circuit.
- *Transmission Line Propagation*—The sample is a part of a nonresonating transmission line circuit.
- *Free-Space Propagation*—The microwave propagation through or reflection from the sample is investigated.

In this section, the basic features of these groups are presented and some useful measurement methods are examined in detail. The theory of the methods can be found in Chapters 3 (lumped circuits, resonators) and 4 (propagation methods). Understanding some of the material presented below requires previous knowledge of radio engineering or that the reader first study Chapters 3 and 4.

Lumped Circuits

The common feature of the methods is that a capacitance is affected (usually increased) by the measured material. At the lowest frequencies, a real lumped capacitor exists, the insulator of which, or part of it, is the measured material. The admittance $Y_s(= Z_s^{-1})$ of the capacitor filled with a lossy material ($\varepsilon_r = \varepsilon_r' - j\varepsilon_r''$) sample, in the case of a plate capacitor, is

$$Y_s = j\omega\varepsilon_0 \cdot \frac{A}{d}(\varepsilon_r' - j\varepsilon_r'')$$

where A is the area of the plates and the sample, and d is the distance between the plates.

The admittance is usually measured with a resonant circuit consisting of the capacitor and a constant inductor. The permittivity of the sample can be determined by measuring the resonant frequency and quality factor of the lumped circuit. For the parallel resonant circuit in Figure 2.29 for which the power transmission response is measured, the resonator equations (3.15), (3.25), (3.26), and (3.35) can be used:

$$\varepsilon_r' \approx \left(\frac{f_{ro}}{f_r}\right)^2 \tag{2.69}$$

The radiation losses are usually small compared to the external losses (caused by the coupling to the measurement equipment) and metal losses, so (3.35) can be used:

$$\varepsilon_r'' \approx \varepsilon_r' \left[\frac{1}{Q_u} - \frac{1}{Q_{mo}} \cdot \left(\frac{f_{ro}}{f_r}\right)^{1/2}\right] \tag{2.70}$$

Here, Q_{mo} (the metal quality factor with no material in the capacitor) is assumed equal to the unloaded quality factor Q_u for the circuit with the empty capacitor. The unloaded quality factor Q_u is

$$Q_u = \frac{Q_l}{1 - \sqrt{a_r}} \tag{2.71}$$

$$Q_l = \frac{f_r}{B_{hp}} \tag{2.72}$$

where Q_l is the loaded quality factor. The resonant frequency f_r, half-power bandwidth B_{hp}, and insertion loss a_r at the resonant frequency are measured with and without the sample, and with the same capacitor plate spacing. If the material losses

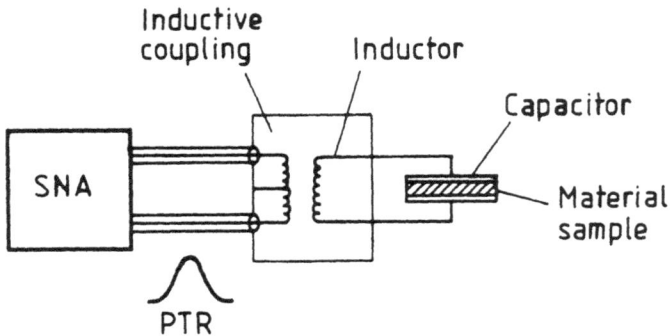

Figure 2.29 The lumped circuit permittivity measurement system for low frequencies. The power transmission response of the parallel LC resonator is measured with a scalar network analyzer (SNA) and an inductive coupling to the resonator.

are high, filling the capacitor only partly may be useful, in which case (2.69) and (2.70) give the effective permittivity ε_{eff}. The permittivity of the sample ε_r' and ε_r'', can then be calculated from the effective permittivity:

$$\varepsilon_r - 1 = \frac{A}{A_s}(\varepsilon_{\text{eff}} - 1) \tag{2.73}$$

Here, A is the area of the capacitor and A_s that of the sample. The sample height is equal to the capacitor plate spacing. The above presentation is simplified and does not take into account, for example, the fringe capacitance of the plate capacitor. This can be done by replacing the geometric area A with an effective plate area A_{eff} being slightly larger than A. For circular plates with a radius r, the ratio A_{eff}/A is presented in Figure 2.30 as a function of the plate spacing [von Hippel, 1954, p. 51]. By using A_{eff} instead of A in (2.73), the correction for samples with $A_s \leq A$ can be made. If the sample is clearly larger than A_{eff}, the electric field of the capacitor is totally filled with the material, and (2.69) and (2.70) can be used without corrections.

The lumped circuit method can be used only at low frequencies. The high frequency limit is set by the minimum value of the inductor, which is about 10 nH. Because of this, the capacitor becomes too small at high frequencies, and the effect of parasitic capacitances increases. Another factor limiting the maximum usable frequency of this method is the appearance of standing waves in the circuit changing the resonant frequency from the value given by (2.69). The radiation losses also increase with frequency. To avoid the standing waves and radiation, the overall size of the inductor-capacitor circuit should be less than 0.05 times the free-space wavelength at the highest working frequency of the circuit. These conditions determine the maximum usable frequency of this method to about 100 MHz.

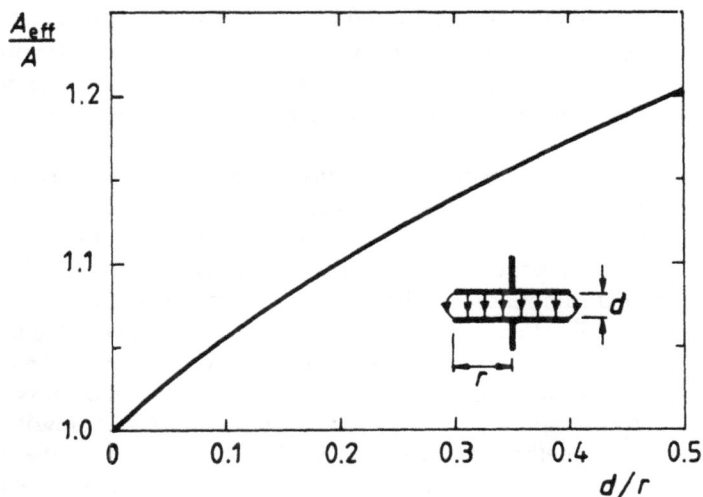

Figure 2.30 The effect of the fringe capacitance of a circular plate capacitor expressed as an increase of the geometric area A of the capacitor plates. (After [von Hippel, 1954, p. 51].)

The quality factor of the resonant circuits for permittivity measurement is usually quite low, typically about 200. In this situation, the $\tan\delta$ measuring accuracy gained by using a scalar network analyzer (see Section 3.4) is about 10^{-4}. Therefore, the method is not suitable for very low-loss materials.

Both low and high ε_r' values can be measured with the method, but with only moderate accuracy due to the low quality factor and the partly unknown parasitic circuit elements. With the highest ε_r' and ε_r'' values the partially filled capacitor and (2.73) can be used to avoid large measurement frequency shift or excessive decrease of quality factor caused by the material.

All material types except gases can be measured with the lumped circuit method. Attention should be paid, however, to the effect of sample holders on the permittivity results. For example, when measuring high-ε_r' liquid with a plate capacitor, the bottom of a plastic sample holder can cause a severe error to the results if not taken into account as a small capacitance in series with the actual measuring capacitance. The shape of the capacitor can be chosen quite freely (different plate shapes, tubular capacitor, *et cetera*) to be suitable for the measured material.

Resonator Method

As presented in Chapter 3, a microwave resonator partly or completely filled with a material can be used for the determination of its permittivity. There are numerous possible resonator structures, from which we present some that are suitable for laboratory permittivity measurements.

For all resonators having an electric field volume that is completely filled with the material to be measured, (3.15) and (3.33) can be used for the determination of the permittivity from the resonant frequency and quality factor. For different measurement methods and situations, some of the more detailed equations for ε_r'', (3.21) to (3.32), can be used.

In some cases, to fill only part of the resonator is useful to decrease the sample size or the effect of the material on the resonator parameters [Cook *et al.*, 1974; Decreton *et al.*, 1975; Weiss *et al.*, 1987; Xu *et al.*, 1987]. In this situation, the connection between the resonator parameters and material permittivity is different for every resonator and sample shape. The exact calculations of the connection can be difficult, or, in the case of complex geometries, practically impossible to perform with adequate accuracy. Therefore, the partially filled resonators are often calibrated with known materials in the permittivity range of interest. Another possibility for certain materials is to use an untuned, large cavity resonator (see below).

If the perturbation method described in Section 3.2.2. is used, the equations are simpler. The method requires that the sample be small, causing a relative resonant frequency shift smaller than 0.001. To solve the connection between the resonator parameters and sample permittivity, only one filling factor S, common to all samples with the same size and shape, is to be determined:

$$\frac{\Delta f_r}{f_r} + j \frac{1}{2} \Delta \left(\frac{1}{Q_l} \right) = C(\varepsilon_r) \cdot S \qquad (2.74)$$

The filling factor S can in some cases be calculated with (3.40), but it can always be determined with a known material whose ε_r' is as close as possible to that of the material we want to measure. The connection $C(\varepsilon_r)$ between ε_r and the perturbations of the resonator parameters depends on the sample shape and direction of the electric field at the sample location (see (3.43) to (3.49)). In many cases, the best results are obtained when the electric field is tangential to the surface of the sample (see (3.43) and (3.44)). In that case, $C(\varepsilon_r)$ is linear and the sensitivity to the changes of ε_r is high:

$$C(\varepsilon_r) = \frac{1}{2} (\varepsilon_r - 1) \qquad (2.75)$$

The changes of the resonator parameters can be measured with several methods based on the phase or amplitude measurement of the reflection or transmission response of the resonator (see (3.19) to (3.26)). The simplest way is to use a *scalar network analyzer* (SNA) to measure the amplitude response. The SNA is also fast and easy to use. In the millimeter-wave range (above 40 GHz), however, a simpler amplitude measurement system with a tunable oscillator and detectors must be used for the measurement of the oscillator output power and the power reflected from or transmitted through the resonator. At all frequencies, the accurate measurement is essential. The required relative frequency measurement accuracy is about one part in $(1000 \cdot Q_l)$. Usually, either a separate frequency counter or an SNA with a crystal stabilized synthesizer oscillator or crystal frequency markers is used. At millimeter-waves, the frequency can be measured with a cavity frequency meter or a phase-locked oscillator can be used. The determination of resonator parameters contains several tuning operations and measurements. Therefore, the measurement system should be as automatic as possible to minimize the measurement time.

With the resonator method, almost any dielectric material can be measured. Gases, with ε_r very close to unity, can be measured with a totally filled resonator if the quality factor is high enough. However, lossy materials with high ε_r', such as saline water, can also be measured quite accurately with the perturbation method. The sample can have a great variety of shapes: sheet, cylinder, cube, sphere, slab, or rod. Even small, arbitrarily shaped samples can be measured, but a calibration routine must then be performed with different known materials of the same sample shape [Rueggeberg, 1971].

Another possibility for the measurement of arbitrarily shaped samples is to use an untuned or "stirred" cavity [Izatt et al., 1981; Backhouse et al., 1985]. This is a large (in wavelengths) cavity where a great number of resonant modes can exist. By using a fan in front of the power input of the cavity and irregular wall surfaces, the time average of the energy distribution in the cavity can be made uniform. In this situation, only the volume of the sample, not its shape, affects the resonant properties of the untuned cavity. By measuring the power level in the cavity, the losses of the sample can be determined. If ε_r' is desired, disk samples can be measured as a function of thickness.

The frequency range for permittivity measurements that can be covered with resonators is from 50 MHz to above 100 GHz. In this range, the free-space wavelenth, which in many cases determines the size of the resonator, changes from 6 m to less than 3 mm. There are, however, ways to keep the size of the resonator as well as the sample at a practical level. At low frequencies, so-called slow-wave structures can be used, where the velocity of microwaves can be more than ten times slower than the free-space velocity. The wavelength is proportional to the wave velocity, and so the resonator structures can be made smaller. Slow-wave structures for TEM or quasi-TEM transmission lines are, for example, the helical coaxial line and the so-called meander line. In the helical coaxial line, the center conductor is a spiral (see Section 2.7.4, Figure 2.39) and in the meander line (for example, strip-line) it is planar and curved.

Another way to shorten a resonator is to use a suitable impedance in one or both ends of the resonator. An inductance can be used in a shorted end, and a capacitance in an open end, as in the so-called re-entrant coaxial resonator of Figure 2.31. The resonant frequency of the empty resonator in the figure is about 900 MHz.

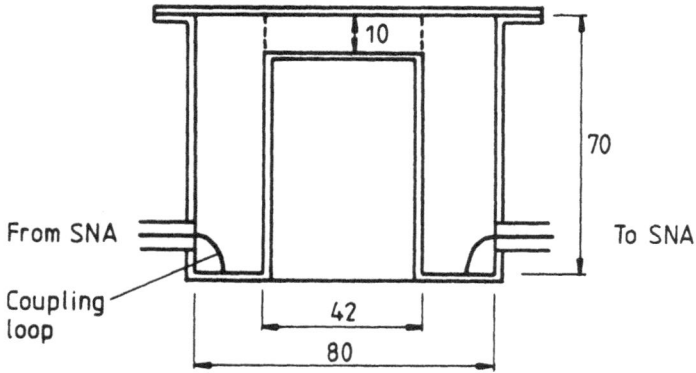

Figure 2.31 A re-entrant coaxial resonator with $f_{ro} \approx 900$ MHz. The coupling for power transmission measurement is made with shorted loops at the shorted end of the resonator.

For a normal quarter-wavelength resonator with the same center conductor length, the resonant frequency would be about 1250 MHz, and so a frequency reduction by a factor 1.4 could be gained. The re-entrant resonator can be used as a totally or partially filled resonator. In the partially filled case, the sample is usually placed in the gap between the center conductor and the end plate of the line.

Above 1 GHz, the size of waveguide cavity resonators becomes suitable for permittivity measurements. The cavities can be used either as totally filled, partially filled [Amato, 1980], or perturbed resonators. In cavity resonators, several resonance modes can exist. With suitable coupling (see Section 3.3.1), the unwanted modes can be attenuated. A useful mode for a totally filled cylindrical cavity is the TE_{011} mode, the electric field distribution of which is shown in Figure 2.32(a). The currents in the end plates of the resonator with this mode are tangential, which is useful in two ways. The first useful property is that there are no currents flowing from the end plates to the walls of the resonator. Therefore, the electric contact of the end plates is not critical, and the opening and closing of the resonator is simple. The other advantage of this mode is that the end plates can be replaced by cylindrical rings (Figure 2.32b) making the resonator quite open and still maintaining a high quality factor, up to more than 7000 [Wenger, 1967]. This property makes the mode suitable for measurements of gases. The gas can flow freely through the cavity and a fast response is gained.

For measurements with a partially filled or perturbed cavity, for example, the TM_{010} mode for the cylindrical cavity and TE_{101} for the rectangular cavity can be used. The TM_{010} mode is especially suitable for the measurement of cylindrical samples because the exact solution of the resonator parameters from the dielectric properties of the sample is known [Agdur *et al.*, 1962]. For a cylindrical sample with

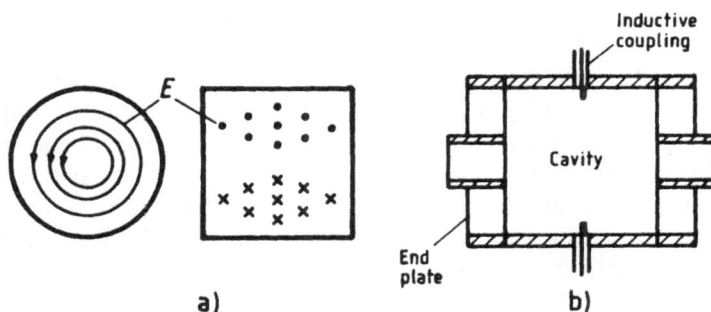

Figure 2.32 The TE_{011} mode of a cylindrical cavity resonator is useful for permittivity measurements: (a) the electric field pattern in the resonator; (b) the end plates of the resonator can be replaced with cylindrical rings.

radius r_s coaxially located in a cavity with radius r_c, we get an implicit equation:

$$\sqrt{\varepsilon_r}\,\frac{J_0'(\sqrt{\varepsilon_r}\,k_0 r_s)}{J_0(\sqrt{\varepsilon_r}\,k_0 r_s)} = \frac{J_0'(k_0 r_s)N_0(k_0 r_c) - J_0(k_0 r_c)N_0'(k_0 r_s)}{J_0(k_0 r_s)N_0(k_0 r_c) - J_0(k_0 r_c)N_0(k_0 r_s)}$$

Here, $k_0 = \omega/c_0$ is the free-space wavenumber, and J_0, N_0, J_0', and N_0' are the zero-order Bessel functions of the first and second kind and their derivatives, respectively.

The sample is placed in the cavity, despite the measuring mode, usually so that the electric field is tangential to the sample surface. The problem with the perturbation method, however, is that the sample often tends to become too small when the first perturbation condition in Section 3.2.2 ($\Delta f_r/f_r < 0.001$) is fulfilled (see Section 2.7.4). The problem can be solved by placing the sample off the maximum of the electric field in the resonator (thus violating the second and third perturbation conditions). The sample can also be located so that the electric field lines are perpendicular to the surface of a thin sample (cylinder, sheet). According to (3.45) and (3.46), the changes of the resonator parameter are then usually smaller than in the case of the tangential field, but the response is also nonlinear with a poor measurement sensitivity at large ε_r' values. The filling factor can also be decreased by using higher-order resonant modes (see Section 2.7.4). The fourth way to deal with the problem is to ignore the first perturbation condition and use the method up to relative frequency shifts of 0.01, with a slightly reduced accuracy as a result, unless we introduce appropriate correction factors for the higher $\Delta f_r/f_r$ values [Berteaud et al., 1975; Chao, 1985].

Above 10 GHz, the size of the fundamental-mode resonators becomes too small to be practical in laboratory measurements. One solution to the problem is to use higher-order resonance modes (see Section 2.7.4). The problem, however, is the great number of different resonances, making difficult the indentification of the wanted resonance mode, especially in the case of a partially filled cavity.

Another possibility at frequencies above 10 GHz is to use open quasioptical resonators. They consist of two mirrors reflecting waves back and forth in the free space between them. Two spherical concave mirrors or one concave and one flat mirror are usually used to form a Gaussian beam (Figure 2.33). The power is connected to and from the resonator through coupling holes or loops in the mirrors. The dielectric sample is normally a plate or slab of even thickness (also, other geometries are possible) placed either in contact with the flat mirror or halfway between the concave mirrors (at the waist of the Gaussian beam). The permittivity of the sample can be calculated exactly from the resonator parameters [Cullen *et al.*, 1971; Cook *et al.*, 1974; Lynch, 1983; Chan *et al.*, 1987]. The quality factor of the resonator is very high, typically greater than 100 000, if the radii of the mirrors are at least four times larger than the so-called scale radius w of the Gaussian beam (the radius where the field strength is $1/e$ times the maximum in the center of the beam, see Section 8.22). This condition gives, for the concave mirror diameter D_c:

$$D_c \geq 4\left[\frac{\lambda l}{\pi}\frac{R/l}{(R/l - 1)^{1/2}}\right]^{1/2}$$

and for the flat mirror diameter D_f (and for the sample diameter D_s):

$$D_f \geq 4\left[\frac{\lambda l}{\pi}(R/l - 1)^{1/2}\right]^{1/2}$$

where λ is the free-space wavelength, R is the curvature radius of the concave mirror, and l is the length of the resonator with one flat and one concave mirror (with two similar concave mirrors, the resonator length is $2l$, see Figure 2.33).

Because of the very high quality factor, this resonator can be used primarily for the measurement of low-loss materials. For medium-loss materials, the quality factor should be lowered, for example, with a lossy sample holder to prevent excessive decrease of the resonance peak, or a very thin sample should be used (see Section 3.4).

Transmission Line Methods

The permittivity can also be measured by using the material sample as the insulator (or a part of it) of a transmission line section. By measuring the effect of the sample on the transmission characteristics of the line or the reflection from the sample, the permittivity can be calculated from known reflection and transmission coefficient equations. Three usual ways to place the sample into the transmission line are presented in Figure 2.34. The voltage (or field) transmission coefficient t_s through a sample can be measured (Figure 2.34a):

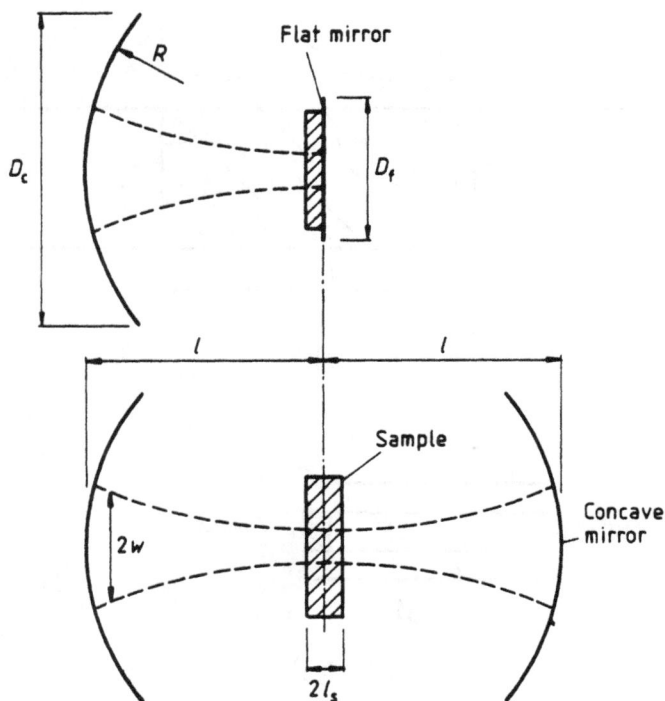

Figure 2.33 Open quasioptical resonators; either one concave and one flat (upper) or two concave mirrors can be used. The wave propagates between the mirrors in the form of a Gaussian beam, the scale width of which is $2w$ (between the dashed lines).

$$t_s = \frac{(1 - \Gamma^2) \exp(- \gamma l_s)}{1 - \Gamma^2 \exp(-2\gamma l_s)} \tag{2.76}$$

Here, Γ is the voltage reflection coefficient of the sample surface:

$$\Gamma = \frac{[1 - (f_c/f)^2]^{1/2} - [\varepsilon_r - (f_c/f)^2]^{1/2}}{[1 - (f_c/f)^2]^{1/2} + [\varepsilon_r - (f_c/f)^2]^{1/2}} \tag{2.77}$$

where f_c is the cut-off frequency of the transmission line without sample ($f_c = 0$ for TEM lines). The propagation factor γ is

$$\gamma = j \frac{\omega}{c_0} \left[\varepsilon_r - \left(\frac{f_c}{f}\right)^2 \right]^{1/2} \tag{2.78}$$

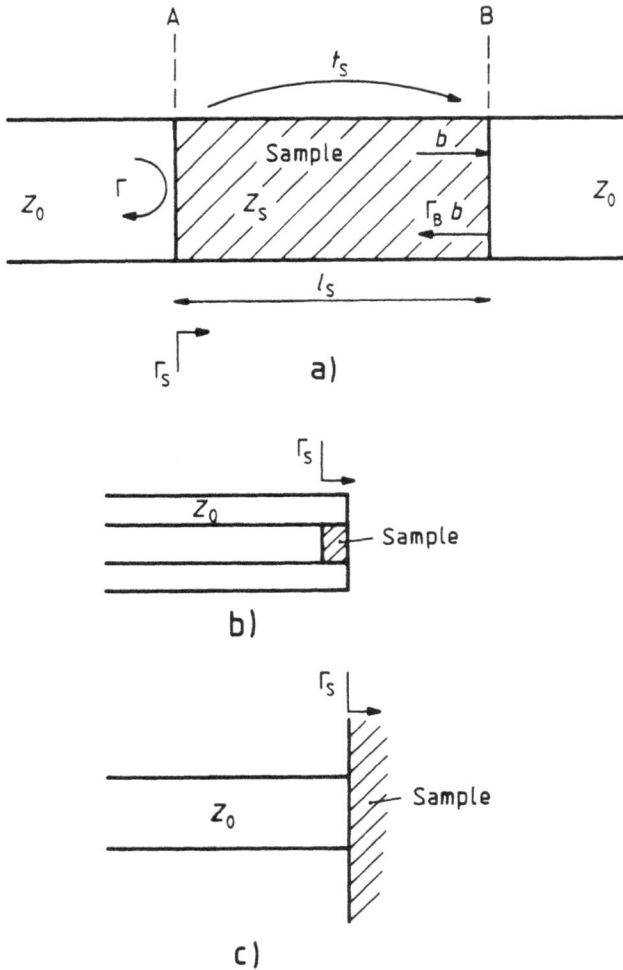

Figure 2.34 Transmission line methods for permittivity measurements: (a) the basic situation when the sample fills a section of an arbitrary transmission line with impedance Z_0; (b) the sample placed between the end of the inner conductor and a short; (c) the sample at the open end of a transmission line.

Another method to investigate the permittivity of a sample in a transmission line is to measure the reflection coefficient Γ_s. The reflection is caused by the sample and the load behind it:

$$\Gamma_s = \frac{\Gamma + \Gamma_B \exp(-2\gamma l_s)}{1 + \Gamma_B \Gamma \exp(-2\gamma l_s)} \tag{2.79}$$

Here, Γ_B is the complex ratio of the returning signal voltage to the transmitted signal, appearing immediately to the left of the interface B in Figure 2.34. There are two generally used possibilities to terminate the transmission line behind the sample. When there is an empty transmission line terminated to a matched load behind the sample, there is no returning signal from the load, and Γ_B is equal to $-\Gamma$. The most common way to measure the reflection is, however, to place a shorted termination next to the interface B. In this situation, the value of Γ_B equals -1. The best accuracy for this method is achieved with a sample length of approximately $\lambda_s/4$, where λ_s is the wavelength in the sample [von Hippel, 1954, pp. 74–75].

To solve the permittivity in the above measurement situations, both the amplitude and phase of the transmission or reflection coefficient are needed. However, by changing either the measurement frequency or the sample length, we can solve ε_r from the amplitude response alone. This can be done by investigating the maxima and minima of the transmitted or reflected amplitude as a function of frequency or sample length [Tinga et al., 1968; Miller et al., 1968]. Further, the cumbersome iterative calculations of the complex transcendental equations (2.76) or (2.79) [Nelson et al., 1973; Ligthart, 1983] can be avoided by using the frequency or sample length variation as shown in the example of Section 2.7.4. By using these variation methods only materials with low or medium losses can be measured because with high-loss materials there are no multiple reflections inside the sample.

The section of transmission line may be partly filled with the sample. For example, a slab of material can be inserted into a waveguide, tangential or perpendicular to the electric field [Altman, 1964, pp. 172–176]. The sample can be inserted on top of a microstrip line as well (see Figure 1.9) [Khalid et al., 1988]. The equations relating the transmission and reflection coefficients and the permittivity are in this case quite complicated.

The sample can also act as a lossy capacitive termination of the transmission line [Stuchly et al., 1980b, pp. 178–179; Stuchly et al., 1987]. In this case, the reflection coefficient Γ_s from the termination is

$$\Gamma_s = \frac{1 - Z_0 Y_s}{1 + Z_0 Y_s} \tag{2.80}$$

Here, Z_0 is the characteristic impedance of the line and Y_s is the load admittance of the line. The admittance Y_s depends on the ε_r and the construction of the end of the line.

In coaxial lines, the sample can be placed between the end of the inner conductor and a short (Figure 2.34b). If the sample diameter is equal to the inner conductor diameter, the admittance Y_s is

$$\frac{Y_s}{j\omega} = \frac{\pi a^2}{l_s} \varepsilon_0 \varepsilon_r + 4a\varepsilon_0 \left[\ln\left(\frac{b-a}{l_s}\right) \right]^{-1} = C_0 \varepsilon_r + C_f \tag{2.81}$$

where b is the outer conductor radius of the coaxial line, l_s is the length of the sample, C_0 is the parallel plate capacitance at the end of the line without sample, and C_f is the fringe capacitance assumed to be independent of ε_r. Now, ε_r can be solved:

$$\varepsilon_r = \frac{Y_s - j\omega C_f}{C_0} = \frac{1}{C_0}\left(\frac{1}{Z_0} \cdot \frac{1 - \Gamma_s}{1 + \Gamma_s} - j\omega C_f\right) \tag{2.82}$$

The sample size is small with this method, which can be either an advantage or a disadvantage (see Section 2.7.3).

The open end of any transmission line can be terminated to a large sample (Figure 2.34c). In this case, Y_s depends on the construction of the transmission line [Decreton *et al.*, 1974; Stuchly *et al.*, 1987].

All liquid and solid materials can be measured by using transmission line methods. The permittivity of gases is, however, too low for these methods, as they are based mainly on one or two transmissions of the wave through the sample. There are problems with the sample preparation of solid materials. Transmission line methods usually demand a certain sample shape, which, for example, in a coaxial line can be somewhat difficult to prepare. Also, in the shorted line end method, the optimum sample length depends on the ε_r' and many different sample sizes are needed.

There are no limits on the measurement of permittivity by use of these methods (excluding those with gases). Both low and high loss materials can usually be measured with the same equipment. Materials with very high losses can be measured by the so-called *infinite sample reflection method*, wherein the attenuation in the sample is so high that the termination reflection Γ_B in (2.79) can be considered to be zero, so $\Gamma_s = \Gamma$. For materials with a high ε_r', there is the problem of higher-order modes having their cut-off frequency exceeded. These modes can cause unpredictable errors to the "correct" reflection and transmission coefficients. However, these modes are not easily excited in normal measurement situations.

The measurement frequency range achieved with a single equipment is usually larger with transmission line methods than with resonators. Coaxial cable methods can be used from VHF to almost 20 GHz. A single coaxial transmission measurement has covered the frequency range of 0.2–18 GHz [Kent *et al.*, 1983] (see Section 2.7.4). However, the size of the coaxial line is most suitable for measurements at the VHF and UHF ranges (up to 3 GHz). At higher frequencies, waveguides are used. Over 20 GHz, the size of waveguides can be too small for some applications, whereupon measurements with partially filled microstrip lines or dielectric waveguides offer solutions up to 100 GHz.

Both phase and amplitude measurements are needed for the basic transmission line methods. The measurement is made on a single frequency, so a swept vector network analyzer is not necessary, but a slotted line or vector voltmeter can be used (in the millimeter-wave range, they are the only possibility). The frequency or sample length variation methods, however, require a greater number of measurements, such that we recommend all means of increasing the measurement speed and level of automation. These methods need only amplitude measurement, however, at the expence of an auxiliary frequency measurement or an increased number of samples. Absolute frequency measurement accuracy is not needed by the frequency variation method, but only the difference between two frequencies. Neither is the accuracy demand nearly as strict as with the resonators (especially with the perturbation method). Respectively, the permittivity measurement accuracy is not as good as with resonators, but as mentioned earlier, accuracy is not always the most important feature when choosing a laboratory method for industrial sensor development.

Free-Space Methods

At frequencies above 3 GHz, measurement of the effect of a material sample on a wave propagating in free space from one antenna to another is possible. Again, both reflection [Garg *et al.*, 1965; Rose *et al.*, 1972], and transmission [Hallikainen *et al.*, 1986] (see Section 2.7.4) methods are possible. The measurement is very similar to the transmission line method. The same relations (2.76) to (2.80) can be used for the transmission (Figure 2.35a) and reflection (Figure 2.35b) cases with $f_c = 0$.

The materials measured with the free-space method must be sufficiently lossy. The standing waves in low-loss materials cause errors in the results. A practical condition for the losses is that the one-way attenuation inside the material layer be at least 10 dB. This condition gives for the thickness l_s and ε_r of the sample:

$$\frac{f l_s \varepsilon_r''}{\sqrt{\varepsilon_r'}} \geq 110 \, \text{MHz} \cdot \text{m} \qquad (2.83)$$

The samples measured with this technique are quite large because the width of the sample must be large enough to cover the signal beam between the antennas. Typically, the diameter of the sample is of the same order as the antenna separation. For example, in a measurement system for frequencies from 3 to 37 GHz, the sample diameter was 30 cm and antenna separation 23–37 cm [Hallikainen *et al.*, 1986].

The frequency range of the free-space method reaches from the lowest microwave frequencies up to over 100 GHz. The only restriction on the frequency is the 10 dB attenuation condition, which can be difficult to fulfill at the lowest frequencies. For these measurements, both amplitude and phase are needed. At the lowest frequencies, a vector network analyzer can be used. At millimeter-wave frequencies, the phase measurement is difficult and some frequency variation may be necessary (see Section 2.7.4), and then a frequency difference measurement with a moderate accuracy is required as well.

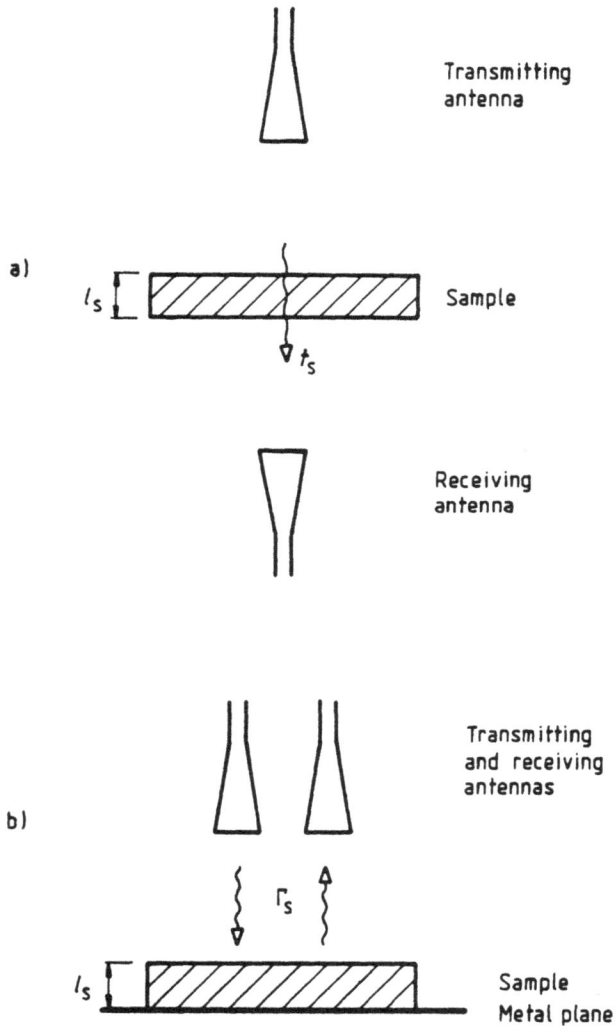

Figure 2.35 Free-space permittivity measurement methods: (a) transmission through the sample; (b) reflection from a sample on a metal plate ($\Gamma_B = -1$ in (2.79)).

2.7.3 Measurement of Different Materials

In this section we investigate practical arrangements for the measurement of the permittivity of various materials as a function of different parameters.

Different States of Materials

With gases the measurement chamber or the resonator should be tight with a sufficient circulation. Also, a reliable and continuous reference measurement method should be available.

For the measurement of liquid solutions, the sample shape can be chosen quite freely. However, the capillary effect of uncovered liquid surfaces can make the surfaces of small cylindrical samples curved. If the liquid sample is an emulsion of two insoluble liquids (for example, oil and water) or contains gases or solid particles, it is usually unstable. Therefore, the sample must be mixed continuously, which is usually possible only outside the measurement applicator. This means that the liquid has to be circulated in a pipe through the applicator and the sample is usually cylindrical.

For homogeneous solid materials, the shape of the sample is important. In some cases, the shape cannot be changed and the measurement method must be chosen accordingly (e.g., resonator perturbation, untuned cavity). Most of the time, however, the sample is machined to fit into the applicator. The shape of the sample thus should be chosen to minimize the preparation time of the samples. A tight fit of the sample is often necessary. To achieve this (except for coaxial lines), the transverse dimensions of the sample can be made slightly oversized, and it can be inserted into the applicator by cooling [Sucher *et al.*, 1963, p. 499]. If there is anisotropy in the material (i.e., fibers, laminations), the sample should be placed into the applicator so that the direction of these is either tangential or perpendicular to the electric field.

For the measurement of granular solid materials, the sample should be large enough to contain at least, 100 granules, for example. Otherwise, the number of granules in different samples causes additional deviation of the results. Also, granular materials (and other semisolid materials) may have many densities. Therefore, we should ensure that the density is as constant as possible in the whole sample. This may be difficult to achieve when the shape of the sample is complicated, as, for example, in a totally filled re-entrant cavity (Figure 2.31).

Control of the Measurement Parameters

The temperature of the sample must be changed in many permittivity measurements. However, the need for measurements at different temperatures is often limited, and the information on the temperature behavior of the material, for example, at about five moisture levels is sufficient. Usually, heaters or thermometers cannot be used during the microwave measurement due to their interference. However, heating and temperature measurement with infrared waves is possible with open applicators. Simultaneous microwave heating and permittivity measurement has also been performed [Jow *et al.*, 1987; Thiebaut *et al.*, 1988]. In normal conduction heating, the sample is placed in contact with some of the applicator metal walls, and the entire applicator near the sample is heated or cooled. Therefore, the applicator should endure the temperature (cables, connectors) and its temperature behavior should be known. The heating and cooling can be performed, for example, by using a jacket on the applicator for water circulation [Kent, 1983] or placing the whole applicator in a test chamber. The sample can be inserted in a gas inlet to control only the sample temperature [Rzepecka, 1973].

The moisture of the material is another commonly varied parameter for permittivity measurements. When preparing moist samples, we should remember that the drying of organic materials may change their dielectric behavior due to changes in the ability to bind water molecules (see Figure 2.24b). Therefore, if water is added to control the moisture of the sample, it ought to be done before the sample has dried too much.

With small or especially thin samples, drying during measurement may be a problem if a closed sample holder cannot be used. To prevent drying, the system should be planned for fast measurements, and the moisture should be checked both before and after the measurement. A thin plastic film cover can be used if the sample is not very small or the applicator can be placed into a test chamber with controlled humidity. The latter method works only with relative low moisture contents and condensation may be a problem.

Permittivity and Applicator

Table 2.1 presents a short summary of the recommended permittivity measurement ranges for different methods. The methods can in some cases be used beyond these limits, but the best accuracy is approximately reached in the ranges presented.

Table 2.1
The Recommended Permittivity Measurement Ranges with Different Measurement Methods

Method	$\varepsilon_r' - 1$	$\varepsilon_r''/\varepsilon_r'$
Lumped Circuits	$10^{-3} \ldots 100$	$(100 \cdot Q_l)^{-1} \ldots 0.05$
Completely Filled Resonators	$10^{-5} \ldots 100$	$(100 \cdot Q_l)^{-1} \ldots 0.05$
Partially Filled Resonators	$10^{-3} \ldots 100$	$(10 \cdot Q_l)^{-1} \ldots 0.05$
Perturbated Resonators	$1 \ldots 10$	$10^{-3} \ldots 10$
Quasioptical Resonators	$10^{-7} \ldots 100$	$10^{-6} \ldots 10^{-2}$
Transmission Line Measurements	$1 \ldots 100$	$10^{-2} \ldots 10$
Free-Space Transmission	$1 \ldots 100$	$10^{-2} \ldots 10$

2.7.4 Examples

In this section we give some examples of useful permittivity measurement techniques at different frequencies. The first four examples deal mainly with low-frequency (below 3 GHz) methods. The next three examples are for the frequency range from 3 to 20 GHz, and the final two are from the millimeter-wave range.

Measurement in the Shorted End of a Coaxial Line with Frequency Variation

At low frequencies, a wideband measurement can be achieved with a reflection measurement in a shorted end of coaxial line (Figure 2.36a). By using a long sample and frequency variation, the complex permittivity can be solved from the amplitude of the reflection ((2.79) ($\varepsilon_r'' \ll \varepsilon_r'$)) [Tinga *et al.*, 1968]:

$$\varepsilon_r' = \left(\frac{c_0}{2l_s \Delta f_{min}}\right)^2$$

$$\varepsilon_r'' = \frac{\alpha c_0 \sqrt{\varepsilon_r'}}{\pi f_{max}}$$

$$\alpha = \frac{1}{2l_s} \log_e\left(\frac{1 - \Gamma|\Gamma_s|_{max}}{|\Gamma_s|_{max} - \Gamma}\right)$$

$$|\Gamma_s|_{max} = 10^{A_{max}/20}$$

$$\Gamma = \frac{1 - (\varepsilon_r')^{1/2}}{1 + (\varepsilon_r')^{1/2}}$$

Here, l_s is the length of the sample, Δf_{min} is the difference in frequency between two minima on the reflection curve of Figure 2.36(b), f_{max} is the frequency at the maximum between the two minima, and A_{max} is the return loss in decibels at the maximum. With this method, the cut-off frequency of the lowest coaxial waveguide mode should be kept in mind:

$$f_c = \frac{c_0}{\sqrt{\varepsilon_r'}\,\pi(a + b)}$$

Here, a and b are the radii of the inner and outer conductors of the coaxial line. In Figure 2.37, a result is presented for water measured by this method. The sample length in the measurement was 0.9 m, and the coaxial line radii $a = 7.5$ mm and $b = 19$ mm. The cut-off frequency for water in this coaxial line is about 400 MHz.

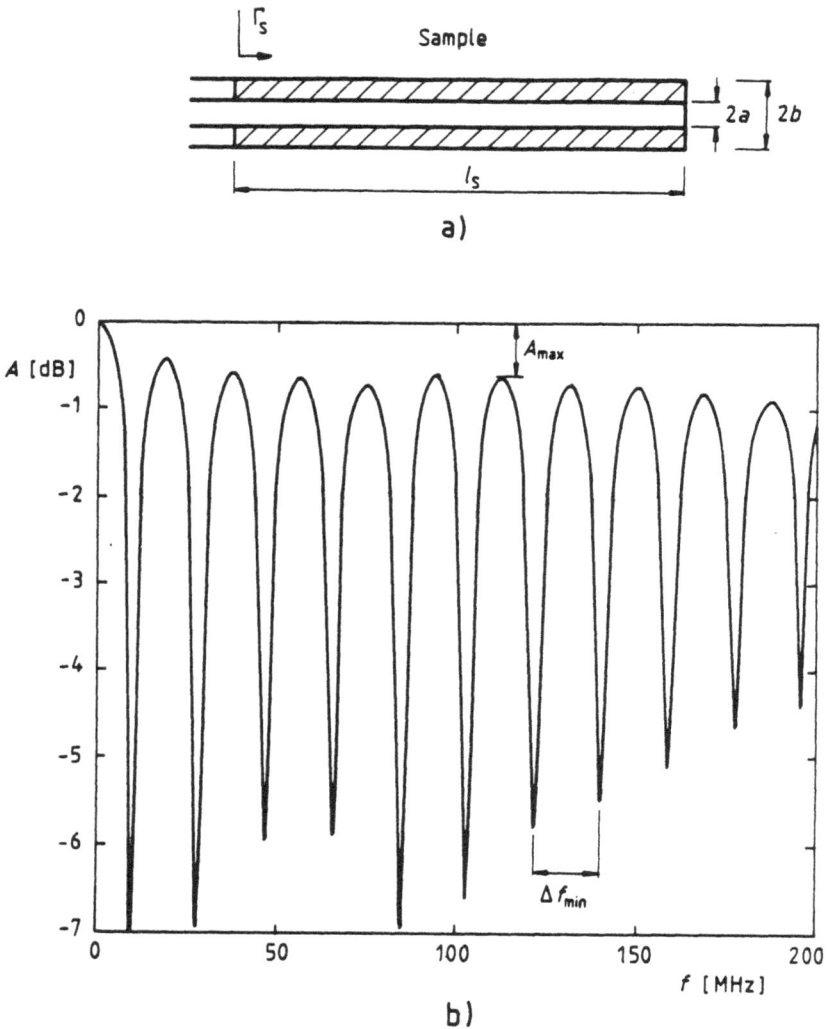

Figure 2.36 Measurement in the shorted end of a coaxial line using frequency variation: (a) the measurement configuration; (b) an example of the reflection loss pattern for tap water ($\sigma \approx 17 \cdot 10^{-3}$ S/m, $l_s = 0.9$ m).

The effect of the waveguide mode can be noticed from the increased inaccuracy of the ε_r'' results above 400 MHz. The reflection curve of Figure 2.36(b) was measured with a scalar network analyzer connected to the 50 Ω coaxial applicator via a tapered transition.

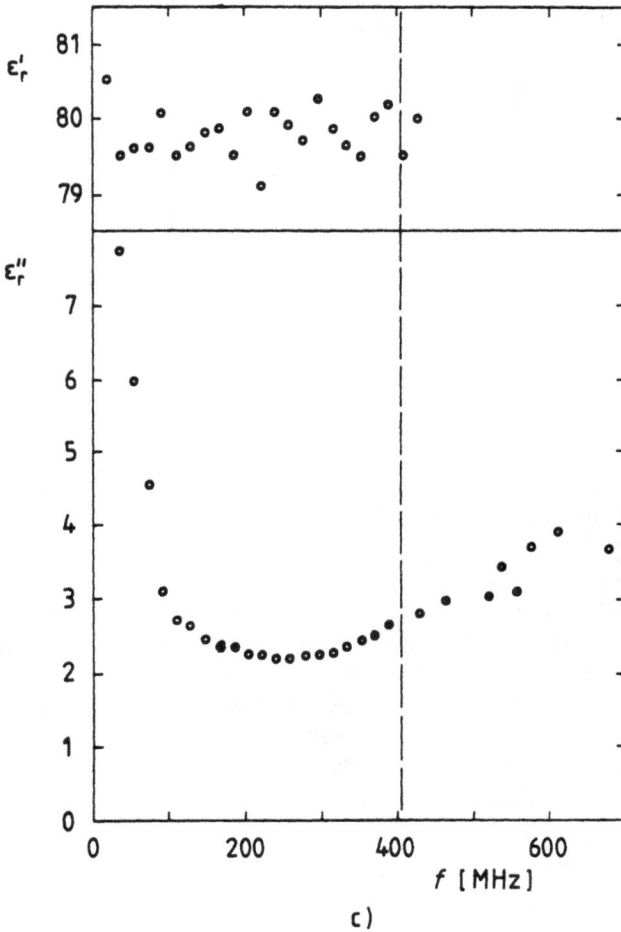

Figure 2.37 The results for the ε_r of tap water with the measurement system of Figure 2.36.

Sample at the End of the Inner Conductor of a Coaxial Line

The measurement configuration of Figure 2.38 is suitable for permittivity measurements at frequencies from 0.1 to 5 GHz [Stuchly *et al.*, 1980b; Stuchly *et al.*, 1987]. For the best accuracy, there is an optimum value C_{opt} for the empty-end parallel-plate capacitance C_0 (see (2.81)):

$$C_{opt} = \frac{1}{\omega Z_0} \cdot \frac{1}{(\varepsilon_r'^2 + \varepsilon_r''^2)^{1/2}}$$

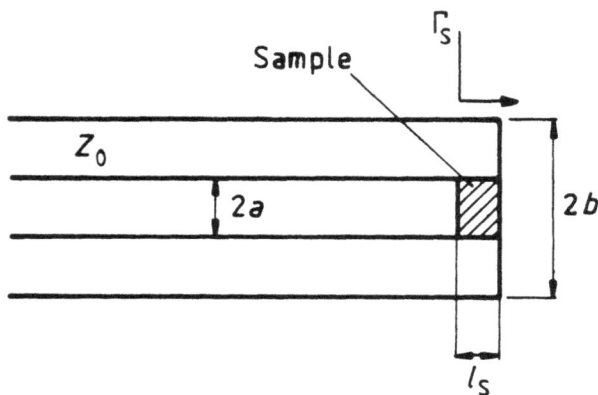

Figure 2.38 The sample at the end of the inner conductor of a coaxial line.

This condition is not critical, but fairly good accuracy is achieved at frequencies one decade downward and upward from that at which C_{opt} is selected. The permittivity can deviate significantly from the optimum value. The main problem with this method arises at higher frequencies, where C_0 becomes small, and changes in the fringe capacitance C_f due to the sample ε_r produce errors in the results. For this reason, the value of C_0 should be over 0.1 pF, this condition thus restricting the measurement frequency and permittivity range. In addition, at the lowest frequencies, the value of C_{opt} can become so high that the sample length becomes too small. The complex reflection coefficient Γ_s in (2.79) can be measured by a vector network analyzer or corresponding equipment. Connection from the applicator (usually $Z_0 = 50 \ \Omega$) to the meter can be made with a tapered transition.

Re-entrant Coaxial Resonator

At frequencies below 1 GHz, the re-entrant resonator (Figure 2.31) is a simple and quite accurate method for permittivity measurements. For gases, liquids, and granular solids, the resonator can be used completely filled. In this situation, ε_r is (see Section 3.2):

$$\varepsilon_r' = \left(\frac{f_{r0}}{f_r}\right)^2$$

$$\frac{\varepsilon_r''}{\varepsilon_r'} = \frac{1}{Q_d} = \frac{B_{hp}}{f_r}(1 - 10^{A/20}) - \frac{B_{hp0}}{f_{r0}}(1 - 10^{A_0/20})\left(\frac{f_0}{f}\right)^{1/2}$$

(2.84)

Here, Q_d is the dielectric quality factor, f_r and f_{r0} are the resonant frequencies, B_{hp} and B_{hp0} are the half-power bandwidths, and A and A_0 are the peak insertion loss values (in dB) of the power transmission curve of the resonator (see Figure 3.6), with and without the sample, respectively. The coupling is assumed equal at the input and output of the resonator. The coupling can be arranged with shorted loops at the shorted end of the resonator (as in Figure 2.31), or with open probes at the open end of the resonator (see Section 3.3.1). The coupling elements should be placed at the opposite sides of the resonator to prevent direct coupling between the elements so that the dynamic range would be as large as possible. The unloaded quality factor of the resonator (brass) in Figure 2.31 is about 1600, which is a typical value.

The measurement frequency for this method depends on the material. However, this is usually not a great disadvantage because the frequency dependence of ε_r is not often very steep. A frequency correction to the results is possible by making additional measurements at the $3\lambda/4$ mode of the resonator.

In (2.84) the condition $\varepsilon_r'' \ll \varepsilon_r'$ was assumed. This is actually necessary because the quality factor of the filled resonator is less than $\varepsilon_r'/\varepsilon_r''$. With high-loss materials or when a small sample is preferred, the resonator can be partly filled. The sample is placed between the center conductor end and top cover (between the dashed lines in Figure 2.31).

The resonance frequency is given below for an empty transmission line section terminated with a short at one end and a reflection coefficient Γ_s at the other [Stuchly *et al.*, 1978]:

$$f_r = \frac{c_0}{2\pi l} \left\{ \left[n\pi - \left(\frac{\pi + \theta_s}{2} \right) \right]^2 - f_c^2 \right\}^{1/2} \qquad (2.85)$$

Here, θ_s is the phase angle of Γ_s, which can be approximated with (2.80) and (2.81), l is the length of the empty transmission line section (the length of the center conductor), f_c is the cut-off frequency of the mode used in the measurement (for TEM lines, $f_c = 0$), and n is an integer.

This equation can be used with a partly filled re-entrant cavity in the following way. Determine first the fringe capacitance C_f of (2.81) without the sample from (2.85) by iteration (the value for C_f in (2.81) is too imprecise). After inserting the sample, calculate θ_s by complex iteration given the assumption that C_f is independent of ε_r of the sample. The quality factor of the resonator depends on the magnitude $|\Gamma_s|$ of the reflection coefficient (see also (3.19)):

$$\log_e|\Gamma_s| = -n\pi \left\{ \frac{1}{\left[1 - \left(\frac{f_c}{f_r} \right)^2 \right] Q_u} - \frac{1}{\left[1 - \left(\frac{f_c}{f_{r0}} \right)^2 \right] Q_{u0}} \right\} \qquad (2.86)$$

Here, f_r and f_{r0} are the resonance frequencies, and Q_u and Q_{u0} are the unloaded quality factors, with and without the sample. The permittivity can now be solved from Γ_s by using (2.82).

Helical Resonator

Another useful low-frequency device is the helical resonator in Figure 2.39(a) [Meyer, 1981]. The resonator is partially filled with the cylindrical sample inserted inside the helical center conductor, where a quite constant axial field exists (Figure 2.39b). The connection between the sample ε_r and resonator parameters is quite complicated. In Figure 2.39(c) [Meyer, 1981], the resonance frequency shift has been presented for the first three resonance modes of the resonator of Figure 2.39(a). It supports about 10 harmonic resonances up to the cut-off frequency of the higher-order modes of the line. The ε_r'' of the sample is ($\varepsilon_r'' \ll \varepsilon_r'$):

$$\varepsilon_r'' = \frac{\varepsilon_r' - 1}{2} \cdot \frac{f_r}{f_{r0} - f_r} \cdot \frac{K}{Q_d} \tag{2.87}$$

where the factor K can be approximated as

$$K \approx 1 + 2 \cdot \frac{f_{r0} - f_r}{f_r} \tag{2.88}$$

The dielectric quality factor Q_d can be measured according to (2.84). The most useful characteristics of this resonator are the small sample size compared to wavelength at the fundamental resonance frequency and the possibility of using 10 resonance modes for wideband measurements. The power coupling is similar to the re-entrant resonator. The unloaded quality factor of the resonator is typically 1000–2000 (or higher for the higher-order modes) [Meyer, 1981].

Sample in a Waveguide Matched Load

For a transmission line reflection measurement, a liquid or powder sample can be inserted in a standard waveguide matched load, which consists of a shorted waveguide section with a lossy plastics wedge (Figure 2.40) [Stuchly, 1970]. In this situation, the returning signal Γ_B from the back end of the sample in (2.79) is zero, resulting in a sample reflection coefficient Γ_s equal to the front surface (as if the sample were very long or lossy) reflection Γ. This gives for the sample permittivity:

$$\varepsilon_r = \left(\frac{1 + \Gamma_s}{1 - \Gamma_s}\right)^2 \left[1 - \left(\frac{f_c}{f}\right)^2\right] + \left(\frac{f_c}{f}\right)^2 \tag{2.89}$$

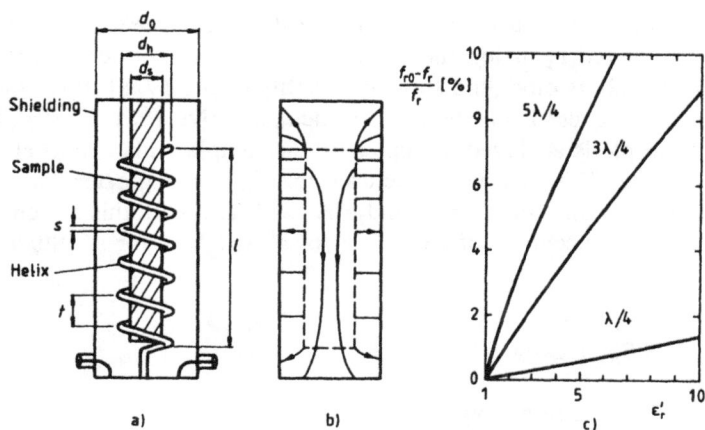

Figure 2.39 The helical resonator: (a) configuration for a fundamental frequency of 200 MHz, where $d_0 = 20$ mm, $d_h = 10$ mm, $d_s = 6$ mm, $l = 50$ mm, $s = 1.5$ mm, $p = 3$ mm; (b) electric field pattern; (c) the frequency shift caused by the sample at the three lowest resonance modes for the 200 MHz resonator. (After [Meyer, 1981].)

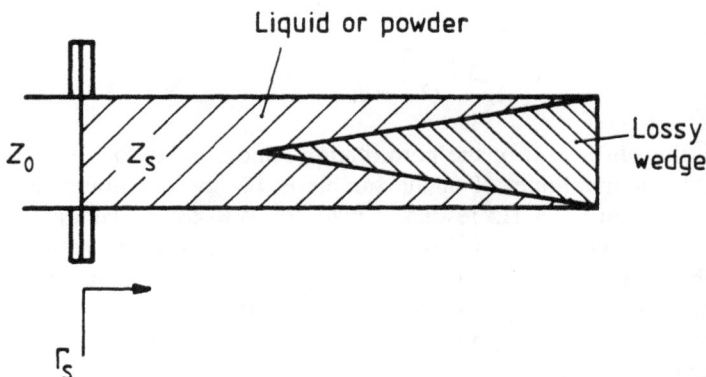

Figure 2.40 The sample in a waveguide matched load. The reflection coefficient Γ_s can be measured either with a reflection measurement (low or medium $|\varepsilon_r|$) or by using the load to terminate a resonator (high $|\varepsilon_r|$).

This method is simple but rather sensitive to measurement errors of the phase of Γ_s. The phase error can be caused, for example, by the capillary effect in liquid samples, making the front surface of the sample curved. Therefore, a thin dielectric film should be used at the surface.

In the case of high ε'_r and high losses, best measurement accuracy for this method can be achieved by using the "infinite" sample as the termination of one end of a resonator with its other end shorted [Stuchly *et al.*, 1978]. Equations (2.85) and (2.86) can then be used (no iteration needed) to solve Γ_s in (2.89). The calibration without sample is achieved by replacing the sample with a short at the plane of the sample surface. The method can be used for $|\Gamma_s|$ values greater than about 0.9 ($Q_u > 30$); that is, typically for $\varepsilon'_r \geq 40$ and $\varepsilon''_r/\varepsilon'_r \geq 0.25$. With this system, Stuchly and others achieved an accuracy of about 5% for ε'_r and 10% for ε''_r [Stuchly *et al.*, 1978].

Perturbation of a TE$_{10n}$ Cavity

As an example of the perturbation method, we have the case of a thin cylindrical sample in a TE$_{10n}$-mode rectangular cavity. The sample is placed in the middle of the cavity tangential to the electric field (Figure 2.41). In this case, the permittivity is

$$\varepsilon'_r = 1 + 2 \left(\frac{\Delta f_r}{f_r} \right) \cdot \frac{1}{S} = 1 + 2 \left(\frac{\Delta f_r}{f_r} \right) \frac{la}{\pi d_s^2}$$

$$\varepsilon''_r = \Delta \left(\frac{1}{Q_l} \right) \frac{la}{\pi d_s^2}$$

(2.90)

Now, the fulfillment of the perturbation condition ($\Delta f_r/f_r$) ≤ 0.001 can be investigated. If the maximum ε'_r of the sample is 10, the maximum sample diameter d_s at TE$_{101}$ mode in a 3 GHz resonator made of WR-284 waveguide ($a = 72$ mm, $b = 34$ mm, $l = 69.5$ mm) is about 0.6 mm and the problem of the small sample size with the perturbation method is clearly seen. The problem can be avoided by using a long resonator at higher-order modes [Rzepecka, 1973]. For example, at the TE$_{109}$ mode, the resonator is nine times longer ($l = 625.5$ mm) and the maximum sample diameter is 1.8 mm.

Transmission Measurement in a Coaxial Line

This is an example of a very wideband permittivity measurement performed in a coaxial line [Kent *et al.*, 1983]. The measurement was made in the frequency range of 0.2–18 GHz with the same sample in a coaxial line with $2b = 7$ mm and $2a = 3.04$ mm. A transmission measurement according to (2.76) through (2.78) was performed with an automatic vector network analyzer. The sample temperature was controlled with a covering jacket, through which water was circulated. The ε'_r of the measured material was in the range of 1.4–2.7 and the ε''_r was in the range of 0.03–0.3. Below 2 GHz, the accuracy of the measurement of ε''_r was only moderate, but

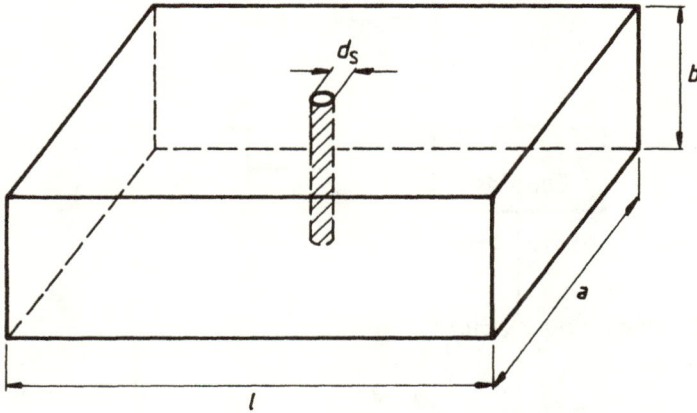

Figure 2.41 The cavity perturbation method for permittivity measurements. The cylindrical sample in the middle of the rectangular cavity working at TE_{10n} mode.

at the higher frequencies the estimated accuracy was $\pm 3\%$ for ε_r' and $\pm 5\%$ for ε_r''.

Free-Space Transmission at 37 GHz

Free-space transmission permittivity measurements at millimeter-wave frequencies (at 37 GHz) have been realized by Hallikainen and others ([Hallikainen *et al.*, 1986]. The measurement equipment is presented in Figure 2.42. The sample diameter in the measurements was 30 cm and the antenna separation was 37 cm. A frequency sweep of 1 GHz was used to eliminate the effect of multiple reflections inside the sample in the attenuation measurement. The average attenuation of the sample was measured between 37 and 38 GHz. This average attenuation is now the absolute value of the numerator of (2.76).

To obtain the $Re\{\sqrt{\varepsilon_r}\}$, they observed the change caused by the sample to the interference pattern occurring at the 1 GHz frequency range. Due to the sample, a maximum in the pattern was shifted from the frequency f_1 to f_2, and the $Re\{\sqrt{\varepsilon_r}\}$ is

$$Re\{\sqrt{\varepsilon_r}\} = 1 + \frac{\lambda_{02}}{l_s}\left[l_r\left(\frac{1}{\lambda_{g2}} - \frac{1}{\lambda_{g1}}\right) - l_t\left(\frac{1}{\lambda_{02}} - \frac{1}{\lambda_{01}}\right)\right]$$

$$- \frac{s}{l_s}[Re\{\sqrt{\varepsilon_{rsh}}\} - 1] \tag{2.91}$$

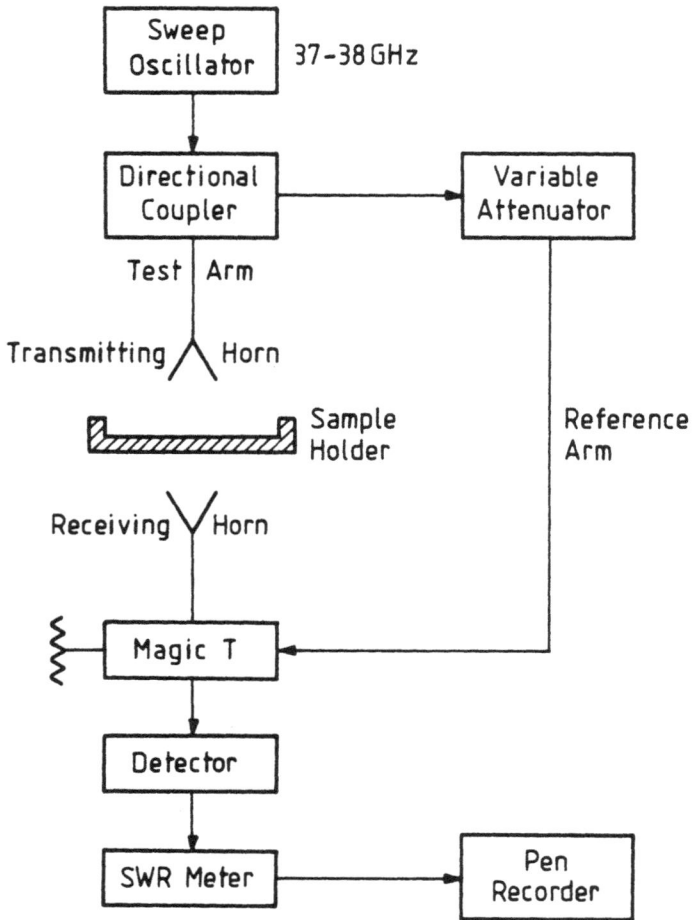

Figure 2.42 A free-space permittivity measurement system for 37 GHz. (After [Hallikainen *et al.*, 1986].)

Here, the subscripts 0 and g refer to the free-space and guide wavelengths, and the subscripts 1 and 2 denote the measurement frequencies. The sample length was l_s, the test arm length was l_t, and the reference arm length was l_r. The effect of the sample holder, (thickness s and permittivity ε_{rsh}) was also taken into account. The permittivity was calculated with an iterative process from the attenuation and the results for $\text{Re}\{\sqrt{\varepsilon_r}\}$. The accuracy of the method was tested with metacrylate ($\varepsilon_r = 2.57 - j0.0082$) and distilled water ($\varepsilon_r = 19.6 - j29.5$). For methacrylate, the error in the measured ε_r' value was only 1.5%, but 25% in the ε_r'' due to the attenuation being too low (about 1 dB) in the sample. For water, both the ε_r' and ε_r'' measurement errors were small, 1.0%.

Comparison of Cavity and Quasioptical Resonator Permittivity Measurements at 35 GHz

In this example, the results obtained for low-loss materials with two millimeter-wave range permittivity measurement systems are compared [Cook *et al.*, 1974]. The systems are a large cylindrical cavity and a quasioptical resonator. The cavity (diameter 50 mm) was constructed of the socalled *helical waveguide* [Barlow, 1960]. With this construction, the number of different resonant modes decreases drastically from that for a normal cylindrical resonator. The cavity was used on the $TE_{01,23}$ mode, which had a quality factor of about 200 000. The low-loss samples were placed at the other end of the cavity.

The open resonator (see Figure 2.33) was constructed of one spherical concave and one flat mirror, with diameters $D_c = 175$ mm and $D_f = 80$ mm, respectively. The separation l of the mirrors was 130 mm and the radius of curvature R of the concave mirror was 141.6 mm. The sample diameter was 70 mm and length l_s was 6–14 mm. The quality factor of the empty resonator was about 170 000.

Comparisons were made of results for PTFE ($\varepsilon'_r = 1.95$, $\varepsilon''_r \cong 10^{-4}$) and polyethylene ($\varepsilon'_r = 2.36$, $\varepsilon''_r \cong 4 \cdot 10^{-4}$) samples. The deviations in the results and estimated uncertainties were very small and quite similar for both methods; for ε'_r, ± 1–$5 \cdot 10^{-3}$ and for ε''_r, ± 1.3–$3.3 \cdot 10^{-6}$. The largest errors were estimated to arise from the uncertainty of the sample length, which should be less than ± 25 μm.

This example shows that very precise permittivity measurements are possible at the millimeter-wave range with both oversized cavities and quasioptical mirror resonators.

REFERENCES

Agdur, B., and B. Enander, "Resonances of a Microwave Cavity Partially Filled with a Plasma," *J. Appl. Phys.*, Vol. 33, No. 2, 1962, pp. 575–581.

Altman, J.L., *Microwave Circuits*, Princeton, NJ: D. Van Nostrand, 1964, 462 p.

Amato, J.C., "Design of Partial Dielectric Cavities for Wide-band Microwave Measurements," *Rev. Sci. Instrum.*, Vol. 51, No. 9, 1980, pp. 1231–1233.

Backhouse, P., N. Apsley, and A. Smart, "Measurement of Dielectric Loss in Thin Parallel-Sided Samples Using an Untuned Cavity," *IEE Proc.*, Vol. 132, Pt.A., No. 5, September 1985, pp. 280–284.

Barlow, H.E.M., "The Use of a Wire-Wound Helix to Form a Circular H_{01} Wavemeter Cavity," *IEE Proc.*, Vol. 107, Pt. B, No. 1, January 1960, p. 66.

Berteaud, A.J., F. Hoffman, and J.F. Mayalut, "Complex Frequency Perturbation of a Microwave Cavity Containing a Lossy Liquid," *J. Microwave Power*, Vol. 10, No. 3, September 1975, pp. 301–313.

Birchak, J.R., L.G. Gardner, J.W. Hipp, and J.M. Victor, "High Dielectric Constant Microwave Probes for Sensing Soil Moisture," *Proc. IEEE*, Vol. 62, No. 1, January 1974, pp. 93–98.

Chaloupka, H., O. Ostwald, and B. Schiek, "Structure Independent Microwave Moisture-Measurements," *J. Microwave Power*, Vol. 15, No. 4, December 1980, pp. 221–231.

Chan, W.F.P., and B. Chambers, "Measurement of Nonplanar Dielectric Samples Using an Open Resonator," *IEEE Trans. Microwave Theory Tech.*, Vol. MTT-35, No. 12, December 1987, pp. 1429–1434.

Chao, S.H., "Measurements of Microwave Conductivity and Dielectric Constant by the Cavity Perturbation Method and their Errors," *IEEE Trans. Microwave Theory Tech.*, Vol. MTT-33, No. 6, June 1985, pp. 519–526.

Cook, R.J., R.G. Jones, and C.B. Rosenberg, "Comparison of Cavity and Open Resonator Measurements of Permittivity and Loss Angle at 35 GHz," *IEEE Trans. Instr. Meas.*, Vol. IM-23, No. 4, December 1974, pp. 438–442.

Cullen, A.L., and P.K. Yu, "The Accurate Measurement of Permittivity by Means of an Open Resonator," *Proc. Roy. Soc. Lond.*, Vol. A325, 1971, pp. 493–509.

Decreton, M.C., and F.E. Gardiol, "Simple Nondestructive Method for the Measurement of Complex Permittivity," *IEEE Trans. Instr. Meas.*, Vol. IM-23, No. 4, December 1974, pp. 434–438.

Decreton, M.C., and M.S. Ramachandraiah, "Nondestructive Measurement of Complex Permittivity for Dielectric Slabs," *IEEE Trans. Microwave Theory Tech.*, Vol. MTT-23, No. 12, December 1975, pp. 1077–1080.

Garg, S.K., H. Kilp, and C.P. Smyth, "Microwave Absorption and Molecular Structure in Liquids. LXV. A Precise Michelson Interferometer for Millimeter Wavelengths and the Dielectric Constants and Losses of some Low-Loss Liquids at 2.1 mm," *J. Chem. Phys.*, Vol. 43, No. 7, 1965, pp. 2341–2346.

Gudmandsen, P., G.H. Jakobsen, "Radio Echo Sounding of the Greenland Inland Ice," *Europhysics News*, Vol. 7, No. 5, May 1976, pp. 2–4.

Hallikainen, M.T., F.T. Ulaby, and M. Abdelrazik, "Dielectric Properties of Snow in the 3 to 37 GHz Range," *IEEE Trans. Antennas Propag.*, Vol. AP-34, No. 11, November 1986, pp. 1329–1340.

Hasegawa, S., and D.P. Stokesberry, "Automatic Digital Microwave Hygrometer," *Rev. Sci. Instrum.*, Vol. 46, No. 7, July 1975, pp. 867–873.

Hasted, J.B., *Aqueous Dielectrics,* London: Chapman and Hall, 1973, 302 p.

von Hippel, A.R., *Dielectric Materials and Applications,* Cambridge, MA: MIT Press, 1954, 438 p.

Izatt, J.R., and F. Kremer, "Millimeter Wave Measurement of Both Parts of the Complex Index of Refraction Using an Untuned Cavity," *Applied Optics,* Vol. 20, No. 14, July 1981, pp. 2555–2559.

Jackson, J.D., *Classical Electrodynamics,* New York: John Wiley and Sons, 2nd Ed., 1975, 848 p.

Jow, J., M.C. Hawley, M. Finzel, J. Asmussen, H.H. Lin, and B. Manring, "Microwave Processing and Diagnosis of Chemically Reacting Materials in a Single-Mode Cavity Applicator," *IEEE Trans. Microwave Theory Tech.*, Vol. MTT-35, No. 12, December 1987, pp. 1435–1443.

Khalid, K.B., T.S.M. Maclean, M. Razaz, and P.W. Webb, "Analysis and Optimal Design of Microstrip Sensors," *IEE Proc.*, Vol. 135, Pt. H, No. 3, June 1988, pp. 187–195.

Kent, M., "Complex Permittivity of Fish Meal: A General Discussion of Temperature, Density and Moisture Dependence," *J. Microwave Power,* Vol. 12, No. 4, December 1977, pp. 341–345.

Kent, M., and E. Kress-Rogers, "Microwave Moisture and Density Measurement in Particulate Solids," *Trans. Inst. Meas. Control,* Vol. 8, No. 3, July-September 1986, pp. 161–168.

Kent, M., and W. Meyer, "A Density-Independent Microwave Moisture Meter for Heterogeneous Foodstuffs," *J. Food Engineering,* No. 1, 1982, pp. 31–42.

Kent, M., and W. Meyer, "Dielectric Relaxation of Adsorbed Water in Microcrystalline Cellulose," *J. Phys. D: Appl. Phys.*, Vol. 16, 1983, pp. 915–925.

Klein, A., "Microwave Determination of Moisture Compared with Capacitive, Infrared and Conductive Measurement Methods. Comparison of On-Line Measurements at Coal Preparation Plants," *Proc. 14th European Microwave Conf.*, Liége, 1984, pp. 661–666.

Kohler, W.E., and G.C. Papanicolaou, "Some Applications of the Coherent Potential Approximation," in *Multiple Scattering and Waves in Random Media,* P.L. Chow, W.E. Kohler, and G.C. Papanicolaou, eds., New York: North-Holland, 1981, pp. 199–223.

Ligthart, L.P., "A Fast Computational Technique for Accurate Permittivity Determination Using Transmission Line Methods," *IEEE Trans. Microwave Theory Tech.*, Vol. MTT-31, No. 3, March 1983, pp. 249–254.

de Loor, G.P., "Dielectric Properties of Heterogeneous Mixtures," Thesis, Leiden, 1956, 93 p.

Looyenga, H., "Dielectric Constants of Mixtures," *Physica*, Vol. 31, 1965, pp. 401–406.

Lynch, A.C., "Measurement of Permittivity Using an Open Resonator," *IEE Proc.*, Vol. 130, Pt. A, No. 7, 1983, pp. 365–368.

Meyer, W., "Helical Resonators for Measuring Dielectric Properties of Materials," *IEEE Trans. Microwave Theory Tech.*, Vol. MTT-29, No. 3, March 1981, pp. 240–247.

Meyer, W., and W.M. Schilz, "Feasibility Study of Density-Independent Moisture Measurement with Microwaves," *IEEE Trans. Microwave Theory Tech.*, Vol. MTT-29, No. 7, July 1981, pp. 732–739.

Meyer, W., and W.M. Schilz, "High Frequency Dielectric Data on Selected Moist Materials," *J. Microwave Power*, Vol. 17, No. 1, January 1982, pp. 67–81.

Meyer, W., and W. Schilz, "Microwave Absorption by Water in Organic Materials," *IEE Conf. Pub. No. 177*, 1979, pp. 215–219.

Miller, G.B., and A. Moore, "Dielectric Parameters of Low Loss Liquids at Microwave Frequencies," *J. Microwave Power*, Vol. 3, No. 3, September 1968, pp. 104–113.

Mätzler, C., and U. Wegmüller, "Dielectric Properties of Fresh-Water Ice at Microwave Frequencies," *J. Phys. D: Appl. Phys.*, Vol. 20, 1987, pp. 1623–1630.

Nelson, S.O., "Electrical Properties of Agricultural Products (A Critical Review)," *Trans. American Society of Agricultural Engineers*, Vol. 16, No. 2, 1973, pp. 1–20.

Nelson, S.O., C.W. Schlaphoff, and L.E. Stetson, "A Computer Program for Short-Circuited Waveguide Dielectric-Properties Measurements on High- and Low-Loss Materials," *J. Microwave Power*, Vol. 8, No. 1, March 1973, pp. 13–22.

Nyfors, E., "On the Dielectric Properties of Dry Snow in the 800 MHz to 13 GHz Range," Helsinki University of Technology, Radio Laboratory, Report S 135, 1982, 17 p.

Polder, D., and J.H. van Santen, "The Effective Permeability of Mixtures of Solids," *Physica*, Vol. XII, No. 5, 1946, pp. 257–271.

Rose, G.C., R.J. Churchill, and K.R. Cook, "Determination of Complex Dielectric Constant of High-Loss Materials," *IEEE Trans. Instr. Meas.*, Vol. IM-21, No. 3, August 1972, pp. 286–287.

Rueggeberg, W., "Determination of Complex Permittivity of Arbitrarily Dimensioned Dielectric Modules at Microwave Frequencies," *IEEE Trans. Microwave Theory Tech.*, Vol. MTT-19, No. 6, June 1971, pp. 517–521.

Rzepecka, M.A., "A Cavity Perturbation Method for Routine permittivity Measurement," *J. Microwave Power*, Vol. 8, No. 1, March 1973, pp. 3–12.

Sihvola, A., and J.A. Kong, "Effective Permittivity of Dielectric Mixtures," *IEEE Trans. Geosci. Remote Sensing*, Vol. GE-26, No. 4, July 1988a, pp. 420–429. (*See also* corrections, Vol. GE-27, No. 1, January 1989, pp. 101–102.)

Sihvola, A., and I.V. Lindell, "Polarizability and Effective Permittivity of Layered and Continuously Inhomogeneous Dielectric Spheres," *J. Electromagnetic Waves and Applications*, Vol. 2, No. 8, 1988b, pp. 741–756.

Sihvola, A., and I.V. Lindell, "Transmission Line Analogy for Calculating the Effective Permittivity of Mixtures with Spherical Multilayer Scatterers," *J. Electromagnetic Waves and Applications*, Vol. 3, No. 1, 1989, pp. 37–60.

Sihvola, A., E. Nyfors, and M. Tiuri, "Mixing Formulae and Experimental Results for the Dielectric Constant of Snow," *J. Glaciology*, Vol. 31, No. 108, 1985, pp. 163–170.

Sihvola, A., and M. Tiuri, "Snow Fork for Field Determination of the Density and Wetness Profiles of a Snow Pack," *IEEE Trans. Geosci. Remote Sensing*, Vol. GE-24, No. 5, September 1986, pp. 717–721.

Stiles, W.H., and F.T. Ulaby, "Dielectric Properties of Snow," The University of Kansas Center for Research, Remote Sensing Laboratory, RSL Technical Report 527-1, 1981, 35 p.

Stuchly, M.A., and S.S. Stuchly, "Dielectric Properties of Biological Substances—Tabulated," *J. Microwave Power*, Vol. 15, No. 1, January 1980a, pp. 19–25.

Stuchly, M.A., and S.S. Stuchly, "Coaxial Line Reflection Methods for Measuring Dielectric Properties of Biological Substances at Radio and Microwave Frequencies—A Review," *IEEE Trans. Instr. Meas.*, Vol. IM-29, No. 3, September 1980b, pp. 176–183.

Stuchly, S.S., "Dielectric Properties of Granular Solids Containing Water," *J. Microwave Power*, Vol. 5, No. 2, July 1970, pp. 62–68.

Stuchly, S.S., A. Kraszewski, and M.A. Stuchly, "Uncertainties in Radiofrequency Dielectric Measurements of Biological Substances," *IEEE Trans. Instr. Meas.*, Vol. IM-36, No. 1, March 1987, pp. 67–70.

Stuchly, S.S., M.A. Stuchly, and B. Carraro, "Permittivity Measurements in a Resonator Terminated by an Infinite Sample," *IEEE Trans. Instr. Meas.*, Vol. IM-27, No. 4, December 1978, pp. 436–439.

Sucher, M., and J. Fox, eds., *Handbook of Microwave Measurements*, New York: Polytechnic Press, Polytechnic Institute of Brooklyn, 1963, 1165 p.

Taylor, L., "Dielectric Properties of Mixtures," *IEEE Trans. Antennas Propag.*, Vol. AP-13, No. 6, June 1965, pp. 943–947.

Thiebaut, J.M., C. Akyel, G. Roussy, and R.G. Bosisio, "Dehydration and Dielectric Permittivity Measurements of a Porous, Inorganic Material (13X Zeolite) Heated with Microwave Power," *IEEE Trans. Instr. Meas.*, Vol. 37, No. 1, March 1988, pp. 114–120.

Thompson, M.C., F.E. Freethey, and D.M. Waters, "End Plate Modification of X-Band TE_{011} Cavity Resonators," *IRE Trans. Microwave Theory Tech.*, Vol. MTT-7, 1959, pp. 388–389.

Thrane, L., "A Second Generation Sea Model. Development and Evaluation of the Radiometric Effects," Technical University of Denmark, Electromagnetics Institute, Report D289, Lyngby, 1976, 37 p.

Tinga, W.R., and E.M. Edwards, "Dielectric Measurements Using Swept Frequency Techniques," *J. Microwave Power*, Vol. 3, No. 3, September 1968, pp. 114–125.

Tinga, W.R., and S.O. Nelson, "Dielectric Properties of Materials for Microwave Processing—Tabulated," *J. Microwave Power*, Vol. 8, No. 1, March 1973a, pp. 23–65.

Tinga, W.R., W.A.G. Voss, and D.F. Blossey, "Generalized Approach to Multiphase Dielectric Mixture Theory," *J. Appl. Phys.*, Vol. 44, No. 9, 1973b, pp. 3897–3902.

To, E.C., R.E. Mudgett, D.I.C. Wang, S.A. Goldblith, and R.V. Decareau, "Dielectric Properties of Food Materials," *J. Microwave Power*, Vol. 9, No. 4, December 1974, pp. 303–315.

Toropainen, A.P., P.V. Vainikainen, and E.G. Nyfors, "Microwave Humidity Sensor for Difficult Environmental Conditions," *Proc. 17th European Microwave Conf.*, Rome, September 1987, pp. 887–891.

Trapp, W., *Das Dielektrische Verhalten von Holz und Zellulose im grossen Frequenz- und Temperaturbereich*, Thesis, Technischen Hochschule, Braunschweig, 1954, 85 p.

Tsang, L., J.A. Kong, and R.T. Shin, *Theory of Microwave Remote Sensing*, New York: John Wiley and Sons, 1985, 613 p.

Vainikainen, P.V., E.G. Nyfors, and M.T. Fischer, "Radiowave Sensor for Measuring the Properties of Dielectric Sheets: Application to Veneer Moisture Content and Mass per Unit Area Measurement," *IEEE Trans. Instr. Meas.*, Vol. IM-36, No. 4, December 1987, pp. 1036–1039.

Wang, J.R., and T.J. Schmugge, "An Empirical Model for the Complex Dielectric Permittivity of Soils as a Function of Water Content," *IEEE Trans. Geosci. Remote Sensing*, Vol. GE-18, No. 4, July 1980, pp. 288–295.

Weiss, J.A., and D.A. Hawks, "Dielectric Constant Evaluation of Insulating Materials: An Accurate Practical Measurement Device," *IEEE MTT-S Int. Microwave Symp. Digest*, Las Vegas, NV, June 1987, Vol. 1, pp. 457–460.

Wenger, N.C., "Resonant Frequency of Open-Ended Cylindrical Cavity," *IEEE Trans. Microwave Theory Tech.*, Vol. MTT-15, No. 6, June 1967, pp. 334–340.

Xu, D., L. Liu, and Z. Jiang, "Measurement of the Dielectric Properties of Biological Substances Using an Improved Open-Ended Coaxial-Line Resonator Method," *IEEE Trans. Microwave Theory Tech.*, Vol. MTT-35, No. 12, December 1987, pp. 1424–1428.

Chapter 3
Resonator Sensors

3.1 INTRODUCTION

We will try to give the reader an idea of what resonance is and how a microwave resonator works without delving deeply into electromagnetic theory. In the next section, we will derive the formulas needed for practical calculations, and in Section 3.3 we will study actual sensor structures for different purposes, their design, and use. In Section 3.4 we present different methods for measuring resonances, and in Section 3.5 we study some examples of realized solutions.

3.1.1 The Resonance Phenomenon

Any structure having a *natural frequency of oscillation* is a *resonator*. Mechanical resonators are familiar to everyone. Their natural frequency or *resonant frequency* is determined by the modulus of elasticity, weight, and structure. During oscillation, the deformation of the body causes a stress force, which causes movement, which causes deformation, which causes a stress force, *et cetera*. The energy alternates between kinetic and stress. One example of a mechanical resonator is the tuning fork (Figure 3.1). If it is excited by an impulse (knock), the fork resonator will ring with the natural frequency for some time. The amplitude of oscillation (maximum displacement) decreases exponentially with time because of viscosity in the material and air resistance. The higher is the quality factor of the resonator, the longer the resonator rings.

If the excitation is a continuous sine wave, the amplitude of oscillation depends on the frequency of the excitation. The amplitude will be small for all other frequencies but the natural frequency. This resonance phenomenon can be understood as a build-up of energy when the excitation is in phase with the natural oscillation. When some part of the resonator is moving in a certain direction, the exciting force pushes in the same direction, thus increasing the movement. If the quality factor is

Figure 3.1 The tuning fork is a mechanical resonator. If hit by an object, the fork will ring for some time at a frequency determined by its dimensions and the material.

high, the amplitude can grow to the point where the resonator breaks apart. The classical example is the opera singer and the glass.

Resonance phenomena are also encountered, for example, in acoustics and electricity. The rest of this chapter will be devoted to electromagnetic resonators. We will study their theory and the different ways in which resonators can be used as measurement sensors.

An example of an electromagnetic resonator is the LC circuit in Figure 3.2. It consists of an inductor L and a capacitor C connected in series through a switch K. Let us assume that the capacitor is initially charged to the voltage V, and then contains the energy:

$$E_C = \frac{1}{2} CV^2 \tag{3.1}$$

stored in the electric field between the plates. At the moment when the switch K is closed, a current begins to run through the inductor, causing a magnetic field to build up. When the capacitor is completely discharged, the current reaches a maximum value I, and all of the energy has been transferred from the electric field in the capacitor to the magnetic field in the inductor. The energy now equals

$$E_L = \frac{1}{2} LI^2 \tag{3.2}$$

Because of the law of conservation of energy, that

$$E_L = E_C \tag{3.3}$$

is evident. The current continues to flow because of the magnetic field. The capacitor, a moment later, will therefore become charged with a reversed voltage. Even-

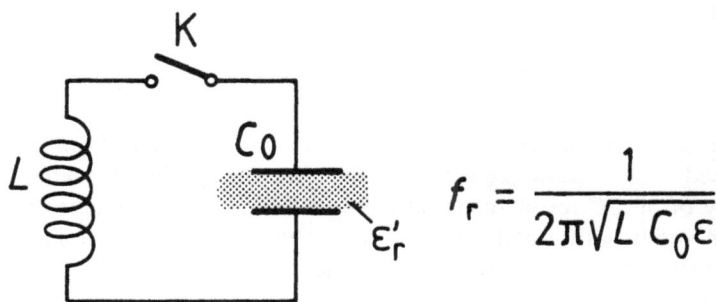

Figure 3.2 The inductor-capacitor circuit is an electromagnetic resonator. The resonant frequency is determined by the inductance and capacitance, which depends on the permittivity of the insulator of the capacitor.

tually, the current will change direction and the oscillation continues. The current oscillates in the resonator until the energy has been dissipated in the resistances of the wires. We can show that the frequency by which the current oscillates (the resonant frequency) is given by

$$f_r = \frac{1}{2\pi\sqrt{LC}} \qquad (3.3)$$

If the capacitor is loaded with a dielectric insulator between the plates, this will change the capacitance and the resonant frequency. The capacitance of a plate capacitor is given by

$$C = \frac{\varepsilon_r'\varepsilon_0 A}{d} = \varepsilon_r' C_0 \qquad (3.4)$$

where A is the area of the plates and d is the distance between them. The resonant frequency is now given by

$$f_r = \frac{1}{2\pi\sqrt{LC_0\varepsilon_r'}} = \frac{f_{r0}}{\sqrt{\varepsilon_r'}} \qquad (3.5)$$

where f_{r0} is the resonant frequency when the capacitor is empty. We see that to use even this resonator to measure ε_r' is possible. If the resonant frequency is first measured with the empty capacitor and then with it filled, the permittivity is given by

$$\varepsilon_r' = \left(\frac{f_{r0}}{f_r}\right)^2 \qquad (3.6)$$

The *LC* circuit is an example of a resonator composed of lumped components. It was used here to show the nature of the electromagnetic resonance. The resonant frequency of an *LC* resonator is, at most, a few hundred megahertz. The plate capacitor in series with a coil inductor can be used in the laboratory for permittivity measurements, but the practical size of the plates limits the resonant frequency to about 100 MHz.

In some cases, *LC* resonators are used in the industry as moisture sensors. The capacitor is made of a planar structure, making contact with the object to be measured. The fringing field penetrates the surface of the object which thus influences the capacitance. Such sensors are usually called *capacitive sensors*.

3.1.2 Microwave Resonators

When we move upward in frequency range, the concepts of voltage and current loops and circuits are replaced by transmission lines and propagating waves. Because the size of the microwave resonators is on the same order of magnitude as the wavelength, they cannot be described by lumped components. They are best described by the following sentence: *A microwave resonator is formed by a section of a transmission line bounded by impedance discontinuities*.

The transmission line can be made of any kind of a structure supporting electromagnetic waves, such as hollow waveguide, coaxial line, slotline, stripline, dielectric waveguide, or two-conductor line (see Sections 1.3.4, 9.1 and 9.2). The role of impedance discontinuities is to cause the propagating wave to be reflected (see Section 1.4.2). Where the transmission line is open-circuited, the reflection coefficient is $\Gamma = +1$, and where the line is short-circuited, $\Gamma = -1$ (Figures 3.3 and 3.4). If the impedance at the discontinuity differs from those mentioned above, the reflection may be partial ($0 \leq |\Gamma| \leq 1$) and the resonator leaky, or the phase angle of Γ may differ from $0°$ ($\Gamma = +1$) or $180°$ ($\Gamma = -1$).

The field in the resonator is excited by the external circuit through some kind of *coupling* (see Section 3.3.1). The coupling can be made for example through an aperture (small hole), coupling loop, or coupling probe. The coupling device radiates a wave into the resonator. The wave propagates along the transmission line and is reflected in alternating directions at the discontinuities.

Resonance occurs if the exciting field is in phase with the reflected components. Hence, they will interfere constructively and destructively to give a standing wave pattern. This will happen only at certain frequencies (resonant frequencies), where the resonance condition is fulfilled. A standing wave with a strong field will

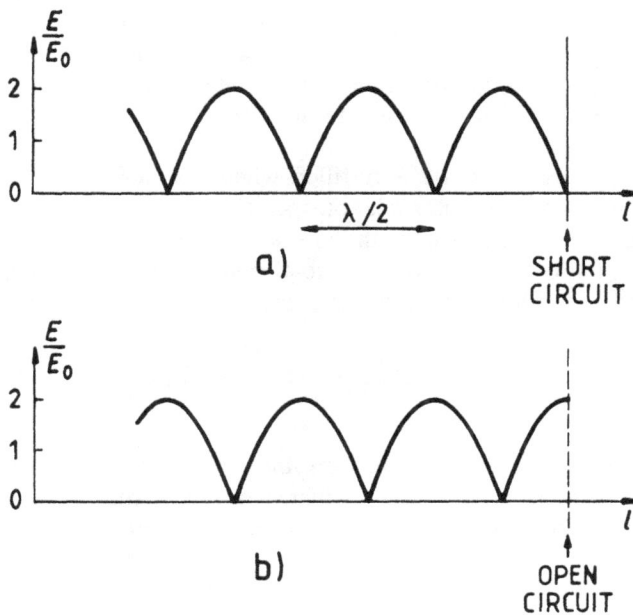

Figure 3.3 A wave of amplitude E_0, propagating in free space or along a transmission line, is reflected from (a) a short circuit, or (b) an open circuit. The wave components, traveling in different directions, combine to form a standing wave with local maxima and zeros.

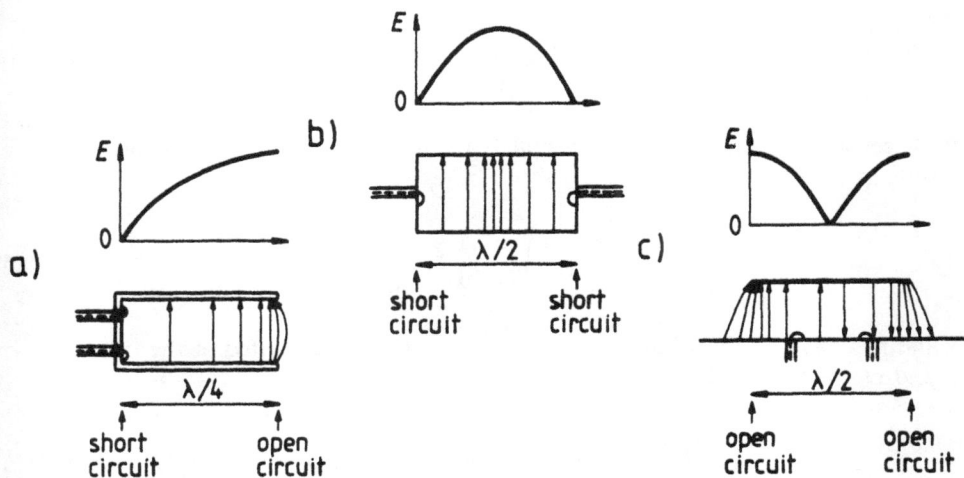

Figure 3.4 Examples of resonators made of (a) two-conductor line, (b) waveguide, and (c) stripline. The electric field lines are indicated. The coupling (input and output for transmission measurement) is made by using coupling loops.

build up, thus storing a great amount of energy. Equilibrium is reached at the level where the loss power in the resonator (in the metal or dielectric, by radiation, or by escaping through the couplings) equals the excitation power. At resonance, the energy alternates between the electric and magnetic field, which also contain the same amount of energy.

The *resonance condition* is fulfilled when the mode wavelength compared to the dimensions of the resonator take on specific values. These values depend on the kind of termination which bounds the resonator. The reflected wave components are in phase with the exciting field if the total phase change experienced by the wave on its way back and forth along the transmission line, is a multiple of 2π:

$$\frac{2\pi}{\lambda} \cdot 2l + \phi_1 + \phi_2 = n \cdot 2\pi \tag{3.7}$$

where l is the length of the transmission line, ϕ_1 and ϕ_2 are the phase angles of the reflection coefficients, and n is an integer. The first term on the left in (3.7) is the phase change of the wave on its way back and forth. Equation (3.7) can be written in the form:

$$l = \left(\frac{n}{2} - \frac{\phi_1 + \phi_2}{4\pi}\right)\lambda \tag{3.8}$$

If the resonator is open-circuited at both ends, $\phi_1 = \phi_2 = 0$, and we obtain for the length:

$$l = \frac{n\lambda}{2} = \frac{1}{2}\lambda, \lambda, 1\frac{1}{2}\lambda, 2\lambda, \ldots \tag{3.9}$$

If the resonator is short-circuited at both ends, $\phi_1 = \phi_2 = \pi$, and we obtain the same results as above:

$$l = \left(\frac{n}{2} - \frac{1}{2}\right)\lambda = \frac{1}{2}\lambda, \lambda, 1\frac{1}{2}\lambda, 2\lambda, \ldots \tag{3.10}$$

If the resonator is short-circuited at one end and open-circuited at the other, $\phi_1 = \pi$ and $\phi_2 = 0$, and we obtain

$$l = \left(\frac{n}{2} - \frac{1}{4}\right)\lambda = \frac{1}{4}\lambda, \frac{3}{4}\lambda, 1\frac{1}{4}\lambda, 1\frac{3}{4}\lambda, \ldots \tag{3.11}$$

For each situation, there is an *infinite series of solutions* satisfying the resonance condition. Therefore, each resonator has an infinite number of resonant frequencies for each wave mode. The lowest resonance is at the frequency for which the wavelength is two or four times the length of the resonator, depending on the terminations.

For exact calculation of the resonant frequency, we need to know the relation between wavelength and frequency. As we have pointed out in Section 1.3, the relation for plane waves and TEM waves is very simple, but it is slightly more complicated for waves in waveguides because the wavelength of a waveguide mode is always longer than for the corresponding plane wave, as is evident from (1.21). For cavity resonators, where both ends are shorted, we have

$$l = \frac{1}{2}\lambda_g, \ \lambda_g, \ 1\frac{1}{2}\lambda_g, \ 2\lambda_g, \ \dots \tag{3.12}$$

where λ_g is the guide wavelength from (1.21).

3.2 THEORY

3.2.1 Resonator Filled with Dielectric

In the previous section, we saw that the resonance condition (3.7) required that the size of resonator measured in wavelengths be constant, for example, $\lambda/4$ or $\lambda/2$. We also know that the wavelength at a specific frequency is shorter in a dielectric material than in vacuum. Therefore, if a resonator (the space where the electric field or the resonator is located) is filled with a dielectric, the resonance condition will be met at a lower frequency than for the hollow resonator. From (1.39), the wavelength is

$$\lambda = \frac{1}{\text{Re}\{f\sqrt{\mu\varepsilon}\}} = \frac{1}{f\sqrt{\mu_0\varepsilon_0}\,\text{Re}\{\sqrt{\varepsilon_r}\}} = \frac{\lambda_0}{\text{Re}\{\sqrt{\varepsilon_r}\}} \tag{3.13}$$

Combining (3.13) and the requirement for the resonance condition that the wavelength be constant, we obtain the change in resonant frequency caused by the dielectric:

$$\frac{1}{f_{r0}\sqrt{\mu_0\varepsilon_0}} = \frac{1}{f_r\sqrt{\mu_0\varepsilon_0}\,\text{Re}\{\sqrt{\varepsilon_r}\}} \tag{3.14}$$

$$\text{Re}\{\sqrt{\varepsilon_r}\} = \frac{f_{r0}}{f_r}$$

where f_{r0} is the resonant frequency of the hollow resonator and f_r that of the filled resonator. For the case of $\varepsilon_r' \gg \varepsilon_r''$, we can approximate (3.14):

$$\varepsilon_r' \approx \left(\frac{f_{r0}}{f_r}\right)^2 \tag{3.15}$$

Equation (3.15) is almost always valid. Because ε_r' is always greater than 1 (except for plasma), the resonant frequency can only become lower when we fill the resonator with a dielectric. As we have shown, obtaining the real part of the permittivity is very simple using a resonator. We need only to measure the resonant frequency of the hollow resonator and that of the filled resonator, and then use (3.15). The value obtained is valid at the frequency f_r.

A resonance has two features, resonant frequency f_r and the quality factor Q. The latter is defined in the following way:

$$Q = \frac{2\pi \times \text{energy stored in the resonator}}{\text{energy dissipated during one cycle}} \tag{3.16}$$
$$= \frac{\omega \times \text{stored energy}}{\text{loss power}}$$

The loss power can be separated into different parts, depending on the source of the loss. Taking the reciprocal of (3.16), we can write the Q-factor as a sum:

$$\frac{1}{Q_u} = \frac{1}{Q_d} + \frac{1}{Q_m} + \frac{1}{Q_{rad}} \tag{3.17}$$

where Q_u is the unloaded Q-factor, Q_d takes into account only the loss in the dielectric, Q_m accounts for only the loss in the metal parts, and Q_{rad} accounts for only the loss through radiation. Q_u is the "real" Q-factor of the resonator, but it is impossible to measure directly. To be able to measure the resonant frequency and quality factor, we must connect the resonator to the measurement circuit through coupling loops, for example. This will "load" the resonator, which means that part of the stored energy escapes through the coupling loops. Consequently, we will measure the loaded Q-factor, Q_l. Therefore, we need to add the reciprocal of the external Q-factor, $1/Q_{ext}$, to (3.17) to obtain Q_l:

$$\frac{1}{Q_l} = \frac{1}{Q_u} + \frac{1}{Q_{ext}} = \frac{1}{Q_d} + \frac{1}{Q_m} + \frac{1}{Q_{rad}} + \frac{1}{Q_{ext}} \tag{3.18}$$

As mentioned above, we can think of a microwave resonator as part of a transmission line bounded by two discontinuities. Expressed in terms of the transmission line parameters (propagation factor $\gamma = \alpha + j\beta$, see Section 1.3.4), Q_u can be shown to be

$$\frac{1}{Q_u} = \frac{1}{\beta l}\left\{2\alpha l - \left[1 - \left(\frac{f_c}{f}\right)^2\right](\log_e|\Gamma_1| + \log_e|\Gamma_2|)\right\} \tag{3.19}$$

where f_c is the cut-off frequency (waveguide), and Γ_1 and Γ_2 are the reflection coefficients at the discontinuities.

We can measure f_r and Q_l in two different ways (see Section 3.4), by the *method of reflection coefficient* or by *insertion loss*. The former method requires only one coupling and the latter requires two. The method of reflection coefficient means that a wave is transmitted along the cable toward the resonator. The reflected power is measured and the ratio between reflected and incident power is the reflection coefficient $|\Gamma|^2$. Based on an equivalent circuit (see Section 3.4.3, Figure 3.27), the amplitude and phase of the reflection coefficient can be approximated in the vicinity of the resonant frequency by the following equations:

$$|\Gamma|^2 = 1 - \frac{4\dfrac{Q_l}{Q_u}\left(1 - \dfrac{Q_l}{Q_u}\right)}{1 + Q_l^2\left(\dfrac{f}{f_r} - \dfrac{f_r}{f}\right)^2} \tag{3.20}$$

$$\phi = \phi_0 - \arctan\left[\frac{2Q_l\left(\dfrac{f}{f_r} - \dfrac{f_r}{f}\right)\left(1 - \dfrac{Q_l}{Q_u}\right)}{2\dfrac{Q_l}{Q_u} - 1 + Q_l^2\left(\dfrac{f}{f_r} - \dfrac{f_r}{f}\right)^2}\right]$$

where ϕ is the phase of the power reflection coefficient and ϕ_0 is a constant that depends on the way of coupling. For coupling loops, apertures and other short-circuit-like devices (inductive coupling) $\phi_0 \cong 180° = \pi$ radians, and for coupling probes (capacitive coupling) $\phi_0 \cong 0$, in the undercoupled case ($Q_e > Q_u$), which is usual for measurement resonators. For overcoupled resonators, ($Q_e < Q_u$) 180° should be added to the values given above. At the resonant frequency, the reflection coefficient has a minimum value because of the loss in the resonator. At other frequencies, the field, and therefore the loss power, in the resonator will be small. The loaded Q-factor determines the sharpness of the resonance. The higher is the loss, the broader is the resonance peak (Figure 3.5). Very high loss smears it out completely. Figure

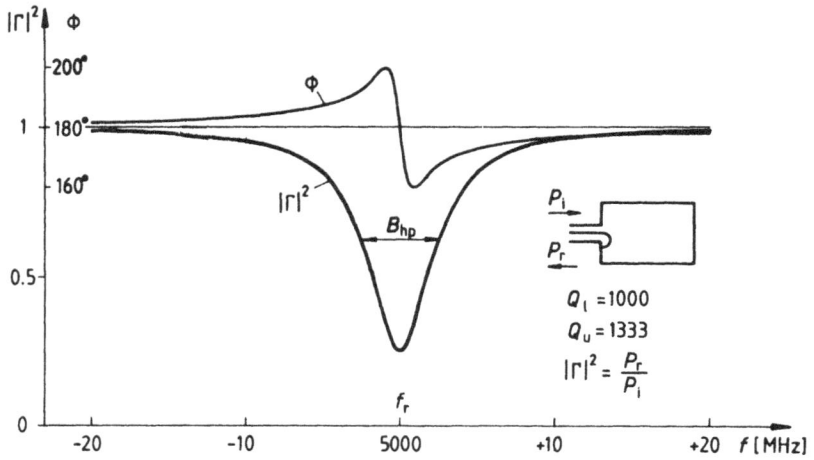

Figure 3.5 Measurement of the resonant frequency and loaded quality factor by the method of reflection coefficient.

3.5 shows the reflection coefficient as a function of frequency. The minimum gives f_r and the half-power width B_{hp} gives Q_l. We can show that

$$Q_l = \frac{f_r}{B_{hp}} \tag{3.21}$$

where B_{hp} is measured halfway down the peak of the power reflection response in (3.20).

The unloaded quality factor can be calculated from Q_l and the power reflection coefficient at the resonant frequency $|\Gamma|_r^2$. From (3.20):

$$Q_u = \frac{2Q_l}{1 \pm |\Gamma|_r} \quad \begin{cases} + = \text{undercoupled} \\ - = \text{overcoupled} \end{cases} \tag{3.22}$$

The method of insertion loss means that the field in the resonator is excited through one coupling, and the field strength is measured through the other coupling. The ratio between the received power and the incident power is the insertion loss a. It depends on Q_l and the coupling coefficients (the size of the loops or probes). If the coupling coefficients are equal, the insertion loss and phase shift can be approximated by the following equations:

$$a = \frac{\left(1 - \dfrac{Q_l}{Q_u}\right)^2}{1 + Q_l^2\left(\dfrac{f}{f_r} - \dfrac{f_r}{f}\right)^2} \tag{3.23}$$

$$\phi = \phi_0 - \arctan\left[Q_l\left(\frac{f_r}{f} - \frac{f}{f_r}\right)\right] \tag{3.24}$$

If the couplings are different, (3.23) must be written in a form containing two coupling coefficients. The constant ϕ_0 is either 0, $\pm 90°$, or $180°$, depending on the locations and types of coupling. There is a phase difference of $180°$ between adjacent lobes in a standing wave pattern (Figure 3.3), so if only one type of coupling is used, only $0°$ and $180°$ are possible. If the types are mixed, probes couple to the electric field, but loops couple to the magnetic field, which is $\pm 90°$ out of phase as compared to the local electric field. This results in $\phi_0 = \pm 90°$. Furthermore, turning the loop $180°$ causes a phase shift of $180°$. The graphs of (3.23) and (3.24) are shown in Figure 3.6. Only at frequencies close to the resonant frequency will any significant field build up in the resonator. Therefore, no signal is transmitted ($a \approx 0$) far from the resonant frequency.

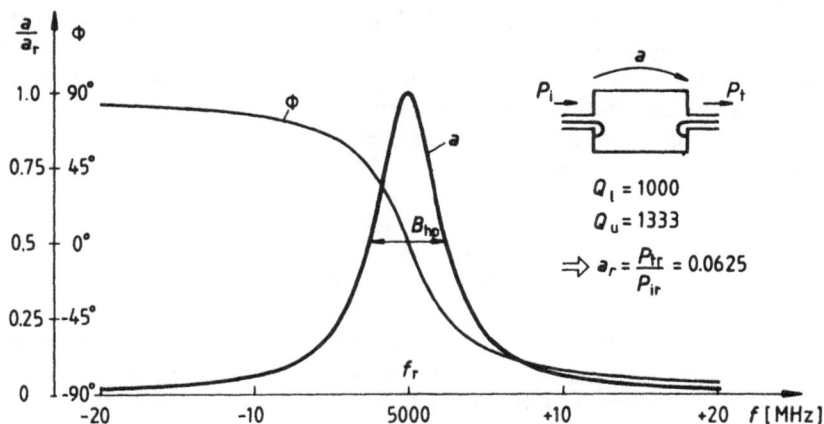

Figure 3.6 Measurement of the resonant frequency and loaded quality factor by the method of insertion loss. The subscript r refers to values at the resonant frequency.

The loaded quality factor can again be calculated from the measurements of resonant frequency and peak width:

$$Q_l = \frac{f_r}{B_{hp}} \tag{3.25}$$

where B_{hp} is the width of the resonance curve (3.23) at the level where the power transmission is half of the maximum value.

If the couplings are different, they must be measured separately by reflection from both sides. If they are equal, the unloaded quality factor can be directly calculated from Q_l and the measured insertion loss at the resonant frequency, a_r. From (3.23), we have

$$Q_u = \frac{Q_l}{1 - \sqrt{a_r}} \tag{3.26}$$

Regardless of which method we use, we always need the resonant frequency, width of the peak, and minimum reflection coefficient or minimum insertion loss (i.e., maximum power transmitted relative to the incident power) to be able to calculate the unloaded quality factor by using (3.21) and (3.22) or (3.25) and (3.26).

To obtain Q_d we still need to know Q_m and Q_{rad}. The loss by radiation is highly dependent on the resonator. "Closed" resonators, like cavity resonators, do not radiate at all, and many "open" ones, like stripline resonators, radiate very little. In these cases, we can approximate $Q_{rad} \approx \infty$. Otherwise, Q_{rad} must usually be calibrated as a function of frequency and loaded quality factor because calculation is difficult and the accuracy would be poor in most cases.

We are still left with the metal quality factor. It can be measured at the resonant frequency of the hollow resonator, f_{r0}. From (3.17):

$$\frac{1}{Q_{m0}} = \frac{1}{Q_{u0}} - \frac{1}{Q_{rad,0}} \tag{3.27}$$

where the subscript 0 refers to the hollow resonator. If $Q_{rad,0} = \infty$, we obtain Q_{m0} directly. In other cases, $Q_{rad,0}$ can be measured by building an identical resonator using another metal with different conductivity. We then have

$$\frac{1}{Q_{rad,0}} = \frac{\dfrac{1}{Q_{u01}} - \dfrac{1}{Q_{u02}} \left(\dfrac{\sigma_2}{\sigma_1}\right)^{1/2}}{1 - \left(\dfrac{\sigma_2}{\sigma_1}\right)^{1/2}} \tag{3.28}$$

where σ_1 and σ_2 are the conductivities of the two metals. The loss in the metal parts, however, is frequency dependent. We therefore need to deduce Q_{m0} from the resonant frequency of the empty resonator, f_{r0}, to the resonant frequency of the filled resonator, f_r. To be able to do so, we return to the definition of the quality factor (3.16). We know that the field pattern in the resonator is constant, the magnetic field energy equals the electric field energy, and the local magnetic field energy content is proportional to the square of the field strength. The total stored energy is therefore proportional to the volume integral of the square of the magnetic field:

$$\text{stored energy} \propto \int_V |H|^2 \mathrm{d}V \tag{3.29}$$

The loss in the metal parts is proportional to the surface integral of the surface resistance times the square of the surface current density. We can show [Collin, 1966, pp. 324–325] that

$$\text{loss power} \propto \sqrt{\omega} \int_S |H|^2 \mathrm{d}S \tag{3.30}$$

The integrals in (3.29) and (3.30) are independent of frequency. From (3.16), we therefore have

$$Q_m \propto \frac{\omega \times \int_V |H|^2 \mathrm{d}V}{\sqrt{\omega} \int_S |H|^2 \mathrm{d}S} \approx \sqrt{\omega} \tag{3.31}$$

This means that the metal loss will increase when the resonant frequency is lowered by filling the resonator with a dielectric. This is important to note because erroneous expressions are found in some textbooks [Sihvola, 1985]. According to (3.31), we can write

$$Q_m = \left(\frac{f_r}{f_{r0}}\right)^{1/2} Q_{m0} \tag{3.32}$$

To acquire the relation between ε_r'' and dielectric quality factor, Q_d, we calculate the energy content of the electric field and loss in the dielectric. We can show [Collin, 1966, p. 325] that

$$\frac{1}{Q_d} = \tan\delta = \frac{\varepsilon_r''}{\varepsilon_r'} \tag{3.33}$$

where tanδ is the loss tangent.

Collecting the results for one equation, we have

$$\varepsilon_r'' = \varepsilon_r'\left[\frac{1}{Q_u} - \frac{1}{Q_{\text{rad}}} - \frac{1}{Q_{m0}}\cdot\left(\frac{f_{r0}}{f_r}\right)^{1/2}\right] \tag{3.34}$$

If the resonator does not radiate, or Q_{rad} is very high, we have

$$\varepsilon_r'' \approx \varepsilon_r'\left[\frac{1}{Q_u} - \frac{1}{Q_{u0}}\cdot\left(\frac{f_{r0}}{f_r}\right)^{1/2}\right] \tag{3.35}$$

where Q_u and Q_{u0} are calculated from (3.22) or (3.26), depending on the measurement method.

If the radiation loss is significant, the metal loss can often be neglected. We then have

$$\varepsilon_r'' \approx \varepsilon_r'\left(\frac{1}{Q_u} - \frac{1}{Q_{\text{rad}}}\right) \tag{3.36}$$

Here, we need to remember that Q_{rad} is frequency dependent, so it must be calibrated by using materials with known ε_r (see, for example, Figure 2.22). If the dielectric is very lossy, we have

$$\varepsilon_r'' \approx \varepsilon_r'\cdot\frac{1}{Q_u} \tag{3.37}$$

In this section, we have assumed that the resonator is filled with the dielectric to be measured or the field of the resonator does not extend beyond the sample. For these cases, the formulas give accurate results. If the assumption is not fulfilled, the formulas can still be used in many cases. The values for ε_r' and ε_r'' given by the formulas will be smaller than the real values. We will call them *effective values*. The relation between the effective and real values of the sample depend on the type of resonator, resonant wave mode, and location, size, and shape of the sample. This relation can be calculated, in some cases; otherwise, it must be calibrated by using samples of known ε_r. These matters will be briefly studied in the next section.

3.2.2 Partially Filled Resonator, Perturbation Formulas

If the sample to be measured fills only a small portion of the resonator, the change in resonant frequency can be calculated from the *perturbation theory* [Waldron, 1960]. It also applies for larger samples if the permittivity of the sample is low or there is only a small change in the permittivity. A detailed presentation of the theory has been given by Harrington, for example, [Harrington, 1961].

The general perturbation formula is

$$\frac{\omega_{r2} - \omega_{r1}}{\omega_{r2}} = -\frac{\int_v [(\varepsilon_{r2} - \varepsilon_{r1})\varepsilon_0 \bar{\mathbf{E}}_2 \cdot \bar{\mathbf{E}}_1^* + (\mu_{r2} - \mu_{r1})\mu_0 \bar{\mathbf{H}}_2 \cdot \bar{\mathbf{H}}_1^*] dV}{\int_v [\varepsilon_{r1}\varepsilon_0 \bar{\mathbf{E}}_2 \cdot \bar{\mathbf{E}}_1^* + \mu_{r1}\mu_0 \bar{\mathbf{H}}_2 \cdot \bar{\mathbf{H}}_1^*] dV} \tag{3.38}$$

Equation (3.38) is the exact formula, and it applies to a change in magnetic permeability (μ) as well as to a change in permittivity. The subscript 1 refers to the situation before the insertion of the sample, or the change in permittivity or permeability. The subscript 2 refers to the situation after the change.

In deriving the formulas from (3.38) for different practical situations, some approximations are to be made. They will impose some limitations on the use of the formulas. The results calculated by using the perturbation theory are generally considered accurate, provided that

- the relative shift in resonant frequency $\Delta f_r/f_r \leqslant 0.001$;
- the electric field is constant in the volume of the the sample to be measured;
- the symmetry is preserved.

For practical applications in the industry, the requirements on the absolute accuracy of the measurements are usually less stringent than in the laboratory. The resonant frequency shift can therefore in many cases be allowed to be at least one decade greater than stated above. The requirement of constant field at the sample means that if the volume of the sample is large and the field strength is different at different points of the surface, the internal field is no longer given by the boundary conditions as assumed. The third requirement means that insertion of the sample must not alter the shape of the wave mode.

Henceforth, we will assume that μ is constant. The denominator in (3.38) is the total energy in the resonator, and can therefore be substituted by twice the value of the first term in the integral. If the change in permittivity is small, the electric field is approximately unchanged:

$$\frac{\Delta\omega_r}{\omega_r} \approx -\frac{\int_v (\varepsilon_{r2} - \varepsilon_{r1})|E|^2 dV}{2\int_v \varepsilon_{r1}|E|^2 dV} \tag{3.39}$$

If $\varepsilon_{r1} = 1$ and ε_{r2} is constant in the sample, we have

$$\frac{\Delta\omega_r}{\omega_r} \approx -\frac{(\varepsilon_r - 1)\int_{vs} |E|^2 dV}{2\int_v |E|^2 dV} = -\frac{(\varepsilon_r - 1)}{2}S \tag{3.40}$$

where vs refers to the volume of the sample and v to the whole resonator. The ratio of the integrals, S, is the filling factor.

Where the permittivity of the sample may be large, but the volume fraction filled by the sample is small, the formula is

$$\frac{\Delta\omega_r}{\omega_r} = -\frac{\int_{vs} (\varepsilon_{r2} - \varepsilon_{r1})\bar{\mathbf{E}}_i \cdot \bar{\mathbf{E}}^* dV}{2\varepsilon_{r1}\int_v \bar{\mathbf{E}}_i \cdot \bar{\mathbf{E}}^* dV} \tag{3.41}$$

where E_i is the electric field inside the sample. If the sample is lossy, the change in angular frequency, given by (3.41), becomes complex (see (1.38)). From the imaginary part, we obtain the change in the loaded quality factor:

$$\frac{\Delta\omega_r}{\omega_r} = \frac{\Delta f_r}{f_r} + j\frac{1}{2}\left(\frac{1}{Q_{l2}} - \frac{1}{Q_{l1}}\right) \tag{3.42}$$

Because the sample is small, we can approximate the internal field $\bar{E}_i$ for some cases with well defined geometry by using the boundary conditions in Section 2.2.1. In the following cases, we assume that $\varepsilon_{r1} = 1$.

When the electric field is tangential to the surface of the sample (e.g., slab, cylinder), the internal field equals the external field, and we have

$$\frac{\Delta\omega_r}{\omega_r} \approx -\frac{\varepsilon_r' - 1 - j\varepsilon_r''}{2} \cdot \frac{\int_{vs}|E|^2 dV}{\int_v |E|^2 dV} = -\frac{\varepsilon_r' - 1}{2}S + j\frac{\varepsilon_r''}{2}S$$

$$\frac{\Delta f_r}{f_r} \approx -\frac{\varepsilon_r' - 1}{2}S \tag{3.43}$$

$$\Delta\left(\frac{1}{Q_l}\right) \approx \varepsilon_r''S$$

If we take the ratio of the relative change of frequency and change of $1/Q_l$, we have

$$\frac{\Delta f_r/f_r}{\Delta(1/Q_l)} \approx -\frac{\varepsilon_r' - 1}{2\varepsilon_r''} \tag{3.44}$$

which is the half of the constant A, used for density independent moisture measurement (see Section 2.5.5, Equation (2.67)). The reader may note that the result is also independent of the filling factor S (i.e., the dimensions of the sample).

In the case of the external electric field E_e perpendicular to the surface of the sample, the boundary condition is $E_i = E_e/\varepsilon_r$. From the perturbation formula (3.41) (assuming that $\varepsilon_r'' < 0.1\varepsilon_r'$), we have

$$\frac{\Delta\omega_r}{\omega_r} \approx -\frac{\varepsilon_r - 1}{2\varepsilon_r} \cdot S \approx -\frac{\varepsilon_r' - 1}{\varepsilon_r'}\frac{S}{2} + j\frac{\varepsilon_r''}{(\varepsilon_r')^2}\cdot\frac{S}{2}$$

$$\frac{\Delta f_r}{f_r} \approx -\frac{\varepsilon_r' - 1}{2\varepsilon_r'} \cdot S \tag{3.45}$$

$$\Delta\left(\frac{1}{Q_l}\right) \approx \frac{\varepsilon_r''}{(\varepsilon_r')^2} \cdot S$$

$$\frac{\Delta f_r/f_r}{\Delta(1/Q_l)} \approx \frac{\varepsilon_r'}{2}\cdot\frac{(\varepsilon_r' - 1)}{\varepsilon_r''}$$

If $\varepsilon_r'' > 0.1\varepsilon_r'$, the real and imaginary parts could not be approximated as above, but should be calculated exactly from the left-hand equation in the first line of (3.45). The same applies to (3.47) and (3.49). In the case of the electric field perpendicular to the axis of a cylinder, the internal field is given by

$$E_i = \frac{2}{\varepsilon_r + 1} E_e \tag{3.46}$$

By substituting (3.46) into (3.41), we have ($\varepsilon_r'' < 0.1\varepsilon_r'$):

$$\frac{\Delta\omega_r}{\omega_r} \approx -\frac{\varepsilon_r - 1}{\varepsilon_r + 1} \cdot S \approx -\frac{\varepsilon_r' - 1}{\varepsilon_r' + 1} \cdot S + j\frac{2\varepsilon_r''}{(\varepsilon_r' + 1)^2} \cdot S$$

$$\frac{\Delta f_r}{f_r} \approx -\frac{\varepsilon_r' - 1}{\varepsilon_r' + 1} \cdot S$$

$$\Delta\left(\frac{1}{Q_l}\right) \approx \frac{4\varepsilon_r''}{(\varepsilon_r' + 1)^2} \cdot S \tag{3.47}$$

$$\frac{\Delta f_r/f_r}{\Delta(1/Q_l)} \approx \frac{(\varepsilon_r')^2 - 1}{4\varepsilon_r''}$$

In the case of a small sphere in the resonator, the internal field is

$$E_i = \frac{3}{\varepsilon_r + 2} E_e \tag{3.48}$$

Substituting into (3.41) gives

$$\frac{\Delta\omega_r}{\omega_r} \approx -\frac{3}{2} \cdot \frac{\varepsilon_r - 1}{\varepsilon_r + 2} \cdot S \approx \frac{3S}{2}\left[\frac{\varepsilon_r' - 1}{\varepsilon_r' + 2} - j\frac{3\varepsilon_r''}{(\varepsilon_r' + 2)^2}\right]$$

$$\frac{\Delta f_r}{f_r} \approx -\frac{3}{2} \cdot \frac{\varepsilon_r' - 1}{\varepsilon_r' + 2} \cdot S$$

$$\Delta\left(\frac{1}{Q_l}\right) \approx \frac{9\varepsilon_r''}{(\varepsilon_r' + 2)^2} \cdot S \tag{3.49}$$

$$\frac{\Delta f_r/f_r}{\Delta(1/Q_l)} = \frac{(\varepsilon_r' - 1)(\varepsilon_r' + 2)}{6\varepsilon_r''}$$

All of the formulas for the change in resonant frequency and quality factor given above contain the filling factor S. When we calculate ε_r' and ε_r'' from the measurement results by using these formulas, our results will also depend on S. Because the integral in the numerator in the expression for S (3.40), is taken over the volume of the sample, we need to know accurately the dimensions of the sample to obtain good results.

There are two ways of using the perturbation formulas that do not depend on S. Both methods are based on the two-parameter technique. One has already been shown. By taking the ratio of $\Delta f_r/f_r$ and $\Delta(1/Q_l)$, we have an expression that equals, or in many cases resembles, the constant A (2.67), used for density-independent moisture measurement. This ratio can often be used in the same way as A.

The other method is the *orthogonality technique* (see Section 2.5.4), where we use two orthogonal wave modes. Orthogonality in this case means that the electric field of one mode is tangential to the surface of the sample and the other mode's field is perpendicular. This is possible, for example, with a slab or a sheet in a rectangular cavity or stripline resonator (see Section 3.5). By taking the ratio of the relative changes in the resonant frequencies of (3.43) and (3.45), we have

$$\frac{\Delta f_{\|}/f_{\|}}{\Delta f_{\perp}/f_{\perp}} \approx \varepsilon_r' \frac{S_{\|}}{S_{\perp}} \qquad (3.50)$$

where the subscript $\|$ stands for tangential (3.43) and $\perp$ for perpendicular (3.45). By using this technique we acquire the value of ε_r' multiplied by the ratio of the filling factors. The ratio $S_{\|}/S_{\perp}$ depends on the wave modes, but much less so on the dimensions of the sample.

Were the volume fraction of the resonator filled by the sample to be large and the permittivity of the sample high, perturbation theory would not give good results. These situations are usually difficult to handle theoretically because the wave modes may be different from those of the empty resonator. In some cases, however, it is possible to calculate the exact result, but unfortunately space does not allow us to study them further. See, for example [Harrington, 1961, p. 158–163].

3.3 RESONATOR STRUCTURES

3.3.1 Coupling Devices

To be able to perform measurements with a resonator we couple it to the electronics. As we pointed out in Section 3.2.1, there are two ways of measuring the resonant frequency and quality factor by using either reflection coefficient or insertion loss. The former requires only one coupling and the latter requires two, but the possible coupling devices are the same. In this section, we describe the devices that are normally used. The discussion is limited to qualitative matters because the exact calculation of the coupled power is difficult, but we give the dependence of the equivalent dipole moment of the coupling structures on the dimensions. The excited field is directly proportional to the dipole moment, which means that, for example, the insertion loss depends on the fourth power of the dipole moment in the symmetrically coupled case.

Probe

We may extend the center conductor of the feeding coaxial cable a slight distance into the resonator to form a *coupling probe* (Figure 3.7). This is a convenient coupling device at VHF and UHF. The coupling probe is short compared to the wavelength, and it therefore looks almost like an open circuit to the feeding transmission line. The current in the probe is consequently very small, but the voltage creates an electric field between the probe and the adjacent wall of the resonator. The field radiates energy into the resonator like a small monopole antenna. The dipole moment is approximately proportional to the square of the length of the probe. It will, however, couple only to resonance modes that have a finite electric field along the probe. The closer to a field maximum that the probe is located and the longer it is, the stronger will be the coupling. An advantage of the coupling probe is that of easy tuning. The probe is initially made longer than necessary, and then trimmed until the desired coupling coefficient is obtained.

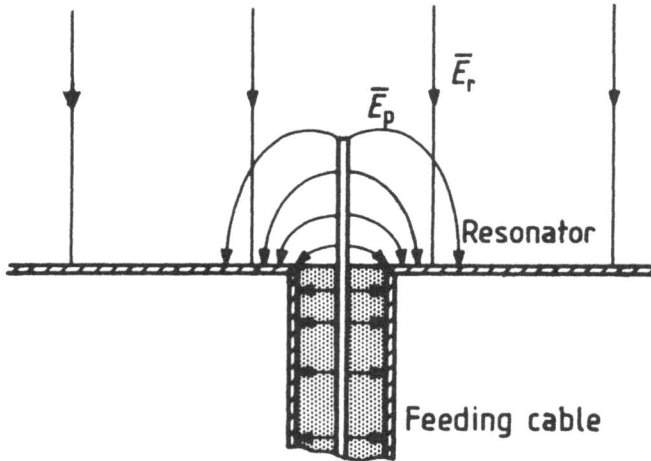

Figure 3.7 The electric field ($\bar{E}_p$) of a coupling probe couples to the resonance mode through the electric field ($\bar{E}_r$) parallel to the probe.

Loop

If we bend the extended center conductor of the feeding coaxial cable and ground it to the wall of the resonator, we have a *coupling loop* (Figure 3.8). The size of the loop is usually made much smaller than the wavelength. Because of the grounding, the loop almost forms a short circuit to the feeding line. The voltage of an

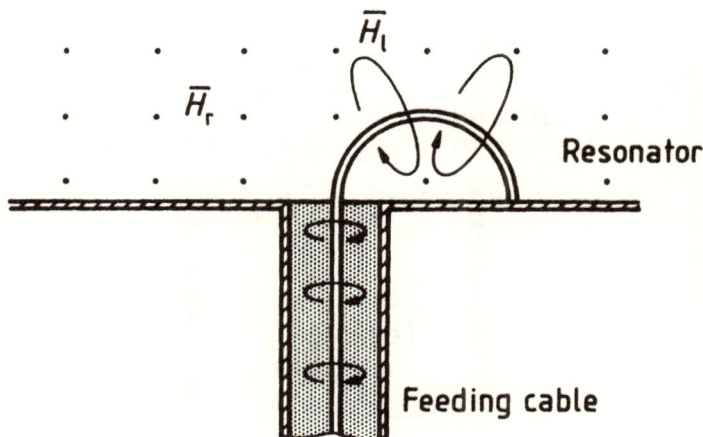

Figure 3.8 The magnetic field ($\bar{H}_l$) of a coupling loop couples to the resonance mode through the magnetic field ($\bar{H}_r$) perpendicular to the plane of the loop.

incident signal becomes negligible, but the current is strong, creating a magnetic field. The field radiates like a magnetic dipole tangential to the wall. The dipole moment is proportional to the loop area. The radiation couples to the tangential magnetic field of the resonance modes. The closer to a magnetic field maximum of a specific mode the loop is located and the larger its area is, the stronger will be the coupling to that mode. The direction of the plane of the loop, however, is also important. For maximum coupling, the field lines of the resonance mode should be perpendicular to this plane.

The coupling loop is more rigid than the probe but more difficult to tune, and the quality of the short circuit must be good.

Aperture

If a waveguide is used as the feeding transmission line, the *aperture* (Figure 3.9) is the natural coupling device. It is usually a small round hole in the waveguide wall or the shorted end. Depending on the location of the aperture in the waveguide, the tangential magnetic field or normal electric field will penetrate the aperture and couple to the resonance mode. The strength of the coupling (the magnetic or electric dipole moment) is proportional to the third power of the aperture radius. The coupling depends, of course, on the location of the aperture with respect to the field of the resonance mode and direction of the field lines in the case of magnetic coupling.

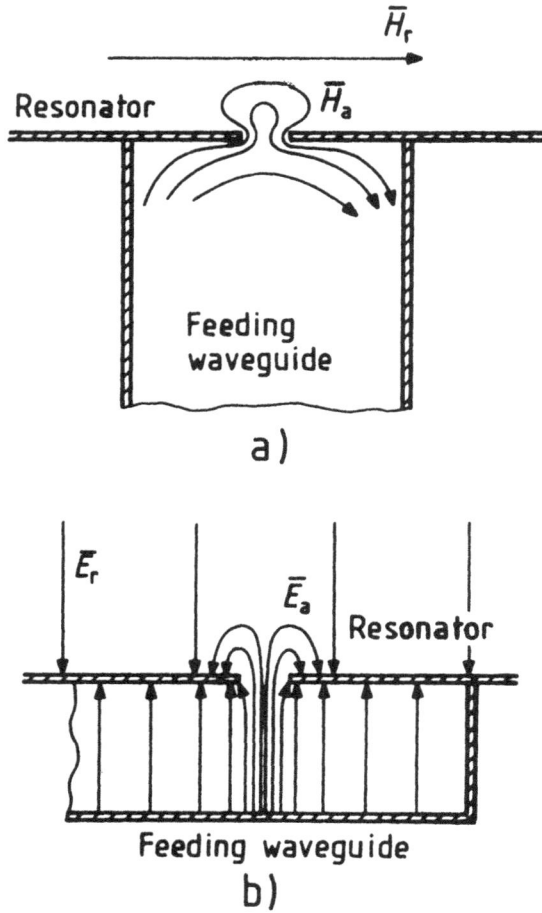

Figure 3.9 (a) Magnetic and (b) electric coupling aperture.

Capacitive Coupling

Especially when the resonator is made of a section of TEM or quasi-TEM transmission line, *capacitive coupling* may be a convenient choice. The signal is then brought, for example, to the center conductor through a lumped capacitance or a capacitive air gap (Figure 3.10).

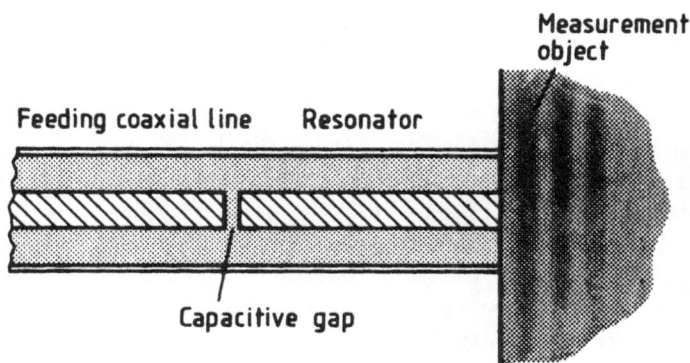

Figure 3.10 Capacitive coupling of a coaxial resonator.

Free-Space Coupling

Open resonators, in which the field extends into the surrounding space (in contrast to cavities) can be excited by a signal traveling through free space. At resonance, a strong field will build up, which is radiated back into the environment and is then picked up by a receiver. This provides a means of measuring resonator sensors from a distance. The same can be achieved with closed resonators if an antenna is connected from the outside to the coupling probe or loop.

Location and Strength of Coupling for Measurement of Insertion Loss

For the characterization of resonator sensors by using the method of insertion loss, we need two couplings. If the coupling coefficients are made equal, the simple expression (3.26) can be used for calculation of the unloaded quality factor. The equality can easily be checked by measurement of the reflection coefficient. The dip at resonance in the reflection coefficient curve (Figure 3.5) should be equally deep when measured through both couplings.

In the design of the sensor, direct coupling between the couplings should be taken into account because it defines the minimum value of a that can be measured. The higher are the losses, the lower is the resonance curve, and if the losses are greater than a critical value, the resonance disappears below the level of direct coupling. If a broad dynamic range in the measurement of ε_r'' were desirable, direct coupling should be minimized. The couplings should be located as far from each other as possible and close to field maxima, keeping the size of the coupling devices small.

3.3.2 Hollow Cavity Resonator

Theory

A resonator made of a piece of hollow waveguide is a *cavity resonator*. It is usually shorted at both ends because of the strong radiation from an open-ended waveguide. The resonant frequencies occur for the TE_{nm} and TM_{nm} modes, when the length of the cavity is a multiple of the half-wavelength for that mode. The resonant modes are called TE_{nml} or TM_{nml}, where the integers n, m, and l refer to the number of electric field maxima in the standing wave pattern along the x, y, and z directions for rectangular waveguides, or ϕ, r, and z directions for circular waveguides. Every cavity therefore has an infinite number of resonances.

The resonant frequency for a mode in a hollow cavity is easily calculated from the wavelength and the condition above. For rectangular cavities, we have

$$f_{r,nml} = \frac{c}{2}\left[\left(\frac{n}{a}\right)^2 + \left(\frac{m}{b}\right)^2 + \left(\frac{l}{d}\right)^2\right]^{1/2} \tag{3.51}$$

where d is the length of the cavity. For TE_{nml} modes in cylindrical cavities, we have

$$f_{r,nml} = \frac{c}{2}\left[\left(\frac{p'_{nm}}{\pi a}\right)^2 + \left(\frac{l}{d}\right)^2\right]^{1/2} \tag{3.52}$$

where, for TM_{nml} modes, p'_{nm} (see Section 9.2) is to be substituted by p_{nm}. A graphical representation of (3.52) is shown in Figure 3.11. In some cases, several modes will have the same resonant frequency. They are called *degenerate modes*.

Because of the great number of resonances, we must design the measurement sensor and couplings in such a way as to avoid the interference from other modes. This is not a problem for cavities completely filled with an isotropic dielectric because all resonant frequencies are shifted in the same way as a function of ε'_r. In partly filled resonators, the frequency shift for each mode depends on the location of the sample with respect to the electric field of that mode.

The fields of the resonance modes can be calculated from the fields of two wave components traveling in opposite directions in a waveguide. If we take into account the boundary conditions at the short circuits, the fields can be solved. They are given for the modes in rectangular and cylindrical cavities in Section 9.2. In Figure 3.12, the fields are indicated for some of the lowest modes.

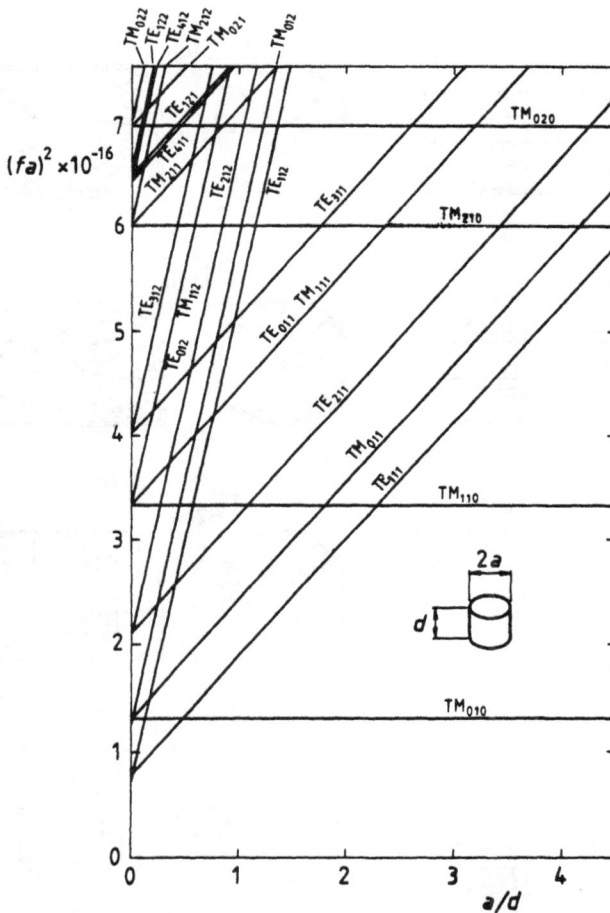

Figure 3.11 Diagram for determining the resonant frequencies of the modes in a cylindrical cavity resonator.

The Cavity Resonator as a Sensor

A cavity can always be used as a completely filled and closed resonator, and in such cases (3.15) and (3.35) are directly applicable. For continuous measurements, we make holes or slots in the walls, where the material to be measured can flow or move through or past the cavity sensor. A dielectric pipe filled with the unknown material, or a thread or rod, can be led through the cavity via holes. Otherwise, a sheet-like material may be led through a pair of slots or between the halves of a split

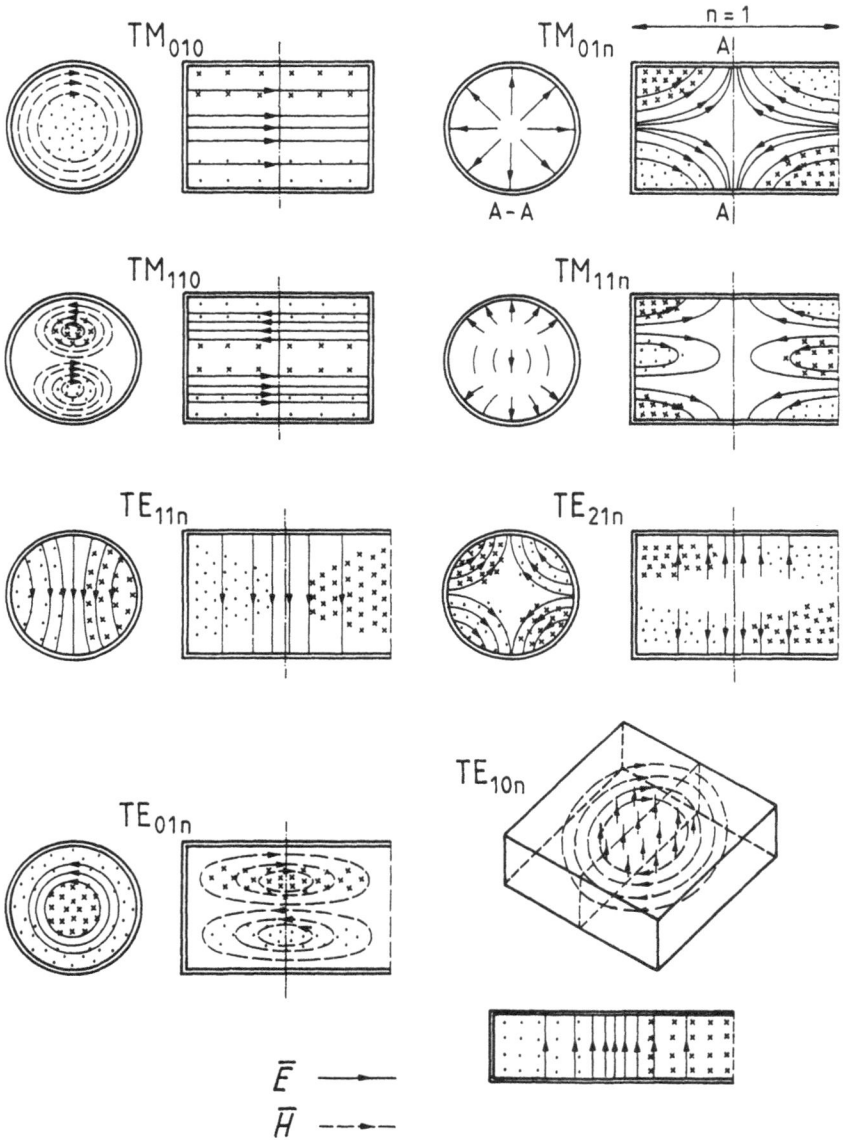

Figure 3.12 A qualitative presentation of the fields of some resonance modes in cavity resonators.

cavity (Figure 3.13). In these cases, the methods of Section 3.2.2 can be applied. The apertures should be designed in such a way that they would disturb the resonance mode used as little as possible. To prevent radiation, the round apertures should be small and preferably located where neither field would be strong. In some cases, a metal net or grid can be used as a cover on the aperture, thus decreasing the disturbance. If necessary, waveguide muffs can be mounted around the apertures on the outside. The diameter should be small enough so that the lowest cut-off frequency in the muff would be higher than the resonant frequency of the used mode under all conditions. Bad electrical contacts in joints should also be avoided, especially where the surface currents were strong. Slots need not obstruct the surface currents of the used resonance mode, but ought to be oriented along the current flow. The surface current is always perpendicular to the tangential magnetic field and is given by (1.16). For example, the TE_{011} mode in cylindrical cavities has an axial magnetic field on the cylindrical surface (see Figure 3.12). The cavity can therefore be cut in two with a slot parallel to the end plates.

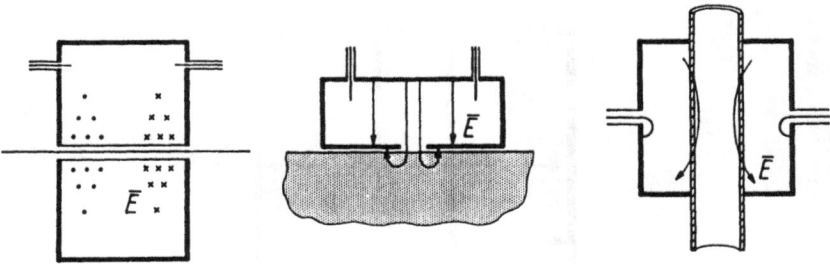

Figure 3.13 Cavity resonator sensors for measuring sheets, solid surfaces, and material inside a dielectric pipe.

If another resonance mode (for example, a degenerate mode) may disturb the used mode, the excitation of the disturbing mode can be prevented by locating the slots so that they obstruct the surface current of that mode. Additional slots can be cut for this purpose, or the couplings can be chosen and located so as to provide the desired selective excitation.

If the cavity is used as a contacting sensor to measure the surface of an object with the electric field protruding through an aperture, it, of course, ought to be located where the normal electric field is strong. In this case and when the perturbation theory is valid, the filling factor is low. Thermal expansion in the metal, or even changes in the humidity or temperature of the air, may therefore cause errors. Because all modes equally sense the air and thermal expansion, another mode that has zero electric field at the aperture or location of the measurement object can be used as a reference for compensation. Cavities are best suited for measurement of liquids, gases, thin rods or threads, sheets, and surfaces.

3.3.3 Coaxial and Helical Resonators

A coaxial resonator is made of a section of coaxial transmission line. Especially if it is short-circuited at both ends, the coaxial resonator can be used in the same way as resonant cavities, but it normally supports only one class of wave modes, being the TEM modes. At sufficiently high frequenices, so-called waveguide modes are also possible, but they are normally not used in measurement sensors. The most useful structure is the quarter-wavelength resonator, which is short-circuited at one end and open at the other. The open end may have a metal cover, in which case the resonator is called a *re-entrant cavity*. Without the cover (Figure 3.14), fringing fields exist at the open end, but, because of the cylindrical symmetry, the loss by radiation is low. The open end can be held against the surface of an object, in which case its ε_r will influence the reflection coefficient at the open end. The formulas for calculation of the reflection coefficient are given in Section 6.2. The resonant frequency can hence be calculated from (3.7).

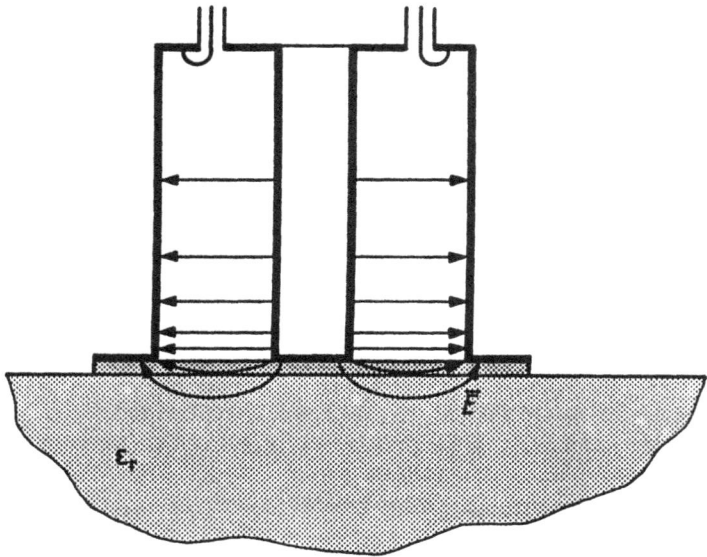

Figure 3.14 A quarter-wavelength coaxial resonator for measurement of solid surfaces. It is equipped with a dielectric window, which reduces the influence of the strong field at the sharp margins.

The coaxial resonator sensor for surface measurements is sensitive to the quality of the contact because the electric field strength is high (theoretically it is infinitely high) at the edge of the inner conductor, and to a lesser degree at the inner edge of the outer conductor. Even a small roughness or curvature of the surface causes large variations in the measured results. The errors can be decreased by holding the sensor at a constant distance from the surface, for example, by mounting a thin dielectric sheet or film on the open end (Figure 3.14).

If, instead of a straight inner conductor, we were to use a helix, we would have a *helical resonator* (Figure 3.15). The use of such resonators for laboratory measurements was described in detail in Section 2.7.4, where the necessary formulas and graphs were given. The helical resonator can be used as a sensor for cylindrical objects of diameter smaller than the inner diameter of the helix. For this purpose, apertures are made in the end plates. The substance to be measured can, for example, be led through the resonator inside a dielectric pipe. An advantage of helical resonators is their low resonant frequency compared to the relatively small size, which means that objects much smaller than a wavelength can be measured. Another advantage is the great number of multiples of the fundamental resonant frequency, a useful property for broadband measurements.

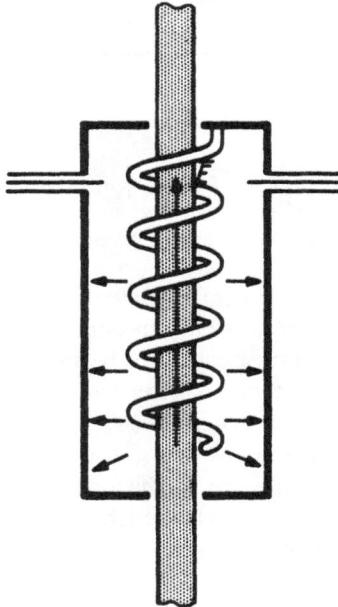

Figure 3.15 The helical resonator is well suited for measurement of rods or material in dielectric pipes when small size compared to the wavelength is desirable.

3.3.4 Strip Resonators

The stripline and microstrip line (Figure 1.9) are closely related and well known quasi-TEM transmission lines, but other, similar structures can be used as transmission lines and resonators. These may consist of one or two ground planes with one or more strip-like center conductors above or between the ground planes. The possible structures are numerous, but we will concentrate here on three useful designs: the *microstrip,* the *stripline,* and the *two-conductor stripline* (Figure 3.16). They are all flat structures that are suitable for measurement of flat surfaces, slabs, or sheet-like materials. The resonators are most conveniently formed of open-ended strips ($\lambda/2$) with larger ground planes. In most cases, the ends could also be shorted, which would usually increase the radiation loss. The unsymmetrical electric fields (microstrip) also radiate to some extent, but use of modes with a symmetrical field may result in radiation quality factors of several thousands.

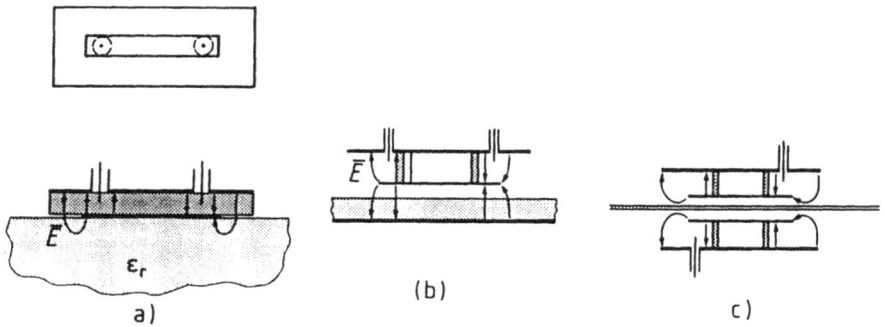

Figure 3.16 Side view of (a) microstrip, (b) stripline, and (c) two-conductor stripline resonator for measurement of solid surfaces, dielectric layers, and sheets.

An advantage of these resonators is their simple structure. An array of resonators with common ground planes is easy to build for profiling purposes. To eliminate mutual coupling, only one resonator can be measured at a time, and the other resonators must be inactivated. Switching on and off can be done with *pin* diodes mounted on the strips [Vainikainen *et al.,* 1987], as shown in Figure 3.17.

The spatial resolution of straight strips is lower along the strip than across it because the electric field maxima are at the open ends. For shaped strips (e.g., H-shaped, butterfly-shaped, bent, coiled, helical), a desired measurement area and lower resonant frequency, compared to the largest dimension of the strip, may be achieved [Vainikainen *et al.,* 1987].

Microstrip Resonator

The *microstrip resonator* sensor is usually made of a strip on a dielectric substrate on a ground plane. The electric field (Figure 3.18) is concentrated between the strip and ground plane, but a weak field also exists above the dielectric. In microstrip design, this is usually taken into account by using an effective permittivity, which is lower than that of the substrate. The smaller is the width-to-height ratio W/h and

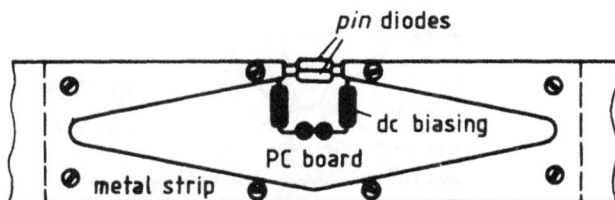

Figure 3.17 A strip switch for activating and deactivating resonators in an array.

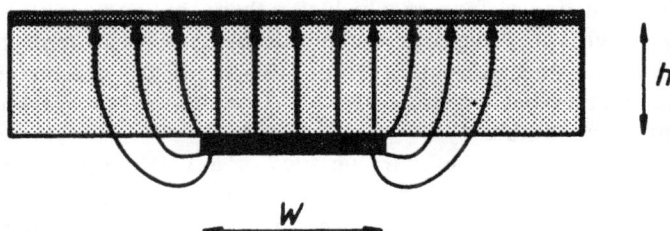

Figure 3.18 A qualitative presentation of the field of a microstrip transmission line.

the lower is the permittivity of the substrate, the lower will be the effective permittivity compared to that of the substrate. If the microstrip resonator is held against a dielectric slab, the fringing field will penetrate the surface of the slab, thus changing the effective permittivity and hence the resonant frequency. The radiation changes, too, making the measurement of ε_r'' dependent on calibration. Measurement of ε_r'' in low-loss materials is difficult. In practice, use of a protective dielectric layer on the sensor is often necessary to prevent wear and corrosion. This, of course, will decrease the sensitivity of the sensor, but nonetheless permits the measurement of substances with higher losses.

The microstrip resonator can be capacitively coupled from a feeding microstrip line, but then the coupling will change with the effective permittivity. Another possibility is coupling through the ground plane with a probe in the substrate.

Stripline Resonator

The *stripline resonator* (Figure 3.19) has two ground planes and a symmetrical electric field. For perfect symmetry, the radiation loss is almost zero, but any unsymmetric loading will upset the balance. Measurement of fairly low losses, however, is possible with an optimized sensor after calibration.

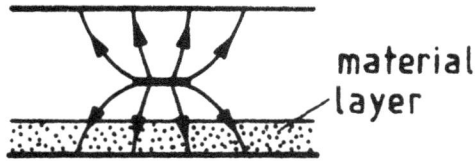

Figure 3.19 End view of the electric field in a stripline resonator.

The stripline resonator is best suited for noncontact measurement of a dielectric layer on a conducting surface, in which case there are two possibilities: measurement of the permittivity of a layer of constant thickness, or measurement of the thickness of a layer of constant permittivity. For two-parameter measurement, with proper calibration, simultaneous measurement of, for example, mass of water and dry mass per area may be possible.

Figure 3.20 shows the characteristics of a VHF stripline resonator used for measurement of the thickness of a layer (or, actually, the mass per area) of wooden chips. The measurement of mass per area is based on the fact that small variations in density are compensated by the resulting variations in permittivity. The accuracy depends on the electric field in the gap between the lower ground plane and the strip. A broad strip gives a more uniform field and a good accuracy. A narrow strip emphasizes the upper layers of the material to be measured, but gives a better spatial resolution in the cross direction.

From Figure 3.20, we can see that the resonant frequency of the empty resonator depends on the distance to the lower ground plane as a result of the bending of the field lines at the open ends. With the dielectric layer present, the effect of the changing filling factor is the opposite. The reader may note that, for a mean layer thickness d, there is a suspension height h around which a moderate variation in the height has very little effect on the measurement of the layer thickness. However, this is not so for local variations. A knob in the lower ground plane (and the sample layer of even thickness) has a different effect at different places in the resonator. If the sensor is suspended at the optimum height and the resonant frequency is integrated over a distance equal to the length of the strip, the differences will cancel out.

A stripline resonator ought to be designed so that the distance between the ground planes does not exceed 0.3λ because of the increasing radiation. A radiating sensor is difficult to keep within the interference limits unless very low power is used. The sensitivity to interference from other equipment is also high because of the reciprocity of radiating structures. The size of the upper ground plane has no upper limit. A small plane, only slightly larger than the strip, suffices. The coupling to the electronics is conveniently arranged through probes in the upper ground plane.

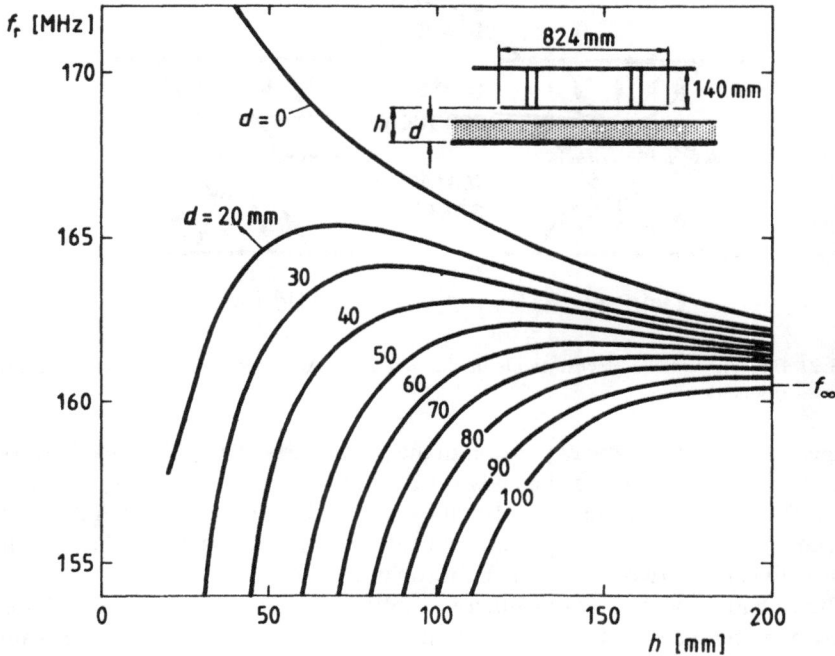

Figure 3.20 The characteristics of a stripline sensor when used for measurement of the thickness of a layer of wooden particles. The breadth of the strip was 71 mm. (After [Nyfors *et al.*, 1984].)

Two-Conductor Stripline Resonator

The *two-conductor stripline* differs from the stripline resonator only in that it has two center conductors. The advantage is that the two-conductor variant can support two different wave modes, even and odd (Figure 3.21). Theoretically, they are degenerate, but in practice the resonant frequencies differ by about 10% because of different fringing of the fields at the ends. Halfway between the strips, the electric fields of the modes are orthogonal in space. This makes possible the noncontact measurement of the permittivity of thin sheet-like materials independent of the thickness using the orthogonality technique (Section 2.5.4).

The two modes can also be used for compensation purposes in measuring thin sheets, like paper [Fischer *et al.*, 1988] when the filling factor is small. If the ε'_r of the sheet is high, the odd mode is saturated and almost independent of small changes in ε'_r (3.45), whereas the even mode is as sensitive as for small values of ε'_r (3.43). Both modes are equally sensitive to the humidity in the air (filling factor = 1) and

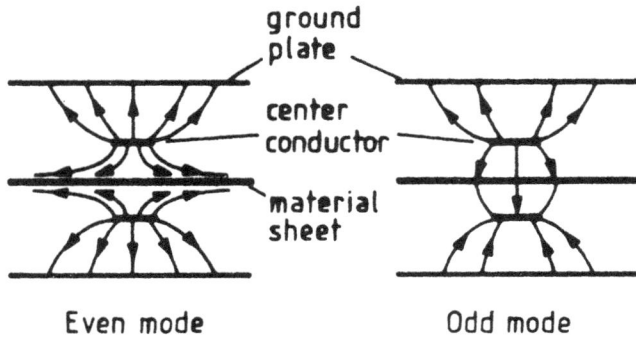

Figure 3.21 End view of the electric fields of the odd and even modes of a two-conductor stripline resonator.

to changes caused by thermal expansion in the metal parts. The difference in resonant frequency $(f_\perp - f_\|)$ therefore depends on ε_r', but not on the humidity or the temperature. The frequency difference also allows much bigger variations in the location of the sample between the strips and reduces the errors caused by changes in the distance between the two halves of the resonator.

The radiation of the two-conductor stripline resonator can be practically eliminated by bending the edges of the ground planes 90° toward the plane of the sample. The edges may at least reach the levels of the strips without altering the normal performance of the sensor. The coupling is conveniently arranged through probes in the ground planes.

3.3.5 Two-Conductor Line Resonator

The *two-conductor line* (Figure 1.8) is a TEM-transmisison line that is well suited for making open resonator sensors. The ends can be left open or short-circuited, depending on the application. The conductors can be implemented on a piece of printed circuit board or they can be made from thin metal rods.

A practical structure is a sensor made of rods that are shorted at one end (Figure 3.22). The first resonance is for $l = \lambda/4$, but fringing effects at the open end will slightly lower the resonant frequency from the theoretical value. The resonator can be pushed into soft, granular, or liquid materials, wherein it is completely filled and the ε_r' of the material can be calculated from (3.15). For accurate measurements, in some cases, we need to use thin wires or to take into account the fact that the wires disturb the medium. In compressible materials, the density at the surface of the rods will become slightly higher than average. This is also the region that will contribute most to the measurement result because the field strength is highest on the surface

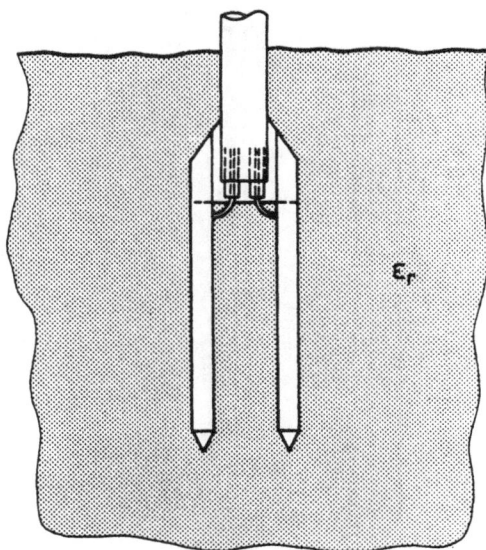

Figure 3.22 The two-conductor line resonator is a sensor that can be pushed into granular materials, powders, or liquids.

of the conductors (Figure 1.8). The value of ε_r'' is impossible in practice to calculate exactly because of the loss by radiation, and is best calibrated with materials of known ε_r (see Figure 2.22). The radiation quality factor (Q_{rad}) is on the order of 100 for a length-to-breadth ratio of 6. Higher ratios give higher Q_{rad} values.

Two-conductor resonators can also be held against the flat surface of the measurement object, in which case the filling factor is less than 1. The contact with the surface and the shape of the conductors will influence the results. We therefore recommend the use of a thin dielectric layer between the conductors and measurement object (or on the surface of the conductors) to reduce the variations in the results. Two-conductor sensors are most easily coupled with loops or capacitively (especially with resonators on printed circuit board).

3.3.6 Slotline Resonator

The *slotline resonator* (Figure 3.23) is closely related to the two-conductor resonator (dual case). The slotline may consist of one or two (or more) slots, which may be separated or joined at the end. The most convenient structure consists of two separate slots, giving a resonator with the electric field maximum in the middle ($l = \lambda/2$). The two-slot structure radiates less energy than a single slot, and can be used as a

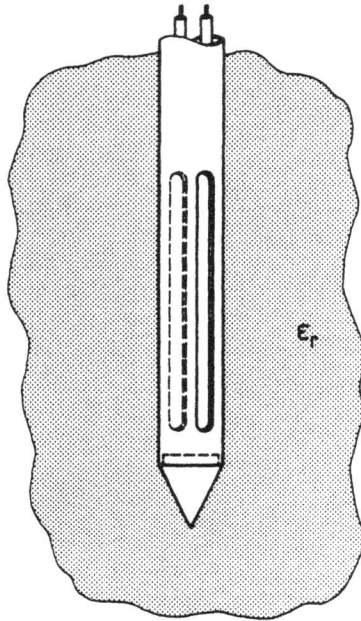

Figure 3.23 The slotline resonator is suitable for measuring noncompressible materials into which the resonator can be pushed. For practical reasons, it is usually filled with epoxy, for example.

flat structure held against the surface of the measurement object, or as a cylinder that can be pushed into the medium. The cylindric resonator can be filled with a dielectric material that keeps the unknown material on the outside, which, at the same time, decreases the sensitivity. The slotline resonator is not well suited for measurement of compressible materials for the reason described in Section 3.3.5. The extension of the cylinder provides a practical shielding for the measurement cables and a rigid handle.

3.3.7 Open Quasioptical Resonator

In cases where the material layer to be measured is thin and a good spatial resolution is needed with a noncontacting sensor, or when a very high frequency is preferred, the *quasioptical resonator* (Figure 2.33) is a possibility. It consists of two spherical concave mirrors, or one concave and one flat mirror. The quasioptical resonator allows for a larger air gap between the sample and resonator structures, as measured in wavelengths, than any other resonator, but is practical only at frequencies above 10 GHz. This resonator was treated in Section 2.7.2 as a means for laboratory measurements, but it can also be used for other applications. One difficulty is the requirement of mechanical rigidity. The exact alignment and a constant distance between the mirrors must be maintained. Neither is the plane of the sample usually allowed to move very much because the field pattern in the Gaussian beam between the mirrors is a normal standing wave pattern. For example, at 35 GHz, the distance between a minimum and the adjacent maximum is only 2.14 mm. However, by moving the sample to find the maximum or minimum value of the shift of resonant frequency, the effect of the errors in the location of the sample can be decreased [Chan *et al.*, 1987].

The formulas for designing the resonator were given in Section 2.7.2. The coupling is most conveniently arranged to a waveguide through an aperture in the middle of the mirror.

3.3.8 Dielectric, Ferromagnetic, and Other Resonators

The *dielectric* and *ferromagnetic resonators* are nonconducting resonators. They are most often used as elements in filters, oscillators, and other microwave components, and seldom used as sensors outside the laboratory. Dielectric resonators are made of a material with very high permittivity. They are usually cylindrical and less than 1 cm in diameter. They are often used for stabilizing the frequency of transistor oscillators. These resonators typically support many different resonant modes, as do cavity resonators. The modes and use of dielectric resonators have been described, for example, in [Kajfez *et al.*, 1986]. One type of ferromagnetic resonator is the YIG (yttrium iron garnet) resonator, which is often encountered in voltage-controlled oscillators. The ferromagnetic resonance frequency is the precession frequency of the electrons, which is tunable with the dc magnetic field from a current loop. The YIG resonator is usually a sphere with a diameter of 0.4–1.0 mm. The modes in dielectric resonators and YIG resonators have an electromagnetic field that extends through their surface into the environment. This field can be used for sensing purposes.

In addition to the most important conventional resonator structures, described above, many others have been used and many more with unprecedented features will be invented. Almost any conducting or dielectric structure will have electromagnetic resonances, just as all mechanical structures exhibit mechanical resonances. We trust, however, that the reader has obtained a general understanding of microwave resonators from the text, which will help understanding and inventing further resonator structures.

3.4 MEASUREMENT OF RESONANT FREQUENCY AND QUALITY FACTOR

3.4.1 Introduction

Equipment that measures the resonance properties and interprets the results fills the last gap in the industrial measurement process for microwave resonators. There are certain properties that are important when we consider the suitability of the equipment for this purpose:

- Either the resonant frequency, quality factor, or both are measured on the basis of the amplitude or phase properties (3.19–3.24) of the resonator;
- Often, there are several resonant modes from which the right one is to be found;
- Often, great relative (sometimes absolute) measuring accuracy is important;
- The changes of resonant frequency and quality factor can be large, in which case both large frequency and power measurement range are demanded;
- Due to rapid changes in the process, great measurement speed may be necessary;
- Sometimes, a large group of sensors is measured with the same equipment;
- The measurement environment may be harsh;
- The measurement equipment usually determines the complexity and thus reliability and price of the measurement system;
- The equipment must be able to manipulate the "raw" results to solve for the wanted material properties or other quantities;
- The equipment must have proper interfaces to the outside world.

The first four of these aspects are characteristic of resonator sensor systems, and the last six are common to all industrial measurement equipment.

3.4.2 The Basic Measurements

Frequency

Depending on the measurement techniques chosen, either frequency, amplitude, or phase measurements are needed to deduce the resonant properties of the resonator sensor.

In almost all resonator applications, frequency measurements are needed. The frequency is the number of oscillations per unit time, so the most natural way to measure is to count the pulses for a certain time determined by a reference oscillator. At low frequencies, up to about 1 GHz, this is possible by using digital counter circuits, realized with the *emitter coupled logic* (ECL). The most usual way, however, is to use a digital prescaler that divides the measured frequency down to the frequency range where TTL or CMOS (up to 100 MHz) circuits can be used (Figure 3.24a). These prescalers are available up to almost 10 GHz, but above about 1.5 GHz they are quite expensive. At these frequencies, a mixer is usually used to lower the frequency. The mixer is a microwave element containing nonlinear semiconductor components (diodes or transistors) and producing (among others) the product of two signals (Figure 3.24b). With sinusoidal signals, this product contains a signal with a frequency equal to the difference of the frequencies of the input signals. By using a stable reference as the other input signal, the measured frequency can be down-converted to a suitable band. At least the other of the mixed signals (the local oscillator (LO) signal) must be at a moderately high power level, typically about 5 mW. The power level of the low frequency output signal (the intermediate or IF signal) is typically about 6 dB lower than that of the weaker input signal. The frequency of the reference signal is sometimes much lower than the frequency of the measured signal, and one of the harmonic components of the reference signal is used. This approach is called *harmonic mixing*.

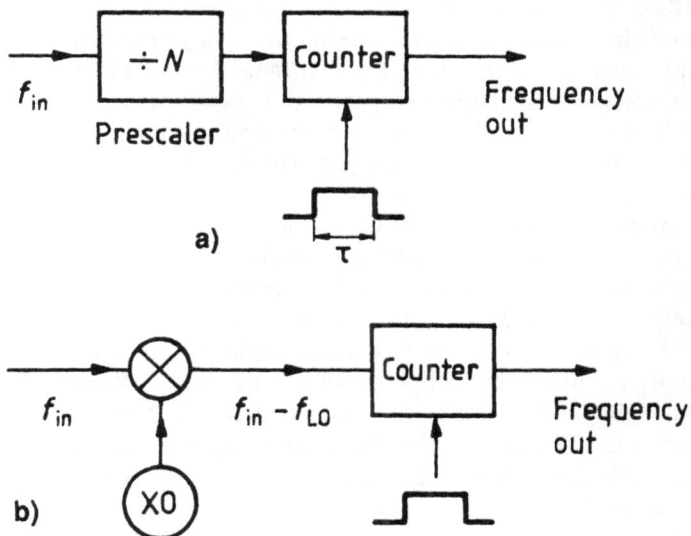

Figure 3.24 Two ways to realize a frequency counter: (a) prescaling ($N = 2–128$) and counting ($f < 1.5$ GHz) (b) mixing with a stable reference (XO) and counting ($f < 20$ GHz).

The long-term accuracy of frequency counting is determined by the reference signal, which is usually based on a crystal oscillator. An overall relative accuracy of about 10^{-5} is quite easy to achieve with crystal oscillators, but 10^{-6} is much more difficult and expensive (a crystal oven or temperature compensation is needed). The short-term errors arise mainly from the instability of the measured signal and the discretization error δf_D determined by the pulse counting period τ and the prescaling ratio N:

$$\delta f_D \leq \pm \frac{N}{\tau} \qquad (3.53)$$

This error determines the time needed for frequency counting; so, if this is critical, as little prescaling as possible should be used (mixing does not affect δf_D).

The same level of accuracy in the determination of measurement frequency as with frequency counters can be achieved by using synthesized signal sources. These are, however, still quite expensive. The sources of error in the determination of frequency are similar to those of frequency counters; long-term errors are caused by the instability of the frequency standard and short-term errors depend on the lock-in speed of the synthesizer.

If the frequency counter is to be replaced with a simpler circuit in the systems described above, a frequency-to-voltage converter can be used. This is a single integrated circuit, the maximum working frequency of which is 0.1–1 MHz. The measured frequency is to be shifted to that range by mixing or prescaling. The linearity of the voltage-to-frequency response is typically about 0.1% and the temperature drift about 0.03%/°C. This circuit offers a simple and cheap solution with moderate accuracy for the frequency measurement in small scale systems such as portable equipment.

At frequencies above 10 GHz, the frequency counter with a stable reference may be too expensive and complicated for simple applications. One solution is to use a circuit with a known amplitude response as a function of frequency (frequency marker) and an amplitude measurement (see below in this section). The circuit can be a stable resonator, or a resonance mode that is not affected by the measured material. By using the difference of the transmission responses of two resonators with a slightly different resonant frequency (see (3.23)), a quite linear voltage *versus* frequency can be accomplished between the resonances (Figure 3.25). The accuracy of this method depends on the stability of the circuits and, above all, the accuracy of the amplitude measurements. Temperature compensation, measurement, or stabilization of the amplitude measuring components is useful. The random error of the frequency measurement of this circuit can be approximated as less than $10^{-3} \Delta f_r$.

At all frequencies, the known connection between the control voltage and frequency of the voltage-controlled oscillator (VCO), can be used as a frequency indicator [Ney *et al.*, 1977; Miyahara *et al.*, 1985]. This connection is, however, usually nonlinear and unstable. Therefore, at least one reference frequency marker must be frequently checked. This marker can be another resonator or resonance mode, or the frequency of the measuring resonator sensor without material.

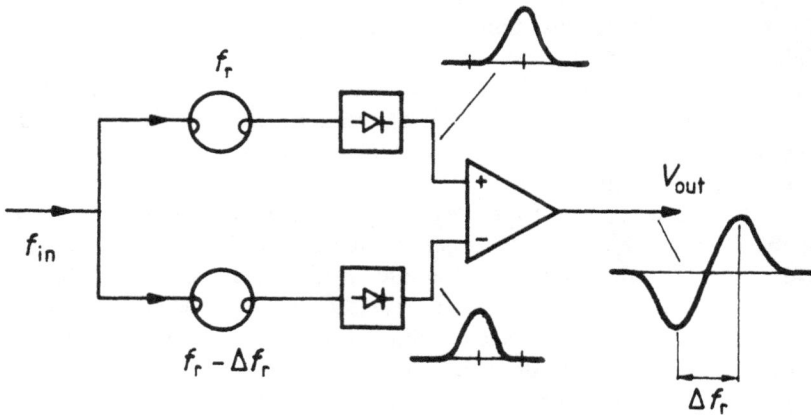

Figure 3.25 Frequency measurement with a frequency-dependent circuit. A linear voltage *versus* frequency is achieved between the resonance peaks.

Amplitude

The amplitude of a microwave signal is usually measured with power indicators. The most suitable for industrial measurements is the Schottky diode detector (the crystal detector). This so-called *square-law detector* rectifies the microwave signal, resulting in a dc voltage proportional to the power of the microwave signal. The component works at power levels typically from about −50 dBm (−50 dB relative to 1 mW) to −15 dBm. The lower limit is set by the noise of the diode and the upper limit by the validity of the square-law behavior. Above the upper limit, the output voltage of the detector is proportional to the microwave signal voltage (linear behavior). The working power range can be shifted down by increasing and up by decreasing the load resistance of the detector. The nominal values are valid for about 10 kΩ load resistance. Respectively, the sensitivity of the detector (typically, about 0.5 mV/μW) increases and the output bandwidth (typically, a few megahertz) decreases with increasing load resistance. The input bandwidth of the detectors is usually large, ranging from less than 10 MHz to 1.5–18 GHz. The frequency response of the component is quite flat, typically <0.5 dB per octave.

The most difficult problem with crystal detectors is the temperature dependence of the sensitivity. It may be halved when the temperature of the detector increases about 20° C. This effect can be avoided by frequent automatic calibration, temperature stabilization, or using a temperature-dependent forward dc bias current for the detector diode. The latter method, however, changes the zero-power dc voltage of the detector, which must be taken into account by calibration or using amplitude modulation of the measurement signal. With some diodes, however, the dc bias current is necessary because without it the output dynamic resistance of the diode is very high.

Another way to measure the microwave power level is to use a local oscillator and mixer to convert the signal to a lower frequency. By using low frequency IF amplifiers with variable gain, a wide power measurement range, -100 to $+10$ dBm, can be achieved. The frequency stability of the local oscillator is not critical and good linearity is gained, but at the cost of a complicated system.

In the microwave signal level measurements, the direction of propagation of the wave is usually important. For example, in the measurement of the reflection loss, the transmitted and reflected power must be separated. Here, directive passive microwave elements are needed, such as directional couplers, hybrids, or circulators. Directional couplers and hybrids couple a certain portion of the signal propagating in a particular direction to the power meter. The most important parameters of directional couplers are the coupling coefficient (-3 dB to -40 dB), which should be constant as a function of frequency, and the directivity, which is the ratio of the coupling coefficients for the signals propagating in the correct and incorrect directions (typically, 15–40 dB).

The circulator is a three-port microwave element in which the signal can propagate only in a certain direction. The attenuation from, for example, port 1 to 2, 2 to 3, and 3 to 1 is low, typically less than 1 dB, and high in the opposite direction, more than 25 dB. The circulator can be used in the reflection measurement (see Section 3.4.3, Figure 3.30c).

Phase

Measurement of phase is a comparison usually performed with a mixer [Dyson, 1966]. As mentioned earlier, a mixer multiplies two signals. If these signals (amplitudes V_1 and V_2) are at the same frequency but have a phase difference $\Delta\phi$, the output voltage V_m of a mixer is

$$V_m = 2KV_1V_2 \sin(2\pi ft) \sin(2\pi ft + \Delta\phi) + V_{m0}$$

$$= KV_1V_2 \cos(\Delta\phi) - KV_1V_2 \cos(4\pi ft + \Delta\phi) + V_{m0} \qquad (3.54)$$

where K is a constant depending on the sensitivity of the device and V_{m0} is a term resulting from the rectifying signals in the mixer. The value of V_{m0} depends on the mixer type and is proportional to $V_1^2 + V_2^2$. With a low-pass filter, we eliminate the second harmonic term and are left with a dc signal proportional to $\cos(\Delta\phi)$. The output signal, however, is also proportional to the input levels. This can be avoided by using high input signal levels to saturate the mixer. The demanded signal levels for both input signals are 1–30 mW, depending on the mixer. The best mixer type for the phase measurements is a double-balanced mixer with an IF response extending to dc (sometimes called a *phase detector*) because $V_{m0} \approx 0$ for it. The reader may note that the output of the mixer phase detector is unambiguous only in a $\Delta\phi$ range of 180°. The temperature drift of the phase detector's output voltage is typically about 0.05%/°C of the full-scale signal. The response time of the phase detector is determined by the IF port bandwidth, which is typically above 10 MHz.

The phase detector can also be realized by using a hybrid junction (the magic tee with waveguides or, for example, 90° stripline hybrid) and two power detectors. By measuring the difference of the detector outputs, a voltage-to-phase response similar to (3.54) can be achieved. We can also accomplish the phase comparison of the signals at a low frequency by using coherent mixing. This kind of phase meter can have a linear response over the whole 360° phase range, but is quite complicated and expensive.

Network Analyzer

The best equipment for all laboratory measurements, production testing, and maintenance operations of microwave sensors is the network analyzer. It measures the reflection and transmission coefficients of linear microwave elements as a function of frequency. Either the amplitude and phase (vector network analyzer) or only the amplitude (scalar network analyzer) can be measured. The frequency range that can be swept is usually large (for example, 0.1–26 GHz) and depends on the signal sources included in the system (Figure 3.26).

The dynamic range of the equipment is high, typically 60–80 dB. The receiver in the vector network analyzer employs narrowband detection with a mixer and constant IF. In scalar analyzers, broadband diode detectors are usually used. Another type of scalar network analyzer can be realized by using a spectrum analyzer with a tracking signal source. In this sytem, the detection bandwidth is also narrow. The narrowband detection usually has a large dynamic range and spurious harmonic signals do not disturb the measurement. However, broadband detection is simple and elements with different input and output frequency (such as mixers) can be measured as well.

The measurement accuracy of the network analyzer depends on the signal levels, frequency, quality of the coupling elements, and especially quality of calibration. The calibration of an *automatic network analyzer* (ANA) is performed as a function of frequency, and corrections are calculated by a computer, which also controls the measurement.

The accuracy of the network analyzer measurement frequency depends on the signal source. If a sweeping synthesizer is used, an additional frequency counter is not needed. With a nonsynthesized source, the accuracy is only moderate and a frequency counter may be needed, as, for example, in resonator measurements.

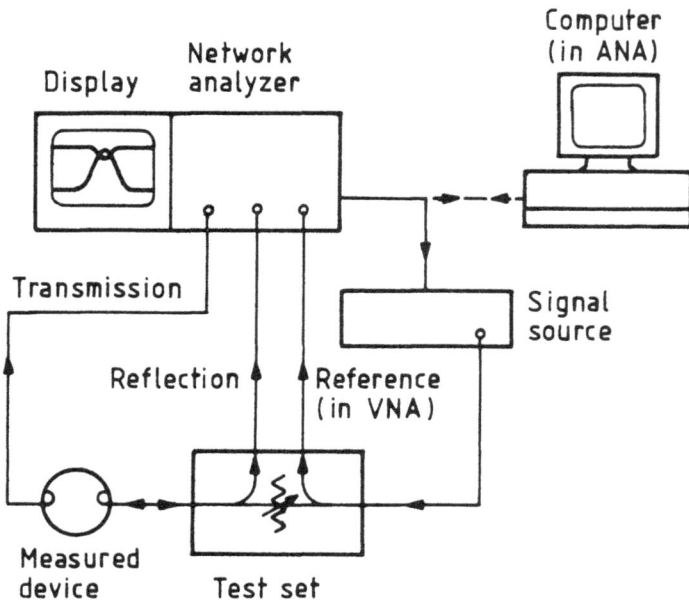

Figure 3.26 A network analyzer system. The signal source is often included in the network analyzer. The system can be automated with a control computer.

There are usually two channels in the analyzer, allowing the reflection and transmission amplitude responses to be simultaneously displayed. Repeated cumbersome measurement routines (for example, laboratory measurements of dielectric properties) can be programmed and automatically executed with the automatic network analyzer.

3.4.3 Different Resonance Measurement Techniques

Near the resonant frequency, the electrical properties of a resonator can be described by a lumped element equivalent circuit. For inductive coupling (see Section 3.3.1), the parallel resonant circuit of Figure 3.27(a), and for capacitive coupling, the series resonant circuit of Figure 3.27(b) can be used if the phase of the reflection response (3.20) of the resonator is also important. Both circuits give the same amplitude response, due to (3.20) and (3.23). Based on the electrical properties of these circuits, several different measurement methods can be used. We can divide these into certain groups based on the connection to the resonator, which response is measured, and the role that the resonator plays in the measurement circuit. We speak of reflection and transmission measurements, phase and amplitude measurements, and active and passive measurement circuits.

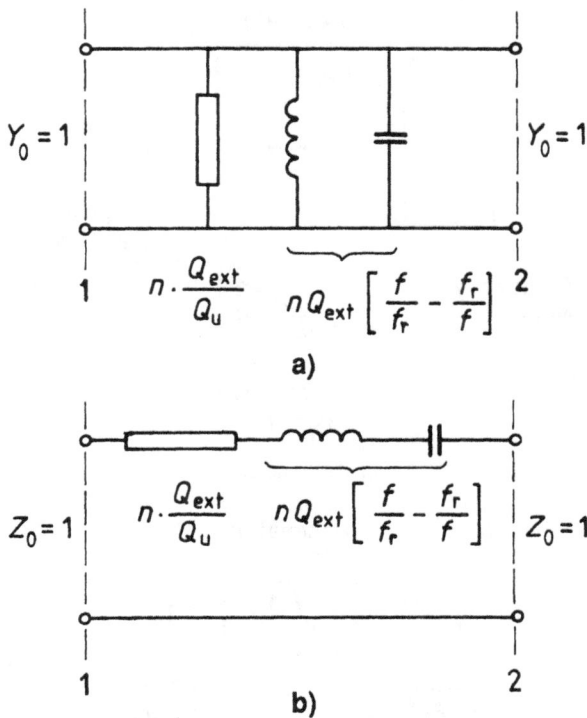

Figure 3.27 The normalized equivalent circuits of resonators. Normally, $(f_r/f - f/f_r) \approx 2\Delta f/f_r$. (a) Admittance with inductive coupling—reflection: port 2 open, $n = 1$; transmission: $n = 2$. (b) Impedance with capacitive coupling—reflection: port 2 shorted, $n = 1$. transmission: $n = 2$.

Resonator as a Part of an Oscillator

Every oscillator contains a resonator that determines the oscillating frequency and stability. This fact is exploited in active resonance measurement methods, where the resonator is a part of the oscillator circuit. As a slight simplification, we can say that the oscillating frequency of the oscillator is determined by the resonant frequency of the resonator, and the amplitude and stability by its quality factor. For example, if the resonator is used as the feedback element of an amplifier (Figure 3.28a) oscillation is possible at a frequency near resonance, provided that

$$\phi + \phi_l + \phi_a = n \cdot 360°, \quad n = 1, 2, 3, \ldots \quad (3.55)$$

Here, ϕ is the phase lag of the resonator, according to (3.24), ϕ_l is that of the connecting transmission lines, and ϕ_a is the phase lag of the amplifier. Another nec-

Resonator

a_r

Frequency
meter

f

Bandpass
filter

$G \cdot a_r > 1$

Microwave
amplifier

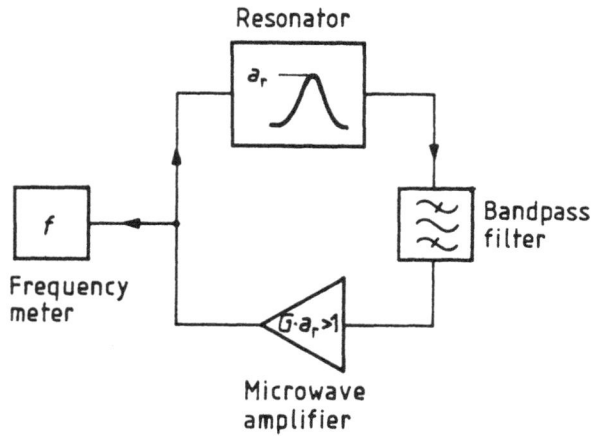

(a) the basic circuit

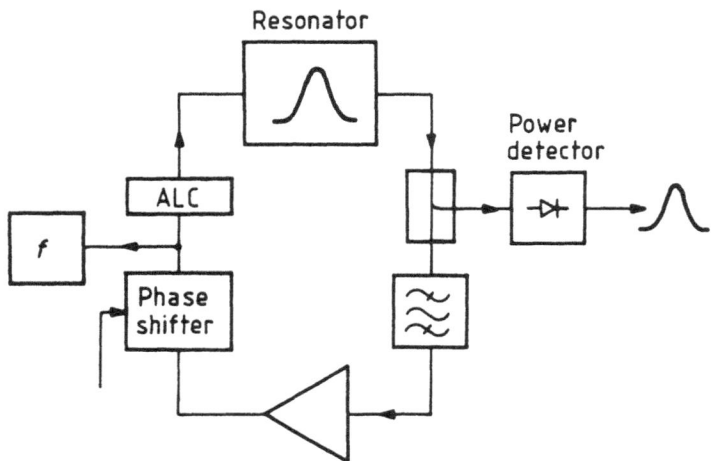

Resonator

Power
detector

ALC

f

Phase
shifter

(b) combination of an oscillator circuit and power measurement

Figure 3.28 Resonator in an oscillator circuit

essary condition for oscillation is that the gain of the amplifier be greater than the insertion loss of the resonator (3.23) at the frequency where condition (3.55) is fulfilled. Because this would not usually occur precisely at the resonant frequency of the resonator, the amplifier should be chosen so that its small signal gain was at least 5 dB greater than the greatest possible insertion loss at the resonant frequency in the measurement situation. The maximum input power of the amplifier, however, should not be exceeded to prevent possible damages. Therefore, use of this method is restricted to measurements for which the quality factor of the resonator does not change very much (low-loss materials, or small filling factor), as we can see from (3.23).

As mentioned above, the oscillation frequency of the circuit in Figure 3.28(a) differs from the center frequency of the resonator. This difference also changes when the resonant frequency changes because ϕ_l depends on the frequency. Changes in the quality factor Q_l of the resonator usually also alter the difference because the frequency derivative of ϕ_r changes. The higher is the quality factor of the resonator, the smaller is the difference between the oscillation and resonant frequencies, and the smaller are the drift and noise of the signal. The drift is caused mainly by the changes of ϕ_a with temperature. To correct these errors may be possible by adding a few circuit elements to the basic oscillator loop (Figure 3.28b). The frequency of the loop can be tuned with a phase shifter. With an ALC circuit at the input and power measurement at the output, the transmission amplitude response of the resonator can be measured to tune the frequency, for example, to the half-power points of the resonator (see below for amplitude transmission measurement). With this improved circuit, the Q_l of the resonator can be accurately measured. With the basic circuit, only a rough estimation of Q_l can be achieved, based on the fact the output power of the oscillator is approximately proportional to the square of the Q_l of the resonator. This may, however, be adequate in the actual measurement situation, especially if Q_l is high [Hoppe *et al.*, 1980]. The active measurement method can be used as well in simultaneous treatment and measurement of the sample via a high-power amplifier in the oscillator loop [Akyel *et al.*, 1985] (see also Section 3.5.1).

The greatest advantage of the active method is the simplicity of the basic circuit of Figure 3.27(a) and the very fast response time. It can be approximated to be about the inverse of the half-power bandwidth of the resonator, and thus the response time is typically a few microseconds. Therefore, very fast changes in the measurement situation can be tracked; the limit is actually set by the frequency meter. The greatest shortcoming of the method is the poor accuracy of the basic circuit, which can be overcome by added complexity of the measuring circuit. The circuit in Figure 3.28(b) combines the fast response time of the active method and accuracy of the passive methods. That circuit also has, however, the basic shortcoming of locking circuits: it does not suit situations where there are several resonances to be measured. A filter is actually used in the oscillator loop to ensure that there is only one resonance into which the circuit can lock.

Passive Amplitude Measurement Methods

In the passive resonance measurement methods, the phase or amplitude response of the resonator to an external stimulus is examined. Usually, the measurement is performed as a function of frequency or, at least, at several frequencies.

The most common way to measure resonator properties is to examine the amplitude response of the resonator. Either the reflection (3.20) or transmission (3.23) coefficient can be measured. As mentioned in Section 3.4.2, the amplitude of microwave signals is usually measured with power detectors, so power equations are also examined (Figure 3.29). In the normal measurement situation, the resonator is undercoupled and the losses caused by the measurement equipment are kept quite low, meaning that $Q_e \gg Q_u$ and $Q_l \approx Q_u$. In Figure 3.30, the basic measurement circuits are presented for reflection and transmission measurements. Instead of using the ALC circuit to stabilize the oscillator output power, it can be measured with a directional coupler and a detector.

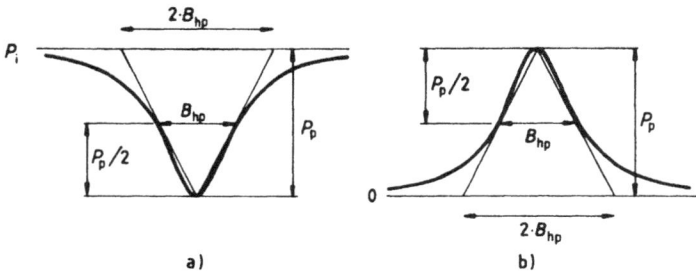

Figure 3.29 The symmetry of the reflection and transmission power responses of resonators. At the half-power points $dP/df = P_p/B_{hp}$.

The problem with the amplitude methods of resonance measurements is to solve three unknowns, Q_l, Q_u, and f_r according to (3.20) and (3.23). The latter are the actual results. To find the values of three unknowns, in principle, at least three independent measurements are needed (see Section 2.5.2). However, in some situations, approximations can be made to decrease the number of measurements and sometimes only the value of either Q_u or f_r is desired. The most demanding measurement situation is the totally filled resonator where both ε_r' and ε_r'' are to be measured. In that situation, three measurements are needed for both the empty and filled resonator [Sihvola *et al.*, 1986].

If the material is lossy, Q_l of the filled resonator can be approximated as equal to Q_d according to (3.37). If this is the case, two measurements are enough to solve Q_d and f_r. For example, the half-power points can be sought, and the following equations may be used to calculate Q_l and f_r (see (3.21) and (3.25)):

$$f_r = \frac{f_{uhp} + f_{lhp}}{2} \tag{3.56}$$

$$Q_l = \frac{1}{2} \cdot \frac{f_{uhp} + f_{lhp}}{f_{uhp} - f_{lhp}} \tag{3.57}$$

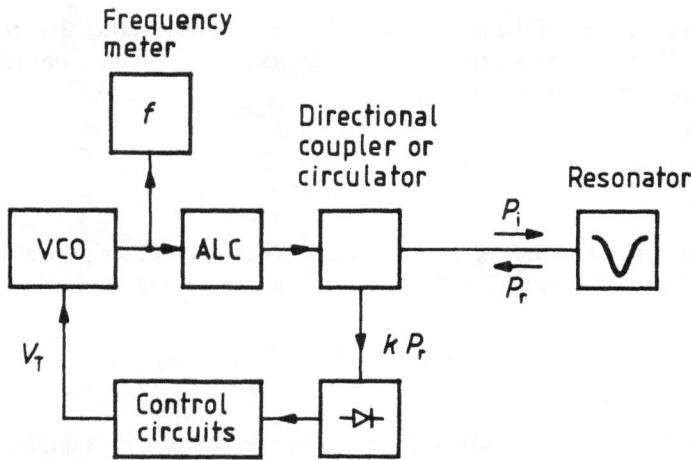

(a) reflection measurement where directional coupler or circulator is needed to separate signals propagating in different directions

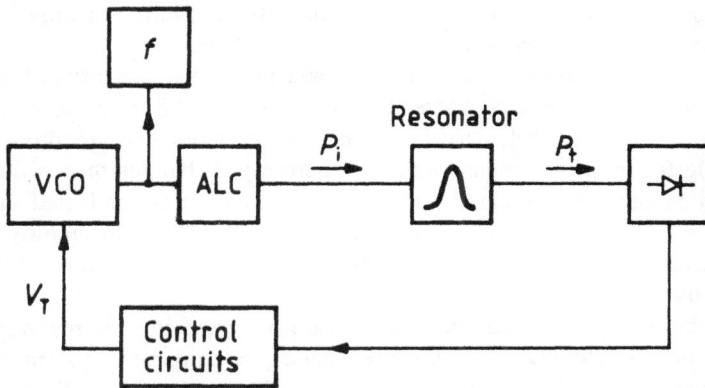

(b) transmission measurement. The ALC in both circuits can be replaced with a measurement of the VCO output power

Figure 3.30 The basic circuits in passive power response measurements of resonators

Here, f_{uhp} and f_{lhp} are the frequencies of the upper and lower half-power points. If the third measurement point is needed, it is usually the peak power of the resonance ($|\Gamma|_r^2$ or a_r). The two-point measurement is sufficient as well for the perturbation measurements, where ε_r'' is solved from $\Delta(1/Q_l)$ (see Section 3.2.2), but the change of Q_l is sometimes very small. In this case, a better accuracy is often achieved by measuring the change of the peak power of the resonance, rather than the half-power bandwidth. When Δf_r is small, Q_{ext} is almost constant. Then, from (3.20), for the reflection measurement, we have

$$\Delta\left(\frac{1}{Q_l}\right) = \frac{1}{Q_{l0}}\left(\frac{1 - |\Gamma|_{r0}}{1 - |\Gamma|_{rd}} - 1\right) \tag{3.58}$$

Here the subscript 0 points at the empty resonator and subscript d at the perturbed situation. Correspondingly, for the transmission measurement, we have

$$\Delta\left(\frac{1}{Q_l}\right) = \frac{1}{Q_{l0}}\left[\left(\frac{a_{r0}}{a_{rd}}\right)^{1/2} - 1\right] \tag{3.59}$$

The quality factor without sample Q_{l0} must be measured, for example, according to (3.57), but the measurement only need be accomplished once because Q_{l0} is constant.

If the measurement of the resonant frequency f_r alone suffices, it can be accomplished by seeking the peak of the resonance curve. However, accurate results are difficult to achieve by seeking the maximum or minimum value of the resonance curve because it is flat at the peak. Therefore, the derivative of the curve, which is zero at the peak, ought to be used [Ney et al., 1977].

Sometimes, especially in laboratory measurements, more than three points are measured from the resonance curve [Pandrangi et al., 1982]. With this method, better accuracy is achieved due to the averaging of errors. The method, however, is not suitable for industrial measurements where rapid changes may occur because all of the points of one resonance measurement ought to be collected approximately from the same curve. In such measurements, a better way to employ averaging is to make several fast measurements in which only one to three points of the resonance curve are used.

The methods described above are based on searching the resonance peak and its certain points. However, tracking resonance measurement systems based on the power response are also often used [Markowski et al., 1977; Tiuri et al., 1974; Kobayashi et al., 1985]. Frequency modulation of the VCO is used in these systems. The reflection or transmission power response of the resonator to the modulated signal is measured, and a synchronous detector is used to create an error signal. Its average value differs in sign on different sides of the resonance peak so that this signal can be used to tune the center frequency of the VCO to resonance.

Such lock-in systems are fast and accurate. They can have a response time on the order of 10 μs [Markowski et al., 1977] and relative resonant frequency measurement accuracy of better than $5 \cdot 10^{-6}$ [Rzepecka et al., 1975]. These systems are best suited to measurements where the changes of f_r and Q_l are small. Only the resonance frequency can usually be measured with the tracking systems.

The dynamic range of the power detector is limited, being only about 35 dB. The maximum possible power in the measurement is received in the reflection response from the resonance and in the transmission response at the resonant frequency of a resonator without sample. In Figure 3.31, the height of the resonance peak P_p is compared to the maximum measurable power P_{max} for both the reflection and transmission response as a function of dielectric losses caused by the measured material. We can notice that total input power range of the power detector is lower for the reflection measurement so that, from the power dynamics point of view, the reflection measurement seems to be superior to the transmission method, especially when Q_d is low. However, there are also other aspects to be examined in this section that influence the choice between reflection and transmission measurement. We can also notice, from Figure 3.31, that the dynamic range demanded in the measurement of lossy materials can be decreased by reducing the quality factor without sample Q_{u0} of the resonator. Q_{u0} can be made smaller by inserting some lossy material in the resonator. The dynamic range can be increased by controlling the transmitted power (with electrically tuned attenuators; see Section 4.4.2) and using a microwave preamplifier in front of the detector.

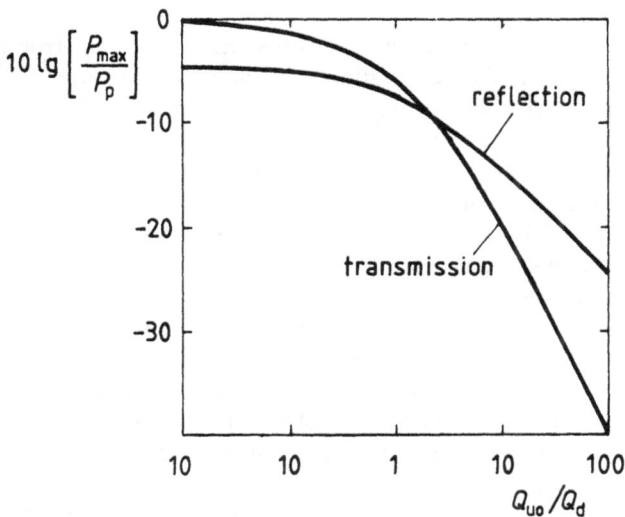

Figure 3.31 The signal levels in typical reflection and transmission measurements as a function of the relative losses of the measured material. For the empty resonator, $Q_{u0}/Q_l = 1.1$. Q_{ext} is approximated to be constant.

The errors of the passive amplitude measurement method are caused by those of either the power or frequency measurement. Short-term random errors (noise) of the power measurement are caused by the oscillator amplitude noise, added noise from the environment, or the power measurement equipment. The latter two are the most common sources of noise and, if proper filtering and other precautions have been taken, the power meter determines the noise level of the power measurement (at least, when a diode detector is used). Systematic errors are caused by the unevenness of the frequency response of either the oscillator (or ALC), connecting circuits, or power meter. The power meter can also have a nonlinear power response.

The power measurement error in a certain measurement point where the frequency is also measured, actually causes a frequency measurement error because the frequency is measured at the wrong point. This error can be calculated in different places of the resonance peak. The absolute value of the result is equal for both reflection and transmission measurement because the shape of the resonance peak is equal for both methods:

$$\frac{\delta f/f_r}{\delta P/P_p} = \pm\frac{1}{4Q_l} \cdot \frac{[1 + (2\Delta f/B_{hp})^2]^2}{2\Delta f/B_{hp}} \tag{3.60}$$

Here, δf is the frequency error caused by the power measurement error δP. The height of the resonance peak is P_p (see Figure 3.29) and the plus sign is valid for the reflection and the minus sign for the transmission measurement. With this equation, we can estimate the effect of different types of power measurement error in different types of resonance measurement. In Figure 3.32, (3.60) is presented. From the curve, we can notice that the frequency error has a minimum at $\Delta f = \pm B_{hp}/\sqrt{12}$, if δP is constant. This is approximately the case when the error is caused by noise, which usually is generated after the resonator.

If the resonance frequency f_r is measured by seeking two symmetrical points on either side of the resonance peak, the error of f_r caused by noise is equal to $(\delta f)_n/\sqrt{2}$, where $(\delta f)_n$ is the measurement error in one point caused by a power measurement error $(\delta P)_n$. This error is the effect of noise on the power measurement and also includes the measurement uncertainty of P_p. Therefore, the points where the effect of noise on the measurement of f_r is lowest are at $f_r \pm B_{hp}/\sqrt{12}$, corresponding to a power level $0.25P_p$ above (reflection) or below (transmission) the peak. At the half-power points, the effect of noise is, in theory, about 30% higher than in the optimum points. In the measurement of Q_l, the error is mostly caused by the measurement error of the bandwidth B_{hp}, which can be calculated to be

$$\delta B_{hp} = \sqrt{2}(\delta f)_n \left|\frac{2\Delta f}{B_{hp}}\right|^{-1} \tag{3.61}$$

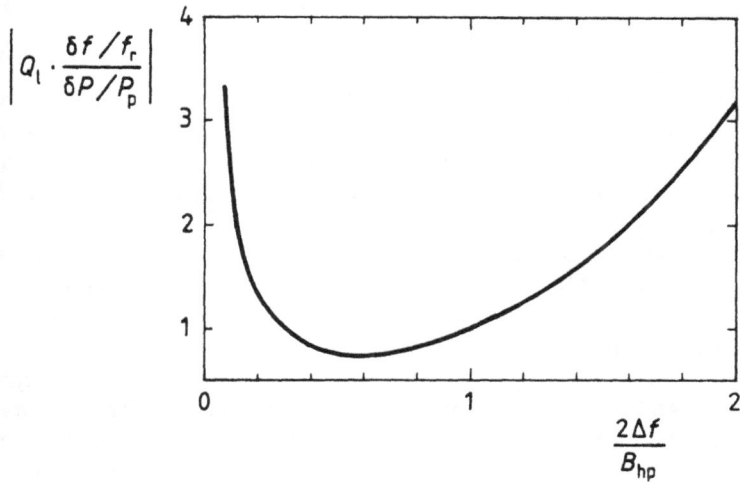

$$\left| Q_l \cdot \frac{\delta f / f_r}{\delta P / P_p} \right|$$

$$\frac{2 \Delta f}{B_{hp}}$$

Figure 3.32 The absolute value of the frequency determination error caused by a power measurement error as a function of relative frequency deviation in the passive power measurements of resonators.

The optimum points are now at the half-power points because of the extra $2\Delta f / B_{hp}$ term so that the optimum is different for f_r and Q_l measurements.

The systematic errors of the power measurement that cause the greatest errors for f_r or B_{hp} measurement are those that create the steepest changes to the frequency response of the measurement system. These are usually the interferences of the actual signal and either multiple reflections in the connecting transmission lines or some leaking signal, such as the wrong-direction coupling of the directional coupler due to finite directivity in the reflection measurement (see Figure 3.29). The interferences create sinusoidal changes to the power response. The greatest error in the measurement of f_r occurs when the first derivative of the response has its maximum or minimum. Respectively, the greatest error in B_{hp} measurement is caused by the maximum or minimum of the second derivative of the "bottom line." The absolute values of the derivatives are greater the longer are the connecting transmission lines in the measurement system. The errors caused by interference are greater the further away are the measurement points from the resonant frequency. The errors are also highly dependent on B_{hp} of the resonator. For the measurement of f_r, according to (3.56), we obtain from (3.60) the worst-case error:

$$(\delta f_r)_s \propto B_{hp}^2 \tag{3.62}$$

and, respectively, for the measurement of B_{hp}:

$$(\delta B_{hp})_s \propto B_{hp}^3 \qquad\qquad (3.63)$$

where the subscript s denotes the systematic errors.

The effect of interference is more severe for the reflection measurement, due to the fact that the height of the measurement peak P_p is often much smaller than the total transmitted and reflected power. The interference, however, is proportional to the total power (see Section 9.5.3), so that the disturbances may be large compared to the measurable signal.

To minimize the interference, the matching of the circuit elements at the ends of long transmission lines should be good and the lines should be as short as possible. The directional couplers or circulators used in the reflection measurement should also have as low leakage as possible.

As an example of the effects of systematic errors, let us consider a resonator with $Q_l = 200$ and $1 - Q_l/Q_u = 0.003$. In transmission measurement, the reflection loss of the ALC output and the detector input (see Figure 3.30b) is 20 dB, and the reflection loss of the resonator is practically 0 dB. The worst-case relative error of f_r measurement by using the average of the optimum noise points is about $\pm(l/\lambda_g) \cdot 20 \cdot 10^{-6}$, where (l/λ_g) is the length of each connecting transmission line in wavelengths. The result does not depend on the height P_p of the resonance peak. For a reflection method with similar parameters and a directional coupler with good (40 dB) directivity, the result would be about $\pm(l/\lambda_g) \cdot 0.8 \cdot 10^{-3}$, which is poor and about 40 times worse than that of the transmission measurement. Actually, in the reflection measurement, the height of the ripple caused by the multiple reflections is many times higher than the resonance peak. The random power measurement errors typically cause relative resonance frequency errors on the order of $5 \cdot 10^{-6}$; so, too, in the transmission measurement are the systematic errors much larger if the connecting transmission lines are long.

The error in the measurement of the peak power is directly caused by the noise amplitude and the total systematic power measurement error. The error of the peak power directly affects determination of Q_u from Q_l, according to (3.22) or (3.26), the measurement of Q_l with perturbation method, according to (3.58) and (3.59), and also indirectly the measurement of B_{hp}.

In fast automatic resonance measurements, a certain reference level is sometimes used, derived from the peak power. The drift of this level, for example, due to the offset drift of low frequency circuits, can cause a systematic error to the measurement of B_{hp}, especially when the peak height P_p is low. This can usually be avoided by choosing circuits with low offset drift to the critical points of the electronics, or by using some kind of base-level checking routine.

The locus of the actual frequency measurement errors is either the signal source or the frequency meter of the system. The noise of the signal source (usually a VCO) is caused by the noise of the tuning voltage or the phase (FM) noise of the oscillator. If the tuning range of the VCO is, for example, one octave, as is quite typical, the signal-to-noise ratio (S/N) of the tuning voltage V_T ought to be about 10^5 (100 dB) to reach the level caused by noise in the power measurement $(\delta f_r/f_r \approx 5 \cdot 10^{-6})$. This is quite difficult, especially if the response time is short and filtering of V_T cannot be used. To avoid substantial noise, the frequency tuning range of the VCO should be as narrow as possible. The FM noise of the VCO is typically around $10 \cdot 10^{-6}$ times the center frequency of the oscillator. This noise is also proportional to the tuning range of the oscillator.

The errors of frequency meters were examined in Section 3.4.2. The discretization error (3.53) of the frequency counter should be chosen smaller than the random frequency measurement errors caused by other parts of the system to enable digital averaging of the result. The stability of the frequency counter, although good, usually determines the stability and repeatability of the resonance frequency measurement with passive power measurement systems. However, in the measurement of $Q_l(B_{hp})$, the stability of the frequency counter has no effect on the total measurement accuracy because the measurement is relative.

For the other frequency determination methods (frequency-to-voltage converter, VCO control voltage, *et cetera*), the problems with the stability of the measurement of f_r are severe. Some kind of stabilization, compensation, or calibration is usually necessary. In the measurement of Q_l, the accuracy of such methods may be adequate.

Passive Phase Measurement Methods

The phase responses of the resonator given in (3.20) and (3.24) can also be used to detect the resonance peak. Phase measurement differs from amplitude measurement in the respect that the connecting circuit has a great effect on the phase response detected as a function of frequency. This is due to the phase lag ϕ_l of transmission lines and other circuit elements:

$$\phi_l = \sum_i \beta_i l_i \qquad (3.64)$$

Here, l_i is the length of the ith circuit element and β_i the phase factor in it (see (1.21)). The phase is a linear function of frequency with TEM lines. Also, with waveguides that are dispersive, the phase response can be approximated as linear in the vicinity of a resonance due to the narrow frequency range. The phase detector measures only phase differences, and so a reference signal is needed as well. The measured phase result ϕ_m is, then,

$$\phi_m = \phi_r + \phi_l - \phi_{ref} \qquad (3.65)$$

Here, ϕ_r is the phase lag of the resonator (see (3.20) and (3.24)) and ϕ_{ref} is the phase lag of the reference signal path that can be calculated in the same way as ϕ_l. If $\phi_l \approx \phi_{ref}$ and the types of transmission line (waveguide, TEM line) are also equal, the effect of the measurement circuit on ϕ_m can be made quite small. However, the measurement is often performed by determining the first derivative of ϕ_m as a function of frequency. When ϕ_l and ϕ_{ref} are practically linear functions of frequency, their derivative is approximately constant. The first derivatives of the reflection and transmission phase responses are shown in Figure 3.33. We can notice that, in the reflection method (Figure 3.33a), the derivative is a steep function of frequency. The height of the peak depends on the coupling (Q_l/Q_u) of the resonator. With the transmission response, the signal does not depend on Q_u, and so solving Q_u from only the transmission phase response is not possible.

The measurement of the resonators' phase response is, in principle, similar to the measurement of the power response. As we can see from Figure 3.34, which presents the basic circuit for the reflection measurement, the power detector is replaced by a phase detector that needs a reference signal. The demand for high signal levels in the inputs of the phase detector (see Section 3.4.2) can be fulfilled in the reflection measurement with loose coupling (P_p is low). In transmission measurement, the signal level at the output of the resonator changes much more, and phase measurement is only possible near the resonance peak by using ALC or power measurement and correction according to (3.54).

When the derivative of the phase response is desired, frequency modulation of the VCO is usually used. The frequency deviation of the modulation should be as constant as possible as a function of frequency. With normal varactor-tuned VCOs, this may be difficult to achieve because the tuning curve of the VCO is usually nonlinear.

When the phase response is used for resonator measurements, the tracking type of resonance measurement circuits is most often employed. The error signal in the tracking may be the actual phase difference caused by the resonator, or frequency modulation may be used. With tracking systems, the effect of the change of the input amplitude of the phase detector is not severe in the measurement of f_r because the "correct" phase response is not essential. However, if both f_r and Q_l are needed, the power is to be measured to be able to lock the circuit to a correct point near the resonance [Linzer *et al.*, 1973].

The random errors in phase response measurement systems are caused by the phase noise of the VCO and noise of the phase measurement system. Systematic errors are caused by the respective sources, as in power response measurements: the connecting circuits, interferences, and phase meter. Their effects on the resonator parameter measurements can be calculated with similar operations to those for the power response measurement. As an example, we can examine the same situation as for power response measurements ($Q_l = 200$, $1 - Q_l/Q_u = 0.003$, $\Delta f = \pm B_{hp}/\sqrt{12}$, reflection losses = 20 dB) to determine the effect of interference.

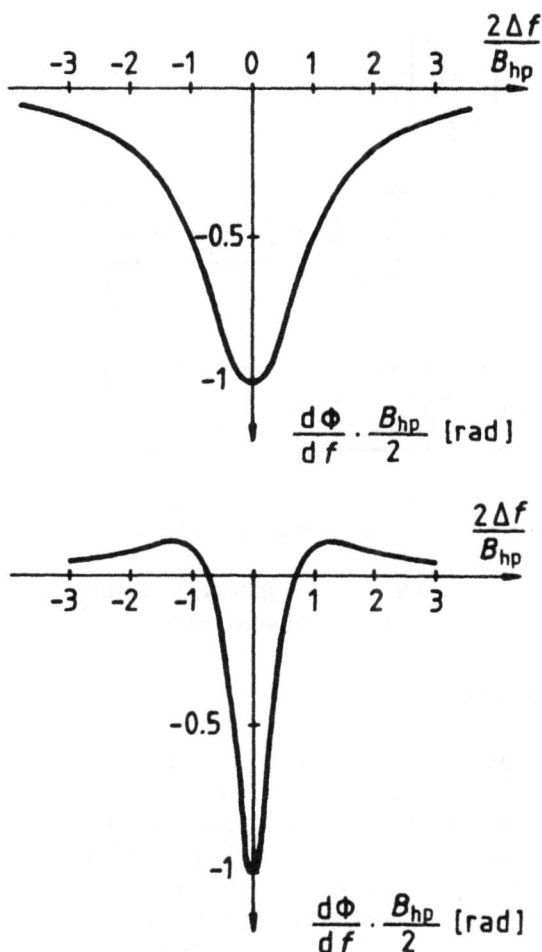

Figure 3.33 The first derivatives of (a) the reflection (3.20) and (b) the transmission (3.24) phase responses for the resonator of Figures 3.5 and 3.6. The peak height in the reflection measurement depends on the Q_l/Q_u of the resonator.

Were we to use the first derivative of the reflection phase response (Figure 3.33a), the result for the relative worst case error of f_r caused by interferences would be about $(l/\lambda_g)^2 \times 20 \times 10^{-6}$. When we compare this result to that for reflection power measurement $[(l/\lambda_g) \cdot 0.8 \cdot 10^{-3}]$, we notice that it is much better, although the effect of the length of the transmission line is greater (due to the use of the derivative). The reason for this is the steepness of the response. We can also notice

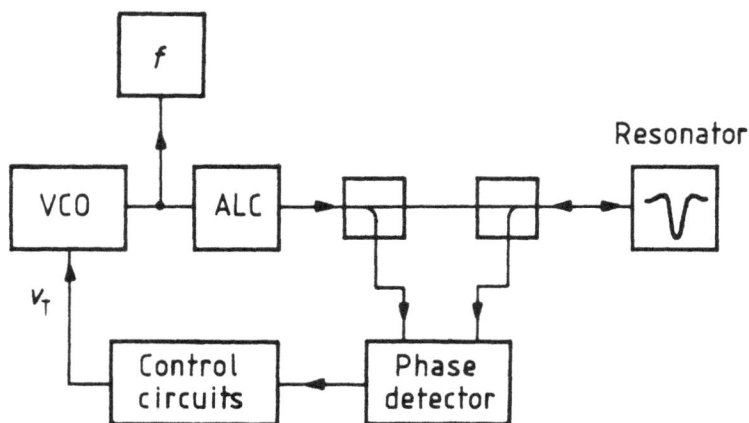

Figure 3.34 The basic circuit for the reflection phase response measurements. The control circuits include a frequency modulator if derivative of the response is measured.

that the phase response method is a significant alternative to the power response measurement, despite being slightly more complicated.

3.4.4 Practical Aspects

In this section, we examine briefly some practical aspects concerning the resonance measurement in different industrial environments and with different types of sensors.

If, in the measurement range, several resonance modes (cavity resonators) can exist, tracking measurement systems are not applicable. If the equipment should also be able to measure several resonances or resonators, some kind of a peak search system, such as a swept measurement, should be used. An example of a measurement system at the UHF range for several (up to 50) resonators with two resonance modes is presented in Figure 3.35. The resonators are connected to the measurement equipment one by one with *pin* diode switches. The measurement is a swept power transmission measurement with fast frequency counting at symmetrical points around the resonance peak. The time of one resonance measurement (f_r and Q_l) is about 10 ms. The speed is limited mainly by the delays in the analog low frequency circuits (preamplifier, sweep control) of the system. The absolute accuracy of the measurement of f_r is about $B_{hp}/100$. The random noise level of a single measurement of f_r is about $10^{-5} \cdot f_r$ [Vainikainen et al., 1987].

Merely relative accuracy is needed in the perturbation measurement. In that case, only those systematic errors with values that change in the measurement range will affect the measurement. By increasing the filling factor of the resonator, the

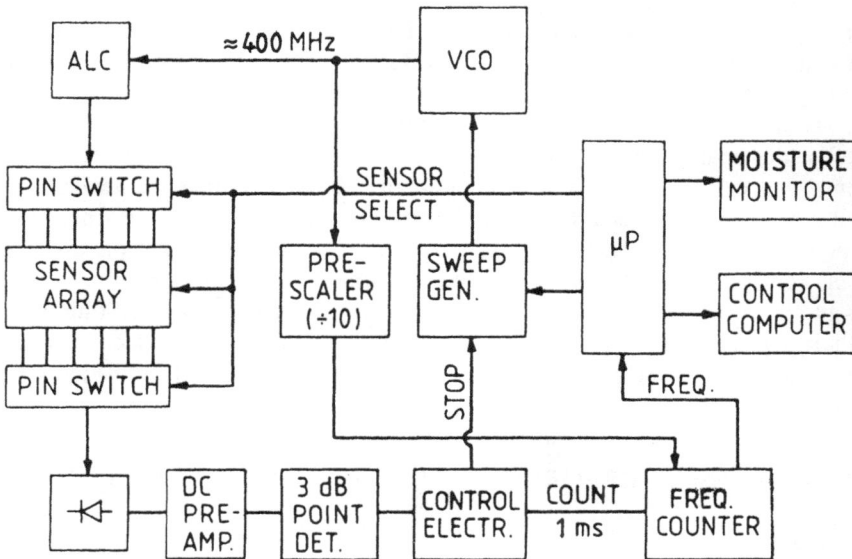

Figure 3.35 An example of a measurement of an array of resonators. The power transmission response is measured to find the half-power points by connecting the resonators one by one to the measuring equipment with high frequency switches. For frequency measurement, a prescaler and counter are used.

measured signals can be increased. However, the effect of the error sources also usually increases due to low Q_l and large change in f_r. This is especially true with high-loss materials. For each measurement method, there are certain ranges in which the resonator parameters are to be kept by choosing the filling factor to ensure proper operation of the equipment. For example, with the passive power transmission method, the change in peak power should not be greater than 20 dB due to the dynamic range of the detector. With all methods, change of resonant frequency greater than one octave or Q_l of less than 10 will cause difficulties.

For example, in the measurements of a fast material flow, a very short response time may be needed [Kobayashi *et al.*, 1985]. In these measurements, a tracking system is usually necessary. The response time of tracking systems is normally on the order of microseconds, whereas with the peak searching systems it is seldom less than 1 ms.

The measurement environment in industry may be harsh. Resonators usually contain only metal and they endure heat, acids, mechanical stress, *et cetera*. The measurement equipment, however, causes problems. The connecting transmission lines closest to the resonator should be either semirigid coaxial cables with PTFE insulator or stainless steel waveguides (low thermal conductivity). In the most difficult situations, wireless connection may be used between resonator and measurement equipment (see Section 3.5). To divide the measurement equipment into parts according to their durability may also be necessary.

The measurement of resonance properties is somewhat complicated. With tracking methods, the analog locking circuits may be complicated and, with peak searching systems, especially when several resonators or resonance modes are measured, some digital control logic is usually needed. The connection between the final results of the measurement (moisture, density, thickness) and the resonator parameters may also be complicated and demand the use of digital computing. However, in many cases the "raw" results are suitable especially for control purposes. Sometimes, the simplicity, price, small size, or low power consumption (with adequate accuracy) of the equipment is the most important property, especially in high-volume applications.

3.5 EXAMPLES OF REPORTED APPLICATIONS OF RESONATORS

3.5.1 Hollow Cavities

On-Line Measurement of Moisture in Paper

The Swedish company, Skandinaviska Processinstrument Ab (Scanpro), has developed a range of both hand-held (Figure 3.36a) and mechanically scanning (Figure 3.36b) cavity resonator sensors for measurement of moisture in paper, pulp, felt, and cellulose. The first model came on market in 1968. As of 1989, its share of the market is substantial.

The sensors employ the one-parameter technique to measure only the resonant frequency, which means that the dry weight of the paper must be known for calculation of moisture. The two-parameter technique would be difficult to use because of the influence of additives on the quality factor. Because of the relatively small perturbation caused by the paper, the sensors use two resonance modes. The reference mode is less affected by the paper than the other mode, thus providing compensation for thermal expansion and humidity variations. The hand-held sensors are contacting one-sided sensors that measure with the fringing field from an aperture. The mechanically scanning sensors and laboratory meters are split cavities for noncontacting measurement. The frequencies of the different modes range through the UHF band. More information can be found in [Agdur, 1969; Bennett, 1973; Steffens, 1983].

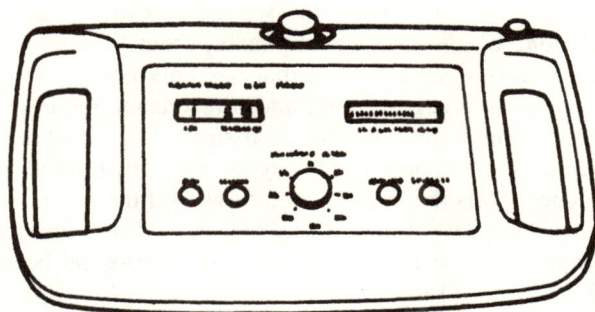

(a)

(b)

Figure 3.36 (a) A hand-held and (b) a mechanically scannable model from the Scanpro cavity resonator sensor family for moisture measurement in paper, pulp, and felt.

Measurement of Flow Rate of Particulate Materials

Several authors have reported flow meters for particulate materials in pneumatic transportation, consisting of a cavity resonator around a dielectric pipe. The resonant frequency gives the amount of material in the cavity at any instant, and the flow rate is calculated considering the speed to be constant. For more accurate results, the speed must also be measured by correlating the outputs from two sensors in series. The speed is the ratio between the separation distance and the delay giving maximum correlation. Kobayaski and Miyahara, from Sumitomo Metal Industries Ltd. in Japan, have developed a system that has only one resonator [Kobayashi *et al.*, 1985]. The cavity is relatively long and a resonance mode is used with $l = 9$ (nine electric field maxima along the material stream). A local density deviation in the stream modulates the resonant frequency as it moves through the maxima. The modulation frequency gives the speed of the flow, and the mean resonant frequency gives the density of the flow. A frequency lock-in system is used, which is able to track the fast changes of the resonant frequency. The sensor has been used to measure the flow of coal dust into a furnace.

Noncontacting Measurement of Thickness of Metal Plates

Sensors for noncontacting measurement of the thickness of metal plates in the steel industry have been reported by several authors [Williams, 1967; Dalton, 1973; Soga, 1973]. Because microwaves do not penetrate through metal sheets, the reported measurement sensors are two-sided. They are all split cavities (Figure 3.37), which together with the metal plate actually form two cavities. Hiromu Soga from Nippon Steel Corporation in Japan has investigated the use of cylindrical cavities using the TE_{011} mode. He found that the resonant frequency of one cavity is a linear function of the distance x (Figure 3.37) to the surface of the plate when $x < \lambda_0/3$. From the two resonant frequencies (f_{r1} and f_{r2}) and the known distance X between the cavities, the plate thickness d can be calculated. Soga also found that tilting the sensor had no effect on the measurement, but a tilt angle of 5^0 decreased the received power by 3 dB. A ring flange was used to reduce the loss by radiation, but for gap widths larger than $\lambda_0/2$, the loss became considerable. To prevent the sensor from making contact with the moving plate, spring suspension was used and an air cushion was formed by a continuous stream of compressed air. An accuracy of ± 0.02 mm was achieved under industrial conditions with a sensor working at 9.4 GHz (see also Section 3.5.5).

Density-Independent Moisture Measurement of Fibrous Materials

The research group for microwave applications at Philips Research Laboratories in Hamburg has developed methods for density-independent moisture measurements based on the ratio $A = (\varepsilon_r' - 1)/\varepsilon_r'$ (see Sec. 2.5.5). In the literature [Hoppe *et al.*, 1980], they describe a cylindrical cavity sensor for measurement of cotton fibres. The material to be measured runs through a pipe mounted axially through the cavity. The TM_{012} mode is used for sensing moisture and TE_{011} for compensation. The measurement system is simple. Two Gunn oscillators are locked through different cou-

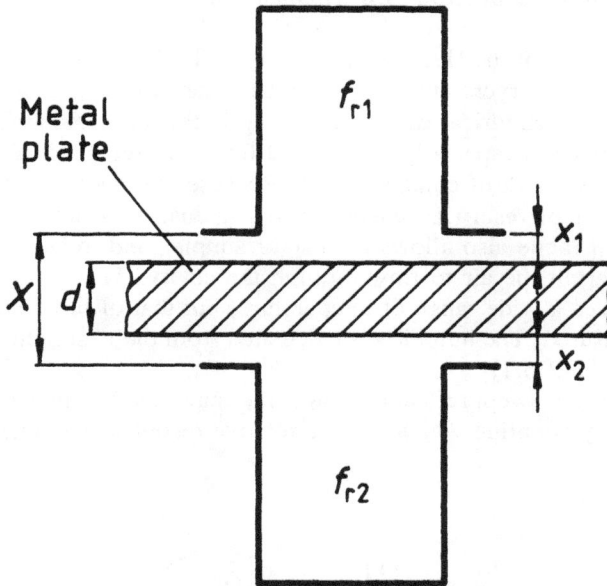

Figure 3.37 A split cylindrical cavity sensor for measurement of thickness of metal plates.

plings to the cavity (active measurement, see Section 3.4.3), one to each mode. The frequency and amplitude of oscillation of the oscillator locked to the TM_{012} mode give ε_r' and ε_r'', respectively.

Measurement of Diameter and ε_r' of Rod-Shaped Samples

At the University of Delaware, a measurement system has been developed for simultaneous measurement of the diameter and ε_r' of dielectric rods [Lakshiminarayana *et al.*, 1979]. The sensor is a cylindrical cavity with the rod inserted axially. Active measurement with a single Gunn oscillator locked to the cavity is used. By temporarily inserting a lossy load (graphite pencil lead), the oscillator can be switched between the modes TM_{010} and TE_{112}. Because of the orthogonality of the electric fields (see (3.43) and (3.47)) at the location of the sample, both the diameter and ε_r' can be calculated. The accuracy was between 1 and 3% for nylon, quartz, and PTFE samples.

Air Humidity Measurement in Harsh Environmental Conditions

At the Radio Laboratory of the Helsinki University of Technology, a meter for measurement of humidity in dryers and ovens has been developed [Toropainen *et al.*, 1987]. The sensor is a cylindrical cavity resonating in the TE_{011} mode (Figure 3.38) at 9.5 GHz. This mode is especially well suited for measurement in dirty environments, where there is a risk of contamination. Because the electric field is zero on all walls, a thin layer of resin (in veneer dryers) or dust does not affect the measurement. The TE_{011} mode also allows the use of simple and relatively open end segments, which permit the air to flow through the cavity. The all-metal structure of the sensor provides a wide range of temperature, but that of both the air and the sensor must be measured. The humidity is calculated from the ε'_r and the temperature of the air (see Section 2.6.1).

The meter uses the swept reflection power measurement technique, a harmonic mixer, and frequency counting which gives a relative resonant frequency resolution of 2×10^{-6}.

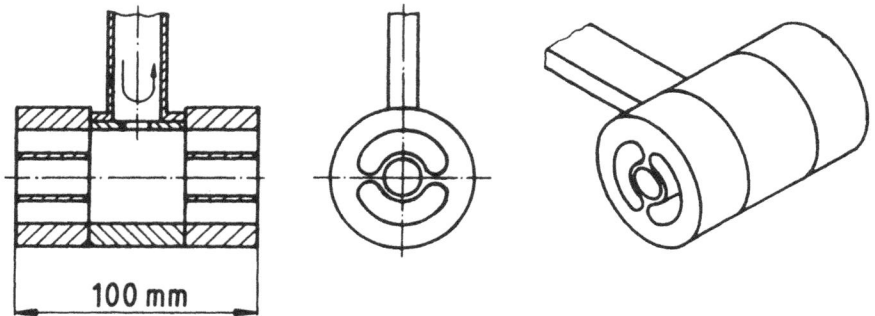

Figure 3.38 A TE_{011} cavity sensor for measurement of humidity in dryers and ovens.

Simultaneous Microwave Treatment and Measurement

Several authors have reported applications where the permittivity of a sample is monitored while the sample is heated or cured (e.g., epoxy) with microwaves [Arenata *et al.*, 1984; Akyel *et al.*, 1985; Jow *et al.*, 1987]. Some use conventional multimode applicators for heating, and perform the measurement at a lower frequency, where the applicator acts like a single-mode cavity. Others use single-mode cavities for heating and measurement on the same frequency.

Konopka and Majewski from the Institute of Physics in Warsaw [Konopka *et al.*, 1980] have studied nonlinear effects in strain-sensitive materials and semiconductors. They have exposed small samples in a cylindrical cavity to strong electro-magnetic fields using the TE_{118} mode at 25.75 GHz, and then studied the nonlinear effects in the samples using the TE_{117} mode at 24.0 GHz. The powerful signal at one frequency modulates the quality factor at the measurement frequency.

Other Applications of Hollow Cavities

S. B. Baumann of Northrop Services, Inc. and others from the Environmental Protection Agency (US) have developed a batteryless miniature temperature transponder [Baumann et al., 1987]. It is a cavity filled with a dielectric of high ε_r', thus providing small size (3 mm in diameter). The cavity is connected to a small antenna, and can be implanted in living animals for research purposes.

The use of microwave techniques for determining the *top dead center* (TDC) of a piston in a combustion engine was proposed in [Merlo, 1970]. The cylinder acts like a variable-length resonator as the piston moves. When measured at a single frequency, the resonator goes through a series of resonances as the piston moves upward. For downward movement, the pattern is reversed, and the TDC can be determined from the point of symmetry. When corrections are made for thermal expansion and pressure changes, an accuracy of $\pm 0.1°$ CA (crank angle) can be achieved [Yamanaka et al., 1985].

The determination of the water content in oil emulsions has been reported, for example, by Doughty of Bartlesville Energy Research Center [Doughty, 1977]. She used a rectangular cavity working in the TE_{102} mode with a thin pipe through the cavity.

Wolff and Schwab from the University of Duisburg (Federal Republic of Germany) have simultaneously measured all components of the permittivity of an anisotropic material using a spherical cavity [Wolff et al., 1980]. The hollow cavity has three orthogonal degenerate components of every resonance mode. When the cavity is loaded with an anisotropic material, the components split and are easily measured. Tiuri and Liimatainen have used a split rectangular cavity with two degenerate modes (TE_{101}, TE_{011}) to measure the anisotropy in paper [Tiuri et al., 1975]. The anisotropy is due to an uneven orientation of the fibers.

3.5.2 Coaxial and Helical Resonators

Measurement of Small Losses with Superconducting Helical Resonators

Very high quality factors can be achieved with cooled, superconducting, closed resonators. Meyer of the Philips Research Laboratories has studied superconducting helical resonators and reported an unloaded quality factor greater than 10^8 [Meyer, 1981], which allows the measurement of very low losses. Helical resonators display many harmonics of the fundamental resonant frequency, and Meyer used the sensors in the frequency range of 100 MHz to 5 GHz. In this range, the small size compared to the wavelength and the small outer diameter compared to the sample diameter are advantageous.

Coaxial Probes and Cavities

Several authors have reported results with open-ended coaxial resonators used as probes for measurement of flat or soft surfaces [Zurcher *et al.*, 1979; King *et al.*, 1983; Moschüring *et al.*, 1984; Xu *et al.*, 1987]. The fields of application range from measuring permittivity of biological substances to moisture in building walls (Figure 3.39). Advantages are portability, nondestructive testing of large objects, and small size of the measurement area. The method is not suitable for rough surfaces.

A group from the Polytechnic University of Madrid has used coaxial cavities for measurement of small deformations [Besada *et al.*, 1980]. Their sensor has sliding junctions in the center and outer conductor so that the length can be varied. The sensor is attached to the measurement object at the location of the short circuits at its ends. This sensor has been used to measure samples of prestressed concrete in the range from room temperature down to cryogenic temperatures. The reader may note that this *extensometer* is not affected by temperature changes in the sensor because the thermal contraction or expansion will automatically be compensated by the sliding junctions.

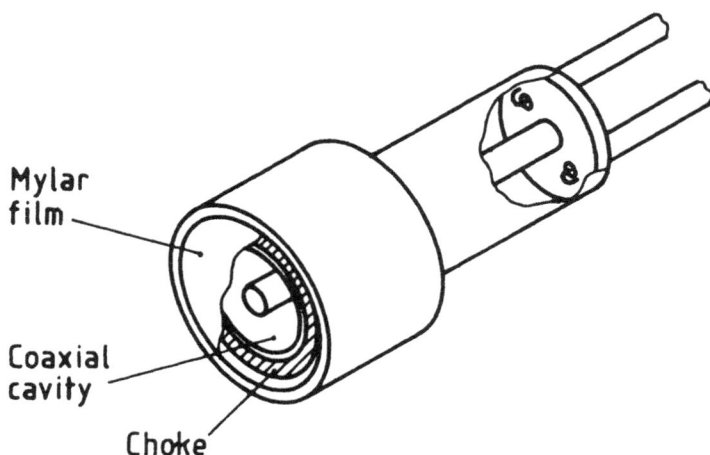

Figure 3.39 An open-ended coaxial resonator probe for measurement of moisture in building walls. The "choke" is a shorted coaxial line, which is a quarter-wavelength long, thus providing an "infinite" impedance at the margin, preventing the fringing of the field. (After [Zurcher *et al.*, 1979].)

3.5.3 Strip Resonators

The research group for industrial applications of microwaves at the Radio Laboratory of the Helsinki University of Technology has developed applications of stripline and two-conductor stripline resonators during the 1980s. Sensors have been developed for noncontacting profiling purposes in the particle board, plywood, and paper industries. Because of the simple and inexpensive resonator structure, the sensors are built in arrays, which can be measured with a single electronics unit. Switches made of *pin* diodes (see Figure 3.17) are used to eliminate mutual coupling between adjacent sensors. A microprocessor-controlled swept insertion loss measurement technique with frequency counting is used. The measurement time for one resonance is about 10 ms, resulting in an almost real-time profiling capability.

A stripline resonator array was developed in cooperation with the company Raute Oy (Finland) for the measurement of the mass per unit area distribution of wooden chips in the particle cake (i.e., the board before pressing) in a particle board factory [Vainikainen *et al.*, 1985]. Only the resonant frequency is used. The characteristics of the sensors are shown in Figure 3.20 and a measured cake in Figure 3.40. If the lower ground plane is wavy or wrinkled (see Section 3.3.4) the sensors must be suspended at the optimum height, which depends on the thickness of the cake and length of the strips.

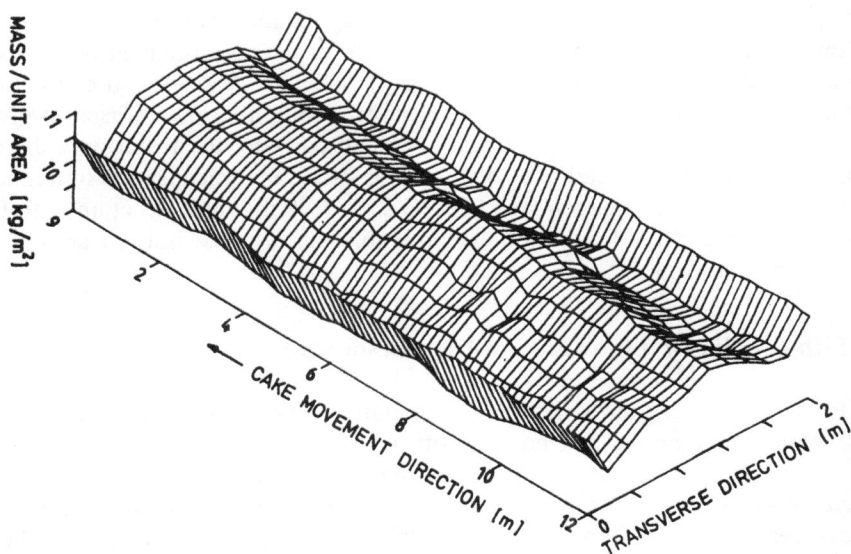

Figure 3.40 A cake of wooden chips as measured with an array of stripline resonator sensors. With proper control of the spreading process and a sensor array, considerable savings in the raw material would be possible in this factory.

A two-conductor stripline array (Figure 3.16c) was developed in cooperation with Raute Oy for the measurement of the moisture distribution in veneer sheets before and after drying [Vainikainen *et al.*, 1987]. Shaped strips (Figure 3.41) are used to give an equal-sided resolution cell. A two-parameter measurement technique based on the even mode is used. Some measurement results are shown in Figure 2.24. The results are independent of the thickness or density of the sheets.

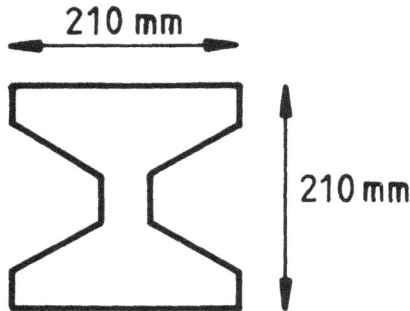

Figure 3.41 Shaped strip for the two-conductor stripline resonator array used for measurement of the moisture distribution in veneer sheets with a spatial resolution of 300 × 300 mm.

Another two-conductor stripline resonator application was developed in co-operation with the company Ivoinfra Oy (Finland) for the measurement of the moisture profile in the wet end of a paper machine [Fischer *et al.*, 1988]. It can be used for the control of, for example, an infrared dryer. Straight strips are used to give a lateral resolution of 10 cm. Both modes (even and odd) are used, but only the resonant frequencies are measured. The odd mode is used for compensation of humidity variations and thermal expansion (see Section 3.3.4). A modular structure with four sensors per module is used, which makes possible the fast assembly of an array of desired width (Figure 3.42).

3.5.4 Two-Conductor Line and Slot-Line Resonators

At the Radio Laboratory of the Helsinki University of Technology two sensors have been developed that can be pushed into soft materials, liquids, powders, *et cetera*. The snow fork [Sihvola *et al.*, 1986] is a two-conductor line resonator (Figure 3.22) designed for field use. The characteristics of the sensor are shown in Figure 2.22. The measurement is based on the snow model in Section 2.5.4, graphically shown in Figure 2.21, from which the dry density (ρ_d) and the volume fraction of liquid water (f_w) in the snow can be read.

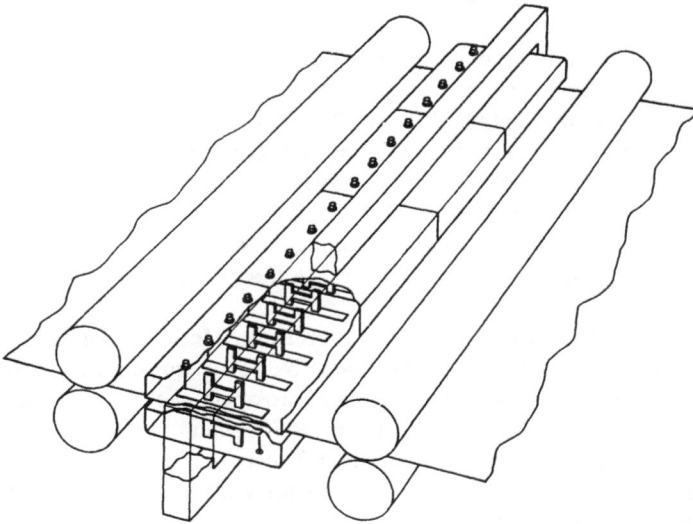

Figure 3.42 A two-conductor stripline resonator array with a modular structure for measurement of the real-time moisture profile with 10 cm resolution in the wet end of a paper machine.

The peat probe developed by Toikka [Tiuri *et al.*, 1983], is a slotline resonator like the one shown in Figure 3.23. It is used together with the peat radar for evaluation of the peat resources in marshes. The probe is mounted in the end of a steel pipe, which is pushed through the peat layers. Using extensions of the pipe, a depth of about 10 m can be reached by using only hand power. The probe measures the ε'_r of the peat, which correlates very well with the energy content expressed in GJ/m^3. A portable, microprocessor-controlled and battery-operated electronics unit for field testing and data sampling with the snow fork and peat probe has been developed at Toikka Engineering, which is also manufacturing the sensors. The measurement is of power transmission. The unit stores the results in a solid-state memory cassette with a capacity of approximately one working day.

Another slotline resonator was developed for the measurement of the surface homogeneity of supercalendar rollers used for polishing paper [Tiuri *et al.*, 1974]. The resonator is a planar structure consisting of two 150 mm long and 4 mm broad slots machined in a plate with the same curvature as the roller. A thin PTFE film was used between the sensor and roller. The resonant frequency is measured with the passive frequency lock-in method.

3.5.5 Open Quasioptical Resonators

Quasioptical resonators have not been used in many cases in industry, probably because of the tight tolerances for the mechanical rigidity and the shape and location of the sample. Chan and Chambers from the University of Sheffield however, have measured at 11.6 GHz the permittivity of curved (convex-concave) dielectric antenna radome samples with various radii of curvature down to 300 mm [Chan *et al.*, 1987]. From perturbation theory, they have derived a correction term for curved samples. In the measurements, they have achieved an accuracy of 1.5% in ε'_r and 35% in $\tan\delta$ (when $\tan\delta \approx 0.005$) using mirrors 216 mm in diameter with radius of curvature of 330 mm.

Miyahara and Kobayashi, from the Sumitomo Metal Industries Ltd. in Japan, have made a thickness meter for metal plates based on quasioptical resonators [Miyahara *et al.*, 1985]. They use one parabolic mirror on each side of the metal plate. The advantage is the larger distance between the resonators and metal plate as compared with cylindrical cavities. The disadvantage is the sensitivity to tilt. Miyahara and Kobayashi used mirrors 195 mm in diameter with a focal length of 70.82 mm. The frequency range was 8.2–9.8 GHz, the measurement range for each mirror was 63–107 mm, and the accuracy ±20 μm. (See also Section 3.5.1.)

3.5.6 Dielectric and Ferromagnetic Resonators

Dielectric resonators have been used as sensors for measurement of the local permittivity of, for example, substrates for microwave integrated circuits (MICs). Clapeau and others have developed a double-sided, shielded (size of the shield is two or three times the size of the dielectric resonator) sensor for measuring the substrate (e.g., alumina, silica, rexolite) plates [Clapeau *et al.*, 1977].

Toutain and others, and Korneta and others have used dielectric resonators for measurement of the permittivity of liquids [Toutain *et al.*, 1976; Korneta *et al.*, 1988]. The sensor is a hollow dielectric resonator, which is filled with the liquid to be measured (Figure 3.43). The method allows the measurement of very small samples. It is also possible to use a solid resonator, which is surrounded by the liquid.

When a ferromagnetic YIG resonator is brought in close vicinity to a metal surface, the resonator will induce eddy currents in the surfaces. Auld and others have developed a miniature sensor using this principle for detection of surface flaws [Auld *et al.*, 1978]. At the location of a flaw, the route of the currents changes, which perturbs the magnetic field at the YIG sphere, thus changing the resonant frequency.

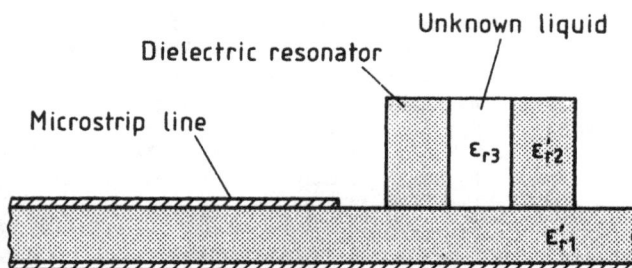

Figure 3.43 Dielectric resonator for measurement of the permittivity of a small quantity of a liquid.

REFERENCES

Agdur, B., "En ny metod för fuktmätning," *Norsk Skogsindustri,* No. 7–8, 1969, pp. 209–219.

Akyel, C., R.C. Labelle, A.J. Berteaud, and R.G. Bosisio, "Computer-Aided Permittivity Measurements of Moistened and Pyrolyzed Materials in Strong RF Fields (Part I)," *IEEE Trans. Instr. Meas.,* Vol. IM-34, No. 1, March 1985, pp. 25–31.

Arenata, J.C., M.E. Brodwin, and G.A. Kriegsmann, "High-temperature Microwave Characterization of Dielectric Rods," *IEEE Trans. Microwave Theory Tech.,* Vol. MTT-32, No. 10, October 1984, pp. 1328–1335.

Auld, B.A., G. Elston, and D.K. Winslow, "A Novel Microwave Ferromagnetic Resonance Probe for Eddy Current Detection of Surface Flaws in Metals," *Proc. 8th European Microwave Conf.,* Paris, September 1978, pp. 603–607.

Baumann, S.B., W.T. Joines, and E. Berman, "Feasibility Study of Batteryless Temperature Transponders Using Miniature Microwave Cavity Resonators," *IEEE Trans. Biomed. Eng.,* Vol. BME-34, No. 9, September 1987, pp. 754–757.

Bennett, P.G., "The Use of a Microwave Moisture Meter for Process Investigations," *Appita,* Vol. 26, No. 4, 1973, pp. 267–272.

Besada, J.L., M. Elices, J. Planas, J. Sanches Miñana, and J.A. García Cachero, "A New Technique of Deformation Measurements Based on Microwave Resonant Cavities," *Proc. 10th European Microwave Conf.,* Warsaw, September 1980, pp. 288–292.

Chan, W.F.P., and B. Chambers, "Measurement of Nonplanar Dielectric Samples Using an Open Resonator," *IEEE Trans. Microwave Theory Tech.,* Vol. MTT-35, No. 12, December 1987, pp. 1429–1434.

Clapeau, M., P. Guillon, and Y. Garault, "Resonant Frequency of Superposed Dielectric Resonators: Application to the Determination of the Local Dielectric Permittivity of M.I.C. Substrates," *Proc. 7th European Microwave Conf.,* Copenhagen, 1977, pp. 545–549.

Collin, R.E., *Foundations for Microwave Engineering,* New York: McGraw-Hill, 1966, 589 p.

Dalton, B.L., "Microwave Noncontact Measurement and Instrumentation in the Steel Industry," *J. Microwave Power,* Vol. 8, No. 3, 1973, pp. 235–244.

Doughty, D.A., "Determination of Water in Oil Emulsions by a Microwave Resonance Procedure," *Analytical Chemistry,* Vol. 49, No. 6, May 1977, pp. 690–694.

Dyson, J.D., "The Measurement of Phase at UHF and Microwave Frequencies," *IEEE Trans. Microwave Theory Tech.,* Vol. MTT-14, No. 9, September 1980, pp. 410–422.

Fischer, M., P. Vainikainen, E. Nyfors, and M. Kara, "Fast Moisture Profile Mapping of a Wet Paper Web with a Dual-Mode Resonator Array," *Proc. 18th European Microwave Conf.,* Stockholm, September 1988, pp. 607–612.

Gardiol, F.E., *Introduction to Microwaves*, Norwood, MA: Artech House, 1984, 495 p.

Harrington, R.F., *Time-Harmonic Electromagnetic Fields*, New York: McGraw-Hill, 1961, 480 p.

Hoppe, W., W. Meyer, and W.M. Schilz, "Density-Independent Moisture Metering in Fibrous Materials Using a Double-Cut-off Gunn Oscillator," *IEEE Trans. Microwave Theory Tech.*, Vol. MTT-28, No. 12, December 1980, pp. 1449–1452.

Jow, J., M.C. Hawley, M. Finzel, J. Asmussen, H-H. Lin, and B. Manring, "Microwave Processing and Diagnosis of Chemically Reacting Materials in a Single-Mode Cavity Applicator," *IEEE Trans. Microwave Theory Tech.*, Vol. MTT-35, No. 12, December 1987, pp. 1435–1443.

Kajfez, D., and P. Guillon, *Dielectric Resonators*, Norwood, MA: Artech House, 1986, 500 p.

King, R.J., and P. Stiles, "Microwave Nondestructive Evaluation of Composites," *Review of Progress in Quantitative Nondestructive Evaluation, Vol. 3, Proc. 10th Annual Review*, Santa Cruz, CA, August 1983, pp. 1073–1081.

Kobayashi, S., and S. Miyahara, "A Particulate Flow Meter Using Microwaves," *Proc. IMEKO*, Prague, 1985, pp. 112–119.

Konopka, J., and J.J. Majewski, "A New Microwave Method for Testing Nonlinear Effects in Solids," *Proc. 10th European Microwave Conf.*, Warsaw, September 1980, pp. 246–250.

Korneta, A., and A. Milewski, "The Application of Two- and Three-Layer Dielectric Resonators to the Investigation of Liquids in the Microwave Region," *IEEE Trans. Instr. Meas.*, Vol. IM-37, No. 1, March 1988, pp. 106–109.

Lakshminarayana, M.R., L.D. Partain, and W.A. Cook, "Simple Microwave Technique for Independent Measurement of Sample Size and Dielectric Constant with Results for a Gunn Oscillator System," *IEEE Trans. Microwave Theory Tech.*, Vol. MTT-27, No. 7, July 1979, pp. 661–665.

Linzer, M., and D.P. Stokesberry, "A Frequency-Lock Method for the Measurement of Q Factors of Reflection and Transmission Resonators," *IEEE Trans. Instr. Meas.*, Vol. IM-22, No. 1, March 1973, pp. 61–77.

Markowski, J., A.D. MacDonald, and S.S. Stuchly, "The Dynamic Response of a Resonant Frequency Tracking System," *IEEE Trans. Instr. Meas.*, Vol. IM-26, No. 3, September 1977, pp. 231–237.

Merlo, A.L., "Combustion Chamber Investigations by Microwave Resonances," *IEEE Trans. Ind. Electron. Contr. Instrum.*, Vol. IECI-17, No. 2, April 1970, pp. 60–66.

Meyer, W., "Helical Resonators for Measuring Dielectric Properties of Materials," *IEEE Trans. Microwave Theory Tech.*, Vol. MTT-29, No. 3, 1981, pp. 240–247.

Miyahara, S., and S. Kobayashi, "A Thickness Meter Using the Resonance of Microwaves," *Trans. IECE of Japan*, Vol. E-68, No. 4, April 1985, pp. 227–232.

Moschüring, H., and I. Wolff, "Inhomogeneous Open-Ended Resonators as Microwave Sensor Elements," *IEEE MTT-S Int. Microwave Symp. Digest*, San Francisco, June 1984, pp. 187–189.

Ney, M., and F.E. Gardiol, "Automatic Monitor for Microwave Resonators," *IEEE Trans. Instr. Meas.*, Vol. IM-26, No. 1, March 1977, pp. 10–13.

Nyfors, E., and P. Vainikainen, "Sensor for Measuring the Mass per Unit Area of a Dielectric Layer," *Proc. 14th European Microwave Conf.*, Liége, September 1984, pp. 667–672.

Pandrangi, R.K., S.S. Stuchly, and M. Barski, "A Digital System for Measurement of Resonant Frequency and Q Factor," *IEEE Trans. Instr. Meas.*, Vol. IM-31, No. 1, March 1982, pp. 18–21.

Rzepecka, M.A., S.S. Stuchly, and M.A.K. Hamid, "High-Frequency Monitoring of Residual and Bound Water in Nonconductive Fluids," *IEEE Trans. Instr. Meas.*, Vol. IM-24, No. 3, September 1975, pp. 205–210.

Sihvola, A., "Note on Frequency Dependence of Quality Factor of Cavity Resonators," *Electron. Lett.*, Vol. 21, No. 17, 1985, pp. 736–737.

Sihvola, A., and M. Tiuri, "Snow Fork for Field Determination of the Density and Wetness Profiles of a Snow Pack," *IEEE Trans. Geoscience and Remote Sensing*, Vol. GE-24, No. 5, September 1986, pp. 717–721.

Soga, H., "A New Microwave Thickness Gauge," *J. Microwave Power,* Vol. 8, No. 3, September 1973, pp. 253–266.

Steffens, D., "Scanpro-Feuchtemessung von der 1. Presse bis zur Aufrollung," *Wochenblatt für Papierfabrikation,* No. 8, 1983, pp. 259–262.

Tiuri, M., and P. Liimatainen, "A Microwave Method for Measurement of Fiber Orientation in Paper," *J. Microwave Power,* Vol. 10, No. 2, June 1975, pp. 141–145.

Tiuri, M., P. Liimatainen, and S. Reinamo, "A Microwave Instrument for Measurement of Small Inhomogeneities in Supercalendar Rollers and Other Dielectric Cylinders," *J. Microwave Power,* Vol. 9, No. 2, June 1974, pp. 117–121.

Tiuri, M., M. Toikka, I. Marttila, and K. Tolonen, "The Use of Radio Wave Probe and Subsurface Interface Radar in Peat Resource Inventory," *Proc. Symposium of IPS Commission I,* Aberdeen, Scotland, 1983, pp. 131–143.

Toropainen, A.P., P.V. Vainikainen, and E.G. Nyfors, "Microwave Humidity Sensor for Difficult Environmental Conditions," *Proc. 17th European Microwave Conf.,* Rome, September 1987, pp. 887–891.

Toutain, S., J. Citerne, P. Gelin, J.P. Parneix, and L. Raczy, "Complex Permittivity Measurements of Liquids Using a Microwave Dielectric Resonator," *Proc. 6th European Microwave Conf.,* Rome, September 1976, pp. 181–185.

Vainikainen, P.V., E.G. Nyfors, and M.T. Fischer, "Radiowave Sensor for Measuring the Properties of Dielectric Sheets: Application to Veneer Moisture Content and Mass per Unit Area Measurement," *IEEE Trans. Instr. Meas.,* Vol. IM-36, No. 4, December 1987, pp. 1036–1039.

Vainikainen, P.V., and E.G. Nyfors, "Sensor for Measuring the Mass per Unit Area of a Dielectric Layer: Results of Using an Array of Sensors in a Particle Board Factory," *Proc. 15th European Microwave Conf.,* Paris, September 1985, pp. 901–905.

Waldron, R.A., "Perturbation theory of resonant cavities," *Proc. IEE,* Vol. 107C, 1960, p. 272–274.

Williams, R.V., "Application of Microwave Techniques in the Iron and Steel Industry," *ISA Proc. Nat. Conf. Inst. for Iron and Steel, 1967.*

Wolff, I., and N. Schwab, "Measurement of the Dielectric Constant of Anisotropic Dielectric Materials Using the Degenerated Modes of a Spherical Cavity," *Proc. 10th European Microwave Conf.,* Warsaw, September 1980, pp. 241–245.

Xu, D., L. Liu, and Z. Jiang, "Measurement of the Dielectric Properties of Biological Substances Using an Improved Open-Ended Coaxial Line Resonator Method," *IEEE Trans. Microwave Theory Tech.,* Vol. MTT-35, No. 12, December 1987, pp. 1424–1428.

Yamanaka, T., M. Esaki, and M. Kinoshita, "Measurement of TDC in Engine by Microwave Technique," *IEEE Trans. Microwave Theory Tech.,* Vol. MTT-33, No. 12, December 1985, pp. 1489–1494.

Zurcher, J.F., and F.E. Gardiol, "Nondestructive Microwave Measurements of Materials' Moisture in Building Walls," *Proc. IMEKO Congress Int. Measurement Confederation,* Moscow, May 1979, pp. 393–398.

Chapter 4
Transmission Sensors

4.1 INTRODUCTION

Most of the early tests with industrial microwave sensors were performed by using some kind of transmission sensor. A microwave signal was directed through a layer of the material to be measured, and the attenuation or phase shift caused by the layer was measured. The simplicity of both the principle and the equipment needed was the most attractive feature. When only the attenuation was measured, the microwave equipment could be reduced to a minimum and interpretation of the data was simple. The attenuation could be calibrated directly, for example, as a function of moisture, which was an advantage because the microprocessor had not yet been invented. Such a simple system however, has several limitations, for example, dependence on layer thickness, density, evenness of surfaces, reflections, and thermal drift. Therefore, modern meters are more complicated and often employ the microwave multiparameter technique combined with γ-ray attenuation measurement. Sensors for many different situations have been developed, and as a result transmission sensors have not become old-fashioned, but today probably form the majority of operational microwave sensors.

The advantage with the transmission sensor is its versatility. The same sensor can be used for different applications. Compared to resonators, conventional transmission sensors (propagation in free space) are better suited to measurement of thick layers of lossy materials in pipes or on conveyors, but are also used with sheet-like materials. Transmission sensors are not suited for measurement of objects with smaller transverse dimensions than a few wavelengths. The term "transmission sensor", however, comprises all devices whereby a microwave signal is transmitted from one point to another in such a way that it is attenuated and phase shifted by the object to be measured in the path. In this chapter, we will study both conventional (called *free-space*) transmission sensors and those based on transmission lines (called *guided wave transmission sensors*). The latter have several features in common with resonators and are used for similar applications (e.g., measurement of small samples in a waveguide).

4.2 FREE-SPACE TRANSMISSION SENSORS

4.2.1 Sensor Structures

In the free-space transmission sensor, the microwave measurement signal ("the wave") propagates from the transmitting antenna to the receiving antenna, passing through the material layer to be measured ("the sample"). Between the antennas, the wave propagates independently of any transmission lines. Because the measurement of phase is comparative, a reference signal is taken from the same oscillator as the measurement signal and fed to the receiver via the reference branch. Free-space transmission sensors are used primarily in two cases: on conveyors and with pipes (Figure 4.1).

The antennas are usually horns because the practical lower frequency limit is a few gigahertz. At lower frequencies, propagation around the sample (diffraction) and antenna near-field effects (nonplanar wavefronts and reactive nonradiating fields) cause difficulties. The most commonly used frequencies fall in the range of 5–15 GHz, but millimeter waves have been used with thin samples because of the higher attenuation. Due to the still very high prices of millimeter-wave components, unnecessarily high frequencies are avoided.

As the wave propagates through the sample, it is attenuated and phase shifted. The thickness of the sample, of course, directly affects the result if it is not kept constant, measured separately, or taken into account in some other way. In a pipe this is not a problem if it is completely filled in all situations. In addition to the desired effects, the wave is also reflected at the surfaces of the sample. If the thickness of the sample is over $\lambda/4$ and attenuation is low, internal multiple reflections will cause a standing wave to appear between the sample surfaces, adding a periodic variation to the attenuation as a function of sample thickness, permittivity (e.g., density, moisture), and frequency. Multiple reflections between the sample and antennas (or other nearby objects) may also appear, causing additional errors in the result. The phase is less affected by the reflections than the attenuation, but, for accurate measurements, all unnecessary reflections ought to be eliminated. The inevitable reflections should be taken into account and possibly even measured, and attenuation in the sample should be high enough. The effect of standing waves can also be eliminated by averaging over a certain minimum frequency range. These matters are studied further in Sections 4.2.3 and 4.4.

4.2.2 Theory

As a first approximation, we may neglect the reflections in our study of the wave propagation through the sample. The antenna transmits a spherical wave, but because we are interested in changes relative to the situation without the sample, we can also

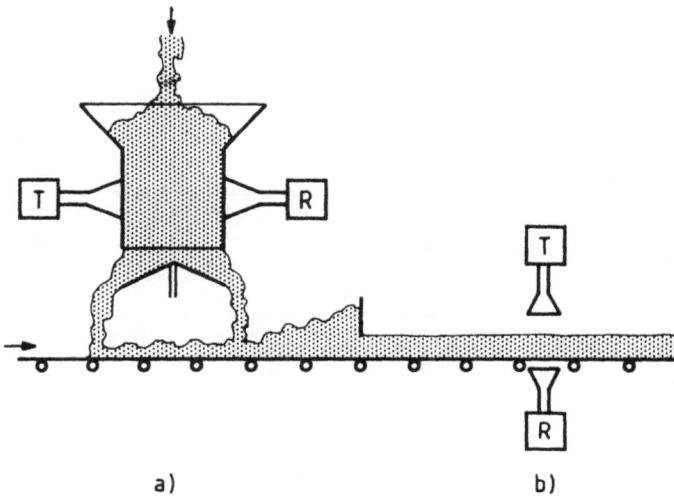

Figure 4.1 Free-space transmission sensors mounted for measurement (a) in a pipe or (b) on a conveyor belt. T is the transmitter and R is the receiver.

neglect the spreading of the wavefront and calculate via plane waves. The freespace attenuation due to spreading of the wave is not affected by the sample, except for possible refraction effects. From (1.33) and (1.35), we have for the field of a plane wave in the sample (see Figure 4.2):

$$E = E_0 \exp(-jk_s x) = E_0 \exp(-jk_s' x) \exp(-k_s'' x)$$

where

$$k_s = k_s' - jk_s'' = \omega \sqrt{\mu\varepsilon} = \frac{2\pi\sqrt{\varepsilon_r'}}{\lambda_0} \left(1 - j\frac{\varepsilon_r''}{\varepsilon_r'}\right)^{1/2} \tag{4.1}$$

is the propagation factor in the sample. For $\varepsilon_r'' \ll \varepsilon_r'$, we can approximate

$$k_s' = \frac{2\pi\sqrt{\varepsilon_r'}}{\lambda_0} = k_0 \sqrt{\varepsilon_r'} \tag{4.2a}$$

$$k_s'' = \frac{\pi\varepsilon_r''}{\lambda_0 \sqrt{\varepsilon_r'}} = k_0 \frac{\varepsilon_r''}{2\sqrt{\varepsilon_r'}} \tag{4.2b}$$

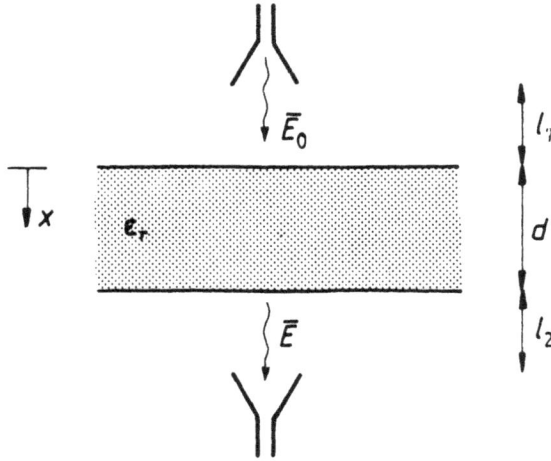

Figure 4.2 A free-space sensor measuring the transmission through a dielectric layer of thickness d.

After transmission we have

$$E = E_0 \exp\{-j[k_s'd + k_0(x - d)]\} \exp(-k_s''d) \tag{4.3}$$

from which we obtain the phase shift $\Delta\phi$, and attenuation Δa compared to the situation without the sample:

$$\Delta\phi = (k_s' - k_0)d \tag{4.4}$$

$$\Delta a = \exp(-k_s''d) \tag{4.5}$$

Expressing the attenuation in decibels, we have

$$\Delta A = (-20 \log_{10} e) \cdot k_s''d = -8.686 \, k_s''d \tag{4.6}$$

From (4.4), (4.6), and (4.2), we can now solve the permittivity of the sample as a function of the measured phase shift and attenuation:

$$\varepsilon_r' = \left(\frac{\Delta\phi}{k_0 d} + 1\right)^2 \tag{4.7a}$$

$$\varepsilon_r'' = \frac{-\Delta A}{k_0 d \, 10 \log_{10} e} \left(\frac{\Delta\phi}{k_0 d} + 1\right) \tag{4.7b}$$

Equation (4.7) contains the thickness of the sample (d), which we must measure separately if we want to measure accurately both ε_r' and ε_r''. In the case of moisture measurements, however, we can measure the ratio $A = (\varepsilon_r' - 1)/\varepsilon_r''$ (see Section 2.5.5) independent of d. In the case of those materials for which A is also independent of density, this is a powerful technique. From (4.7), we have

$$A = \frac{\Delta\phi}{-\Delta A} \cdot 10 \log_{10} e \cdot \frac{\sqrt{\varepsilon_r'} + 1}{\sqrt{\varepsilon_r'}} \qquad (4.8)$$

Because the ratio $(\sqrt{\varepsilon_r'} + 1)/\sqrt{\varepsilon_r'}$ is a slowly changing function, we are justified to use a more convenient ratio A_t:

$$A_t = -\frac{\Delta\phi}{\Delta A} \qquad (4.9)$$

The simple calculations above are fairly useful in cases when ε_r' is moderately low and the total attenuation in the sample is at least 10 dB. In other cases, and for accurate calculations, we must take the reflections into account. This can be done by using the signal-flow graph technique (Section 9.5). For the case of two interfaces, the graph is shown in Figure 4.3, where Γ_{12} and Γ_{23} are the reflection coefficients calculated from (1.46) and (1.47) or (1.53) and (1.54), and θ_2 is the direction of propagation (angle of the k_s' vector) compared to perpendicular in the sample calculated from (1.43) or (1.52). Reduction of the flow graph gives the transmission coefficient:

$$t = \frac{(1 + \Gamma_{12})(1 + \Gamma_{23}) \exp(-jk_s d/\cos \theta_2)}{1 + \Gamma_{12}\Gamma_{23} \exp(-j2k_s d/\cos \theta_2)} \qquad (4.10)$$

If medium 1 is equal to medium 3 (e.g., air), we have $\Gamma_{12} = -\Gamma_{23}$. Because medium 2 is normally lossy, Γ is a complex quantity, which will effect the phase calculations for high-loss materials. For simplicity, however, the following discussion will be limited to cases when Γ is considered real ($\varepsilon_r'' \ll \varepsilon_r'$) and the angle of incidence is perpendicular. Equation (4.10) now simplifies to

$$t = \frac{(1 - \Gamma^2) \exp(-jk_s d)}{1 - \Gamma^2 \exp(-j2k_s d)} \qquad (4.11)$$

where the exponential function in the numerator is equal to the first approximation

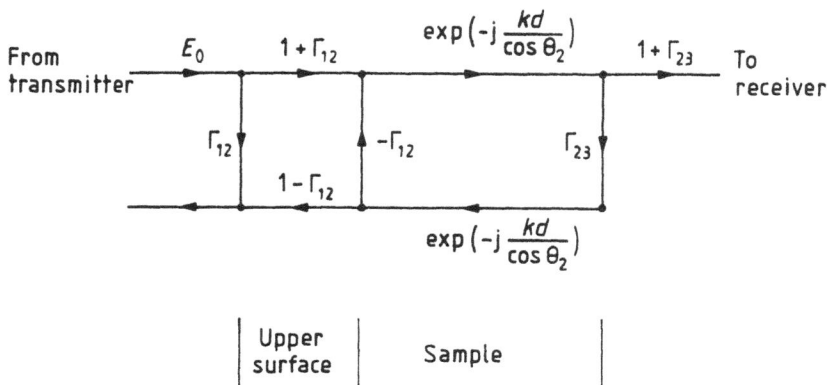

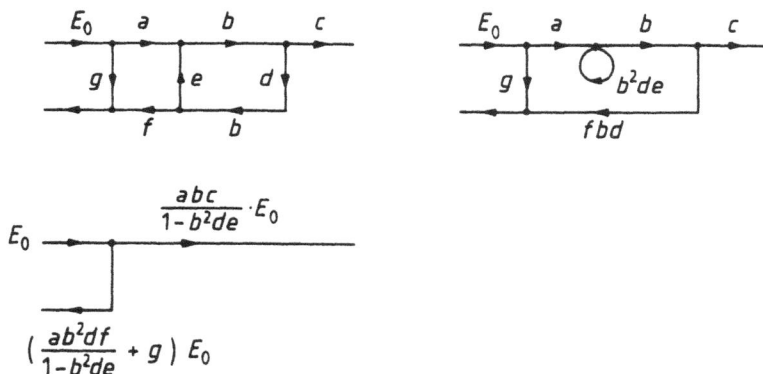

Figure 4.3 The signal flow graph describing the transmission through a double interface. The reduction procedure directly yields the transmission and reflection coefficients. The same procedure can be used with multilayered samples.

above, the $(1 - \Gamma^2)$ term represents the mismatch loss caused by reflections, and the second term in the denominator is caused by the multiple reflections. Under the assumed conditions, the reflection coefficient is

$$\Gamma = \frac{1 - \sqrt{\varepsilon_r'}}{1 + \sqrt{\varepsilon_r'}} \tag{4.12}$$

In analogy with (4.4) and (4.6), the attenuation and phase shift are now

$$\Delta A = \Delta A_0 + \delta A = -20 \log_{10} t \tag{4.13}$$

$$\Delta \phi = \Delta \phi_0 + \delta \phi = \phi_t - k_0 d \tag{4.14}$$

where δA and $\delta \phi$ are the periodic variations caused by the multiple reflections and ϕ_t is the phase angle of t. Figure 4.4 shows the graphs of ΔA and $\Delta \phi$ as functions of sample thickness for $\varepsilon_r = 5 - j0.1$ and $f = 5$ GHz. The periodic variations can be clearly seen for this low-loss material. The variation is much greater in the attenuation than in the phase, but both decrease as the attenuation increases (increasing sample thickness, ε_r'', or frequency). The maximum deviations are

$$\delta A_{\max} = -20 \log_{10}[1 \pm \Gamma^2 \exp(-2k_s'' d)] \tag{4.15}$$

$$\delta \phi_{\max} = \pm \arcsin[\Gamma^2 \exp(-2k_s'' d)] \tag{4.16}$$

The graphs of (4.15) and (4.16) are shown as functions of ΔA_0 in Figure 4.5 for some values of ε_r'. In Figure 4.6 are shown ΔA and $\Delta \phi$ as functions of the moisture of tobacco, based on measured permittivity values [Schilz *et al.*, 1981]. The periodic variation is not visible because of the high losses and relatively low permittivity in tobacco.

A minimum criterion for achieving accurate measurements with a sensor is that the attenuation or phase shift be a monotonic function over the range of measurement conditions. By comparing the amplitudes of the variations with the slopes of ΔA_0 and $\Delta \phi_0$, we obtain the criteria:

$$|\delta A|_{\max} < 3.41 \frac{\varepsilon_r''}{\varepsilon_r'} \tag{4.17}$$

$$|\delta \phi|_{\max} < \frac{\pi}{4} \left(1 - \frac{1}{\sqrt{\varepsilon_r'}} \right) \tag{4.18}$$

From (4.15) and (4.17) we obtain the minimum sample thickness for a given ε_r and frequency f for which the attenuation is monotonic:

$$d_{\min A} = \frac{-\sqrt{\varepsilon_r'} \, c \log_e \left[\dfrac{1 - 10^{-0.1706 \varepsilon_r''/\varepsilon_r'}}{\Gamma^2} \right]}{2\pi f \varepsilon_r''} \tag{4.19}$$

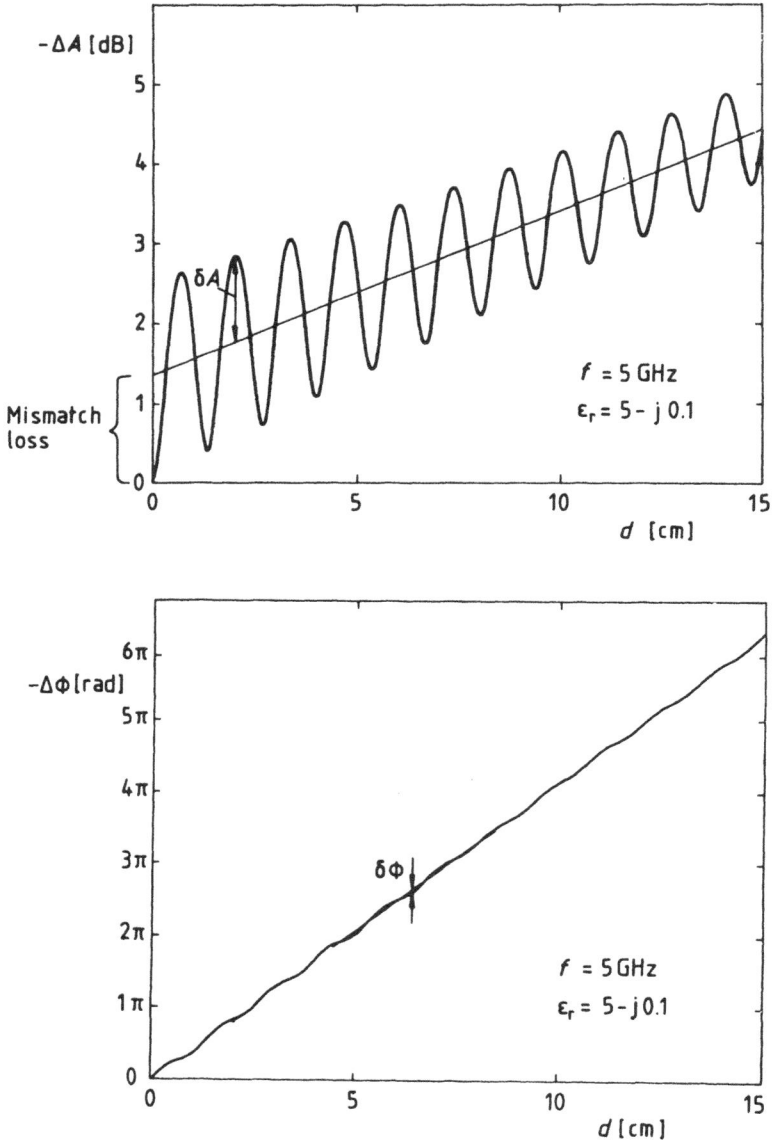

Figure 4.4 The calculated attenuation and phase shift as a function of the sample thickness. The periodic variations are caused by the multiple reflections in the sample. The material losses are too low to be successfully measured by using only the attenuation.

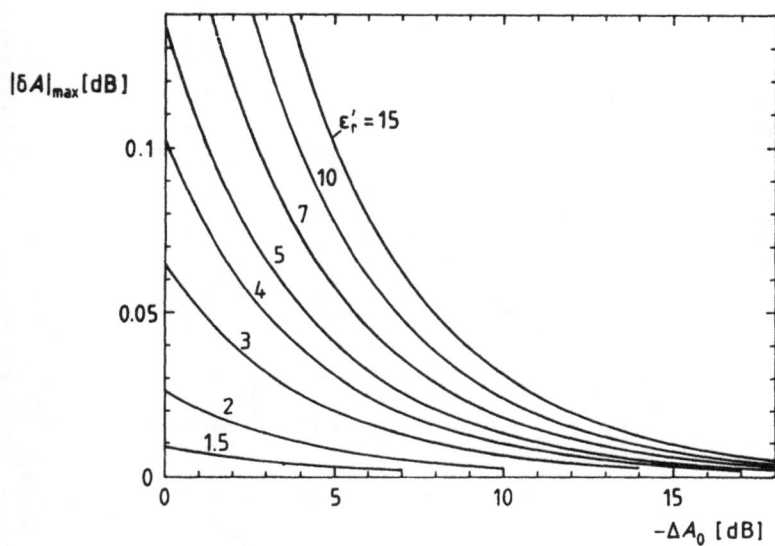

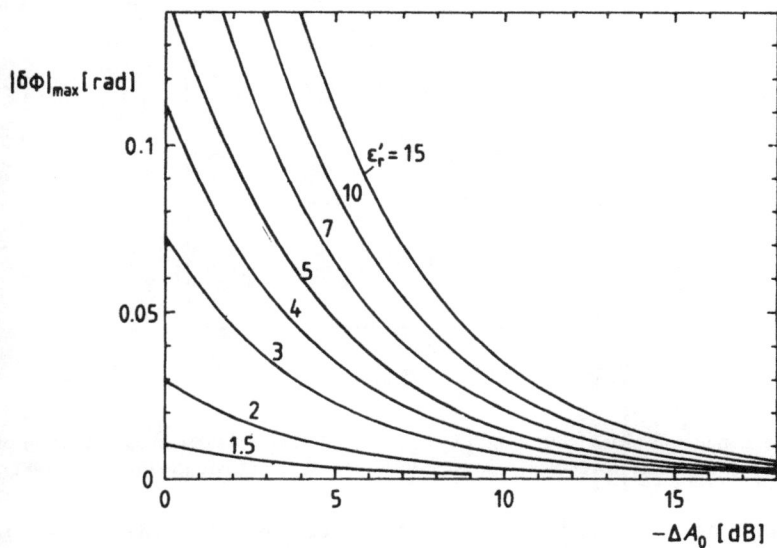

Figure 4.5 The amplitude of the periodic component (caused by the multiple reflections) in the transmission coefficient as a function of ε'_r and mean attenuation in the sample.

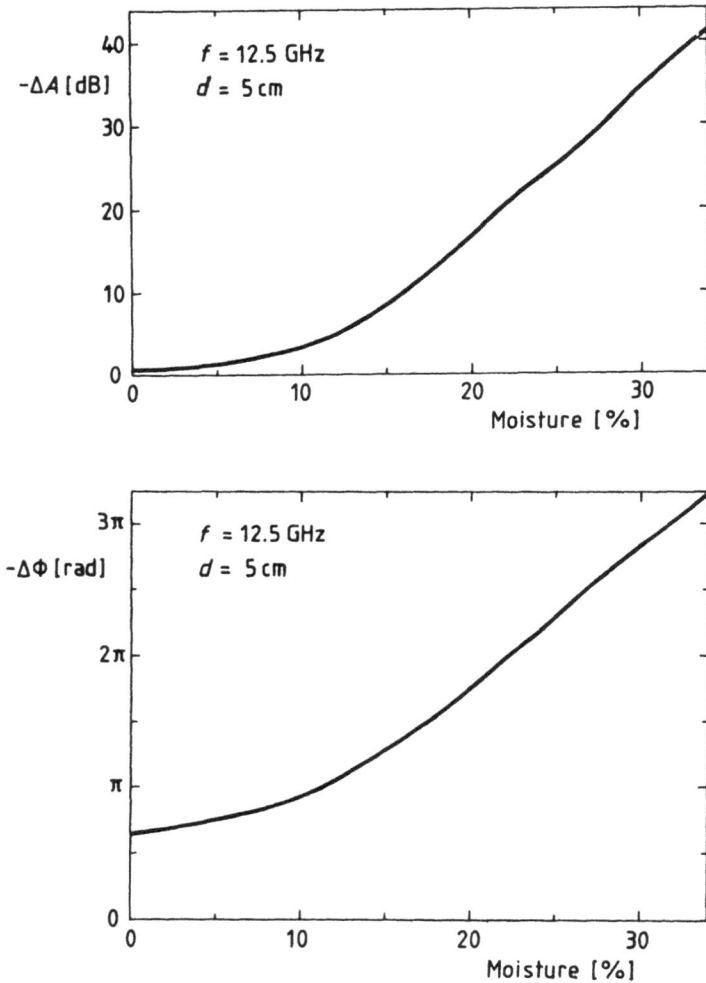

Figure 4.6 The attenuation and phase shift at 12.5 GHz in a 5 cm layer of tobacco as a function of moisture, as calculated by using measured values for the permittivity [Schilz *et al.*, 1981].

A similar formula can be derived for the phase, but comparing (4.18) and (4.16) reveals that the phase shift is always monotonic when $\varepsilon'_r < 83$, which in practice means "always."

Inclusion of multiple reflections between the sample and the antennas in the calculations, is in principle possible with the signal flow graph technique, but the usefulness of such calculations is dubious. We can easily take into account the spreading of the wavefront by introducing a 1/(distance traveled from antenna)-dependence of the field, but the reflection coefficients of the antennas are more problematic. Even a perfectly matched antenna may reflect (or, more correctly, scatter) part of a wave that is incident on the antenna. The scattering properties are generally difficult to treat. They depend primarily on the shape of the metal parts, less so than on the quality of the matching to the receiver or transmitter. For the transmitted signal however, we can assume that the results of the external multiple reflections will qualitatively resemble those for multiple reflections in the sample. In the measured signal, all of the periodic variations arising from both internal and external multiple reflections will be superposed.

4.2.3 Practical Considerations

Multiple Reflections and Mismatch Loss

From the theoretical calculations, we saw that the multiple reflections introduce signals that varied periodically as a function of sample thickness and permittivity. Under industrial conditions, these variations could be regarded as unpredictable and should therefore be minimized. The internal multiple reflections can be minimized by increasing the sample thickness and properly choosing the frequency. As a rule of thumb, the attenuation in the sample should be over 10 dB ($\varepsilon_r' \leqslant 10$, $\tan\delta \geqslant 0.065$).

The effect of multiple reflections in the air gaps can be decreased by increasing the distance between the antennas and sample. The spreading of the spherical wavefront attenuates the reflected signal during each crossing of the air gap. The upper limit is set by the antenna lobes and sample dimensions. The transmitting antenna should not illuminate the borders of the sample by any appreciable amount because that would cause the wave to diffract around the rim. If the attenuation in the sample is high, the diffracted signal may be stronger than the wanted signal. Another way to prevent external multiple reflections from arising is to tilt the sensor with respect to the sample. Then, the reflected wave will be directed away from the transmitting antenna. The antennas can also be partly covered or filled by microwave absorbing materials to reduce the reflection coefficient. An additional way to reduce the variations is to move the pair of antennas along the connecting axis and average the signals. Mechanical movements however, are not preferable under industrial conditions. A better way to produce the same effect for all multiple reflections is to average over a frequency band by sweeping the frequency (FM). The sweep bandwidth Δf should be chosen so that the number of standing waves in the shortest gap between reflecting surfaces would change by at least one during the sweep. For the antenna-to-sample gap, we have

$$\Delta f_{\min} = \frac{c}{2l_i} \qquad (4.20)$$

and for the internal reflections:

$$\Delta f_{\text{min}} = \frac{c}{2d\sqrt{\varepsilon_r'}}$$

(4.21)

Also, in this respect, we should use antennas as widely spaced as possible without causing diffraction. The FM method also decreases the influence of mismatch in the connecting waveguides, cables, and adapters.

When the effects of multiple reflections have been eliminated, the simple formulas presented in the beginning of Section 4.2.2 can be used, with a posible correction for the mismatch loss. Measurement of both the transmitted and reflected signal may be useful for that purpose. The latter can be measured by a receiver and a directional coupler or a circulator in the transmitting branch or by a separate antenna especially in the case of oblique incidence. Problems may arise from poor directivity, mismatch in the antenna, or direct coupling between the antennas.

Attenuation or Phase?

The attenuation is more affected by multiple reflections and mismatch losses than is the phase shift, but attenuation is easier to measure. We are also able to master the reflections to some degree, so, depending on the dielectric properties of the sample, either the attenuation or the phase shift may be the better choice. For two-parameter measurements, both are needed, but other methods are probably better in cases where the accuracy of the other of the two measurements is poor. For example, the combination of measurements of phase shift and γ-attenuation has been shown by Klein to give fairly good results on conveyors without any material stream-forming unit [Klein, 1984]. He also demonstrated the superiority in accuracy of the pipe-mounted sensor, which has clearly defined geometry, as compared with the conveyor sensor.

The internal structure of the sample must also be considered. Inhomogeneous materials cause volume scattering of the wave. The larger are the grains, the more they scatter. The scattered power is subtracted from the measured power to give an erroneous attenuation reading. Again, the phase shift is much less affected than the attenuation [Klein, 1981].

For very thin samples ($d < \lambda/4$), such as paper, the attenuation has a high sensitivity to variations in permittivity or thickness, as we can see from the steep slope near the origin of the ΔA graph in Figure 4.4. The real part of the permittivity has a much greater influence on the attenuation than is the case for thicker samples. The phase shift, however, provides far less sensitivity.

4.3 GUIDED WAVE TRANSMISSION SENSORS

4.3.1 Introduction

In the guided wave transmission sensor, the wave propagates in, for example, a waveguide, coaxial cable, or stripline. The sample is brought into contact with the field on a section of the line, thus affecting the propagation factor (phase and attenuation). The sensing part of the line is equivalent to that of a resonator made from the same transmission line, apart from the terminations. The resonator has reflecting open or shorted ends, whereas the transmission sensor is continuous.

The variety of ways to measure transmission and resonator sensors make them suitable for different applications. In the transmission sensor, the attenuation in the sensing section of the line can be tens of decibels, but in the resonator the maximum is only a few tenths of a decibel. Transmission sensors are therefore more suitable for measurement of high-loss materials and resonators for measurement of low-loss materials.

A guided wave transmission sensor should be designed such that the reflections at the ends of the sensing section are minimized. Otherwise, they will cause multiple reflections and mismatch loss, leading to the same kind of problem that arises in free-space transmission sensors. The minimizing of the reflections can in some cases (e.g., a sensing slot in the waveguide wall) be done with tapered transitions (i.e., slowly increasing and decreasing sensitivity along the line).

The theoretical calculations for free-space sensors are, in principle, also valid for guided wave sensors when jk is changed for γ and the reflection coefficients are calculated from (1.25), unless the transitions have been matched. The values of $\gamma = \alpha + j\beta$ depend on the sensor structure.

4.3.2 Sensor Structures

Waveguide Sensors

Waveguides can be used as transmission sensors in different ways. Measurements with totally filled waveguides were described in Section 2.7.2 as laboratory methods, but can, of course, also be used as sensors outside the laboratory for measurements of liquids, for example. The slotted waveguide (Figure 4.7a) is suitable for measurement of sheetlike materials such as paper, textiles, veneer, or plastic. The same kind of structure is also widely used in power applicators for drying purposes [Metaxas *et al.*, 1983]. The change of γ in the sensor can be calculated for thin samples by using perturbation theory (Section 3.2.2) [Altman, 1964]:

$$\gamma - \gamma_0 = [\varepsilon_r'' + j(\varepsilon_r' - 1)]\, 2\pi(t/a)(\lambda_{g0}/\lambda^2) \qquad (4.22)$$

where ε_r is the permittivity of the sample, λ is the free-space wavelength, γ_0 and λ_{g0} are propagation factor and guide wavelength in the sensor without the sample, and γ the propagation factor with the sample present.

Other ways to use waveguides as transmission sensors are the contacting single-slot sensor (Figure 4.7b), and the waveguide section with a tilted dielectric pipe as sample holder through the broad walls (Figure 4.7c). The former can be used to measure both thin sheet-like and "infinitely" thick samples, whereas the latter is suitable for the measurement of liquids or powders. Both slotted and single-slot waveguides can be used to measure, for example, the moisture profile by mounting output coupling probes at regular intervals along the sensor. The coupling coefficients should be small so that the probes would not disturb the wave.

Microstrip Sensors

Microstrip line can be used as a guided wave transmission sensor in the same way as the microstrip resonator (Figure 3.16a). Microstrip is well suited to the measurement of solid surfaces, liquids, and powders. Microstrip can easily be integrated with the surface of a pipe or trough. The sensitivity per unit length depends on the permittivity of the sample, dimensions of the line and substrate, and protective dielectric layer, which is often necessary to prevent wear and corrosion. For the optimal sensor design, the length of the line must be chosen sufficiently long to give a high enough sensitivity. These factors have been studied by Kent, Khalid, and others [Kent, 1972, 1973; Kent et al., 1984; Khalid et al., 1988].

Other Structures

Practically all types of transmission line can be used as guided wave transmission sensors as well as resonators. The exact calculation of the sensitivity of γ to the variation of ε_r in the sample is usually impossible or very difficult. A rough estimate can often be calculated by using the perturbation formulas in Section 3.2.2, provided that the field is known approximately. The phase factor β is inversely proportional to the resonant frequency and $\alpha = \beta/2Q$. A more detailed discussion is given in [Altman, 1964].

4.4 MEASUREMENT EQUIPMENT FOR TRANSMISSION SENSORS

4.4.1 Introduction

Measurement equipment for industrial microwave transmission sensors has the following general properties:

- The measurement is, in principle, simply that of phase or amplitude at a single frequency (see Section 3.4.2);

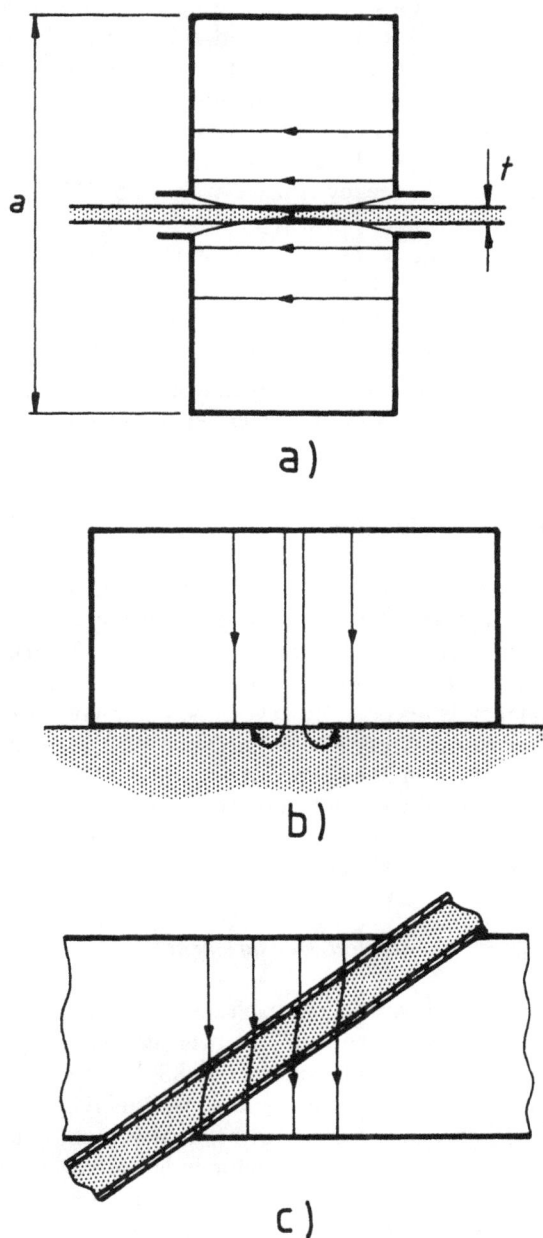

Figure 4.7 Different types of waveguide sensors: (a) slotted waveguide for measurement of sheets; (b) single-slot waveguide for solid surfaces or sheets; (c) waveguide with dielectric sample holder for liquids or powders.

- Due to the nonidealities of the measuring circuits or systems, different corrections or compensations are to be used, which often leads to more complicated systems with modulation, mechanical tuning, or frequency sweep;
- Sometimes, the reflection of the sensor also must be measured;
- The sample thickness is often several wavelengths, and a large dynamic range is needed in the amplitude and phase measurements;
- The measurement frequency is usually above 5 GHz, so waveguide techniques are often used;
- Accurate knowledge of frequency is not usually needed, but the microwave source ought to be stable for phase measurements;
- The measurement can be made in free space or in a transmission line, but the measurement equipment is basically the same in either case.

Additionally, the common properties of industrial meters mentioned in connection with resonator sensors (see Section 3.4.1) apply as well to the measurement equipment of transmission sensors.

4.4.2 Measurement Techniques

As mentioned in the introduction, the basic phase and amplitude measurements are simple. The phase shift or attenuation caused by the sample are measured with techniques described in Section 3.4.2. The best equipment for making the measurements is an accurately calibrated automatic network analyzer, at least, if information as a function of frequency is needed. However, the equipment is expensive and not suitable for the industrial environment because the measurement frequency is usually high, and long transmission lines between the sensor and the measurement equipment cannot be used. Instead of the ANA, several other types of equipment, more suitable for the industrial environment, have been used to achieve accurate measurements with less complexity and cost. The basic types of equipment are described below.

Direct Phase Shift or Attenuation Measurement

The most straightforward way to approach the measurement problem with transmission sensors is to use a phase detector or power meter to detect the phase shift [Ozamiz *et al.*, 1979] or attenuation [Anderson, 1980; Wyslouzil *et al.*, 1979] caused by the sample (Figure 4.8). The measurement is usually performed at a single frequency, so a VCO is not necessary. For the same reason, automatic level control or output power measurement of the generator is not as essential as in the case of resonator measurements. More important is to decrease reflections in the transmission line between the power source and sensor (antenna or transmission line). To do so, an attenuator or isolator can be used. The isolator is the better choice (but also more expensive) because it attenuates only the reflected signal. The isolator often consists

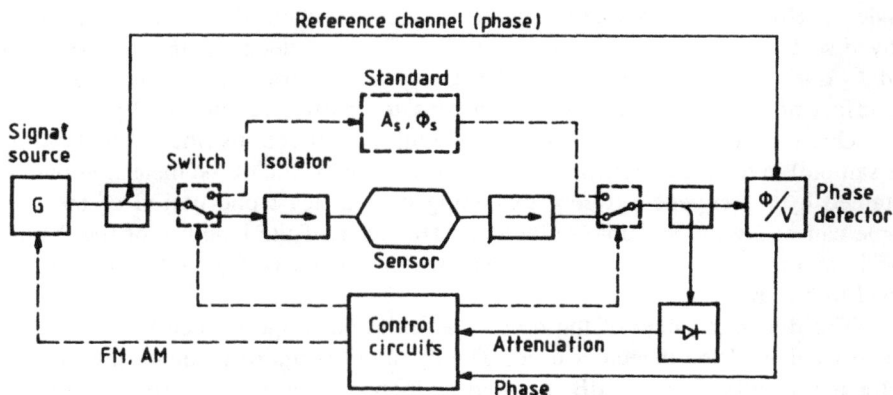

Figure 4.8 Direct measurement of phase shift and attenuation for transmission microwave sensors. Error compensation can be performed with optional (dashed lines) standard load, frequency sweep (FM), or amplitude modulation (AM).

of a circulator (see Section 3.4.2) with one port matched. For the measurement of phase, a reference signal is needed (see Figure 4.8). It is taken from the same signal source as the measurement signal ("the wave"), but is unaffected by the sample.

The biggest drawbacks of direct phase and amplitude measurements are the nonlinearity and drift of the phase and power detectors, and the drift of measurement frequency in phase measurement. The drift is caused mainly by temperature changes, so compensation by temperature measurement or temperature stabilization is needed to decrease the effect on the results. Another way to correct the results is by frequently measuring a standard device with a phase shift or attenuation that is the average of that of the sensor in the measurement situation. With this strategy, the nonlinearity problems can also be avoided if the measurement range of phase or attenuation is sufficiently small. The connection of the standard device can be accomplished by two switches. Mechanical switches (relays) can be used if the checking of the standard is not very often needed because the number of switching operations is typically limited to ten million. If high frequency (greater than 0.1 Hz) switching is needed due to fast drift, solid-state switches (*pin* diode, Schottky diode, FET) must be used.

As in resonator measurements, transmission measurements are affected by the multiple reflections in the equipment's transmission lines. The effect of multiple reflections on the phase or amplitude response of the equipment can be estimated by (4.15) and (4.16) given for the reflections inside the sample. The estimation can be done by replacing the square of the reflection coefficient Γ^2 with the product of the reflection coefficients Γ_1 and Γ_2 at the ends of the transmission line section and by setting the attenuation term $\exp(-2k_s''d)$ equal to one (the line is usually practically lossless). However, a comparison with the sensor without the sample is usually employed so that the actual effect of multiple reflections depends on the changes of Γ_1 and Γ_2 during measurement. With the use of a standard load, as discussed above, the effect of the reflections can be compensated partly, but not totally.

One way to decrease the effect of all multiple reflections (including those inside the sample) in the measurement is to use the average value of the measured parameter obtained by a frequency sweep, extending over at least one period of the slowest ripple caused by the multiple reflections. However, if the length ℓ of the disturbing line is short in wavelengths, the sweep width needed ($\Delta f/f \approx \lambda_g/2\ell$) can be too broad to be practical.

The dynamic range of the direct transmission property measurement is determined by that of the detectors used. The dynamic range of a square-law power detector is typically about 35 dB, but can be increased by filtering (by decreasing the noise bandwidth) after the detector. Filtering, however, increases the response time of the measurement. The detector can also be used above the square-law power limit (typically -15 dBm), but then the nonlinear power response must be corrected. The unambiguous range of the phase detector is usually only $180°$ and the usable range with high frequency phase detectors is about $120°$ (see Section 3.4.2). The number of whole cycles in the phase shift can be solved by slightly changing the measurement frequency and detecting the corresponding change of the phase shift, which is proportional to the total phase shift between the measurement and reference channels in Figure 4.8. For the frequency derivative of the phase difference $\phi_m - \phi_{\mathrm{ref}}$ in the measurement and reference channels (similar transmission lines in both channels), we obtain

$$\frac{\Delta\phi}{\Delta f/f} = \frac{\phi_m - \phi_{\mathrm{ref}}}{[1 - (f_c/f)^2]^{1/2}} \tag{4.23}$$

where $\Delta\phi$ is the phase shift caused by the relative frequency shift $\Delta f/f$, and f_c is the cut-off frequency of the transmission line (for TEM lines $f_c = 0$). From (4.23), we also notice that the greater is the phase difference between the measurement and reference channels, the greater is the demanded frequency stability of the phase measurement. For example, if the phase difference is $9000°$ (50π radians) in a coaxial cable (about 0.5 m at 10 GHz), the phase error per relative frequency shift is $90°/\%$.

The direct measurement of transmission sensor parameters is the fastest method (see below in this section). The measurement response time is limited, in principle, only by the bandwidth of the detectors, which can be several megahertz. Thus, the response time can be as short as a few microseconds, which is enough in practically all industrial measurements.

Balancing Method

Another solution to the problems of the limited dynamic range and nonlinear phase or power response of the detectors is to keep the detector output within certain limits by using a controllable attenuator or phase shifter. The control device can be in the measurement ([Shiraiwa *et al.*, 1980; Okamura, 1981], Fig. 4.9), or in the reference channel (bridge measurement) [Kent, 1973]. The accuracy of this method is determined by both the phase shifter or attenuator and the detector.

The most accurate control devices are electrically tuned mechanical phase shifters or attenuators, but they are slow and can perform only a limited number of tuning operations (typically, $0.2-10 \times 10^6$). Better for fast measurements are totally electrical devices, like diode or ferrite phase shifters [Skolnik, 1981] and solid-state attenuators. The tuning can be either continuous or stepped. If stepped, interpolation can be used to acquire the exact result. A typical tuning range for mechanical step attenuators is $10-130$ dB with good repeatability (<0.05 dB) and temperature stability (<0.01 dB/°C). For solid-state step attenuators, the tuning range can also be up to 100 dB, with repeatability <0.5 dB and temperature drift <0.02 dB/°C. With continuously tunable *pin* diode attenuators, the maximum tuning range is about 60 dB and temperature drift is typically 0.05 dB/°C. Diode phase shifters can be either stepped or continuous. Usually, the tuning range of a stepped phase shifter is 360° in 16 steps of 22.5°. The accuracy of a single point is about $\pm 10°$ but the repeatability is much better. The tuning range of continuous diode phase shifters can also be up to 360°.

The sources of measurement error are the same for the balancing method as with the direct measurement, only the nonlinearity of the detectors is partly eliminated. Therefore, the same methods of error compensation (i.e., standard channel, frequency sweep) can be used. The measurement speed with the balancing method is determined mainly by the tunable devices. Typical tuning time for electronically controlled devices is about 1 μs and for mechanical devices 30 ms. The total measurement time is also affected by the measurement algorithm. In the measurement of a single sensor, continuous tracking can be used to give a short response time. However, if several sensors are used, or if the measurement situation changes considerably in a short time as in the measurement of discrete objects, several tuning steps may be needed to reach the balance.

Simultaneous Measurement of Phase Shift and Attenuation

In the methods described above, different detectors were needed for the phase shift and attenuation measurement. However, measuring both quantities with a single (phase) detector and balancing circuit (Figure 4.10) is possible. In the measurement, an amplitude modulation of the signal in the measurement channel is used to eliminate the

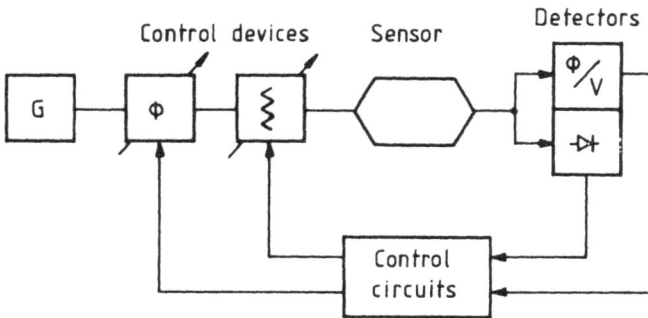

Figure 4.9 The balancing method for measuring the phase shift and attenuation of transmission sensors. The outputs of the power and phase detectors are tuned (in this order) to certain values by the controllable attenuator and phase shifter.

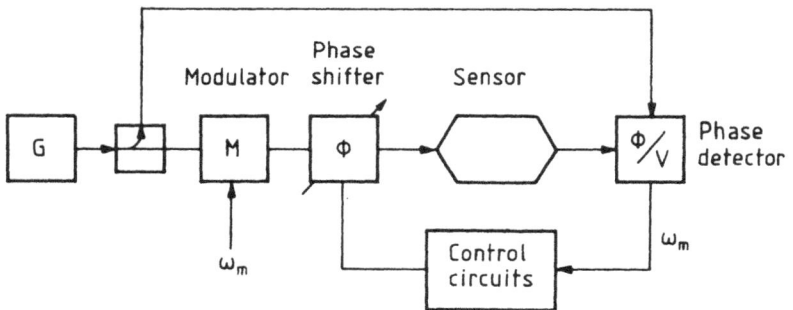

Figure 4.10 The modulated subcarrier method for measurement of the phase shift and attenuation with one phase detector. In the measurement, the minimum (determination of phase) and maximum (determination of attenuation) of the modulated component of the phase detector output are sought by tuning the phase shifter.

dc term of the phase detector output (see (3.55)). The minimum and maximum of the modulated output voltage of the phase detector are found by tuning the phase shifter. If the signal traveling through the sample is considerably attenuated with respect to the reference signal, the value of the phase shifter at the minimum point (90° phase difference at the input of the phase detector) does not depend on the signal level in the measurement channel, but only on the phase shift in the sample. Respectively, at the maximum (0° or 180° phase difference), the value of the detector output voltage depends only on the signal levels. We can notice that the phase measurement with this method is not affected by the attenuation, unlike the previous methods. This method is called the *modulated subcarrier technique* [Schafer, 1960; Schafer *et al.*, 1962].

The sources of error in this method are the same as those mentioned in conjunction with the previous methods plus errors of the phase shifter and modulator. Most of these can be eliminated by an automatic measurement system, called the *chopped subcarrier method* [Kalinski, 1981], where a standard load is used in parallel with the sensor to calibrate the detectors. Furthermore, an additional phase modulator and detector are employed for automatic tuning of the phase shifter. The measurement speed of this method is limited by the tuning time of the phase shifter, modulation frequency, and complexity of the measurement procedure.

4.4.3 Practical Aspects

The information on the signal reflected back from the sample is sometimes important. The reflection properties can be measured with any of the methods described in the previous section by using a directional coupler or a circulator (port 3 of the isolator) in the input of the sensor.

According to the calculated results in Figure 4.4, phase measurement is much less affected by multiple reflections than is attenuation measurement. This has also been confirmed by practical experiments [Klein, 1981]. Unfortunately, the measurement of phase is more complicated than the measurement of amplitude. Even the simplest phase measurement circuit includes either an ALC circuit, a balancing phase shifter or frequency down-converters (mixer and LO), all of which are quite complicated and expensive microwave circuits. Furthermore, a microwave connection is needed between the transmitter and receiver parts of the equipment, which may limit the installation possibilities of the meter.

The sample dimensions or measurement frequency should be chosen such that the maximum attenuation of the signal would not be too high. For the attenuation and down-converted phase measurement, a practical limit is about 60 dB. For phase measurement at high frequency, the limit is lower, 20–40 dB, depending on the method.

There are some measurement methods for transmission sensor arrays. A first method is to use several detectors, one for each sensor, that are fed in parallel. However, calibrating several detectors, and especially maintaining their calibration in a harsh environment, is difficult. Another way is to use switches to connect the sensors to the meter one by one. This method is applicable only at low frequencies, where long coaxial cables can be used. Another method, where neither sensitive semiconductor devices nor many waveguides are needed, is presented in Figure 4.11 [Wyslouzil *et al.*, 1979]. Frequency-selective directional couplers (directional filters) are used to connect the power to the sensors from a single terminated waveguide line. The center frequencies of the filters differ from each other by about 0.5%. A frequency sweep is used for the fast (<0.1 ms per sensor) measurement of the sensors. A typical result obtained for pulp is presented in Figure 4.11.

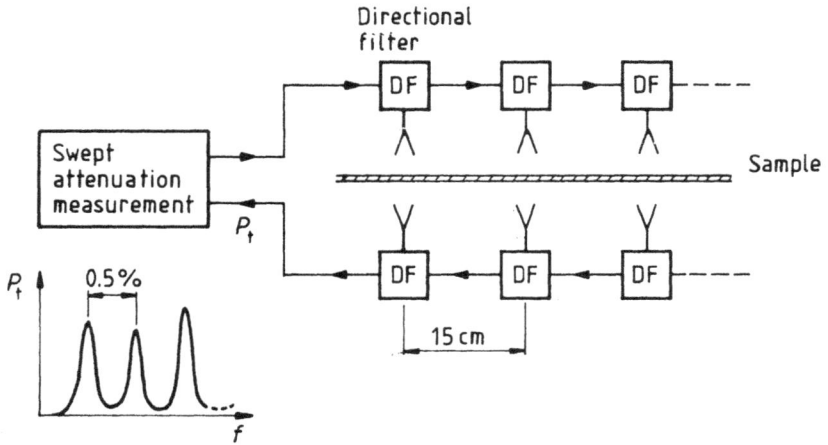

Figure 4.11 A measurement system for transmission sensor arrays. The attenuation of the sensors is measured at different frequencies, determined by the directional filters (DF) using a frequency sweep. The array can be used for the fast mapping of the properties of sheets or plates.

4.5 APPLICATIONS OF TRANSMISSION SENSORS

4.5.1 Free-Space Systems

Measurement of Attenuation Through a Pipe

The Kay-Ray company (US) markets a broad range of industrial radioactive sensors. Among the products is also one that uses microwave attenuation for moisture measurements in powders or granular solids. The meter consists of a section of square pipe (14.5×29.8 cm^2) with a box for sensors and electronics on each side (Figure 4.12). The microwave attenuation is measured with a pair of horn antennas, which make contact with the dielectric windows in the pipe. This arrangement leads to resonance-like multiple reflections in the horns, damped by insertion of microwave-absorbing lossy material [Brodwin *et al.*, 1980]. The γ-ray attenuation through the sample is also measured for the compensation of density variations. The pipe, however, should always be completely filled for proper operation of the meter. A temperature sensor is included to compensate the temperature dependence of the permittivity. The meter is intended for the moisture range 0 to 50% (wet basis). The promised moisture measurement accuracy is very high, but this probably only means the accuracy of the measurement electronics, because the variations in grain size alone will cause a larger error [Klein, 1981]. The manufacturer recommends the meter for measurement of coal, grain, food products, plastic resins, and other chemicals after drying or hydration, or as quality control.

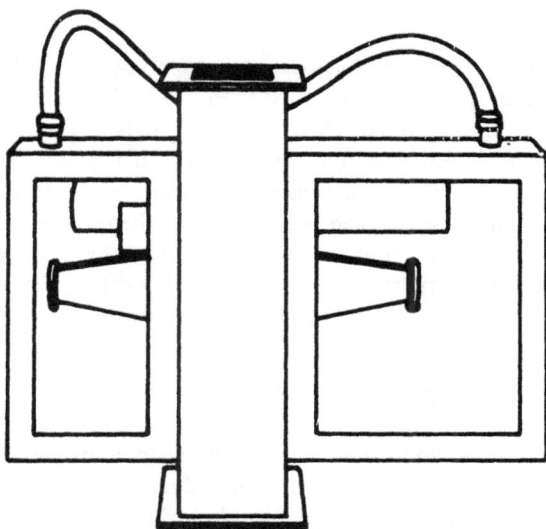

Figure 4.12 Free-space transmission sensor mounted on a section of square tube.

Measurement of Attenuation and Phase on a Conveyor

The Berthold company (Federal Republic of Germany) manufactures a free-space transmission sensor for measurement of moisture, mainly on conveyors. The principle is the same as shown in Figure 4.1(b), with the addition of a γ-attenuation sensor mounted adjacent to the microwave sensor. Either the phase shift, attenuation, or both can be measured and related to the total mass per unit area as measured by the γ-sensor. The manufacturer says that the measurement of relative phase shift results in better accuracy than does the measurement of relative attenuation, but that the best accuracy is achieved with a combination of the two. The measurement range is from 0 to 6 cm water (0 to 60 cm layer thickness for 10% moisture). The meter is recommended for the measurement of a wide range of materials in the building, chemical, food, paper, wood, and basic industries, ranging from sand, coal, and tobacco to particle board.

Shiraiwa and others have reported results of free-space transmission sensors in operation as moisture sensors for limestone [Shiraiwa *et al.*, 1980]. The sensors measure the attenuation through a 16 cm thick layer on a belt conveyor. The measurement frequency is 4 GHz.

Portable Moisture Meter for Grain

A portable moisture meter for grain has been developed by the Institute of Electronics of the Bulgarian Academy of Sciences. The meter is built into a case, which can be carried by hanging on the shoulder. The total weight is 3.2 kg. The grain sample (85 cm^3) is put into a plexiglas sample holder, which fits into a hole in the front panel. Behind the panel is a pair of horns to measure the attenuation through the sample at about 20 GHz. Various kinds of grain, soybeans, peas, lentils, *et cetera* can be measured, but a conversion table must be used for accordingly converting the reading. The measurement range is 5–34% and the accuracy is reported to be ±1% in the lower end of the range and ±2% in the upper end.

Tilted Transmission Sensors for Paper and Pulp Products

The Canadian company, Valmet-Sentrol, markets four microwave sensors for measurement of moisture in pulp and paper products during the manufacturing process. One is a contacting guided wave sensor (see below), but three of them are tilted free-space systems (Figure 4.13). They are tilted so that the multiple reflections will not affect the transmitted signal. In addition, they use the FMCW radar technique (see Section 6.3.2) for further improving the accuracy by separating the reflections from the transmitted signal on the basis of different distance traveled.

In the meter, which is primarily intended for the dry end of a paper machine (Figure 4.13a), the wave passes twice through the sample. Between the antennas on the other side of the paper is only a 40 ns delay line. The correct received signal will therefore have an almost constant frequency shift of 15 kHz compared to the output of the transmitter. After the mixer, the multiple reflected waves can be filtered out because they will appear on different frequencies. The meter uses a central frequency of 22.2 GHz and measures paper with dry weight in the range of 100 to 1000 g/m^2 (moisture typically is 0.5 to 20%).

The meters for thicker products use only a single pass, but they also measure the reflected signal (Figure 4.13b). The version for dry weights between 400 and 3000 g/m^2 works on 22.2 GHz, and that for about 6000 g/m^2 (acoustic tile) works on 2.4 GHz. The moistures range from 55 to 70%.

Other Reported Systems

A team at Philips Research Laboratories has applied the density-independent technique using the ratio $A = (\varepsilon_r' - 1)/\varepsilon_r''$ to transmission sensors [Jacobsen *et al.*, 1980; Kent *et al.*, 1981]. This team has developed a meter which can use both guided wave sensors and a free-space sensor ($f = 9$ GHz). The latter consists of a sample holder with two contacting horns.

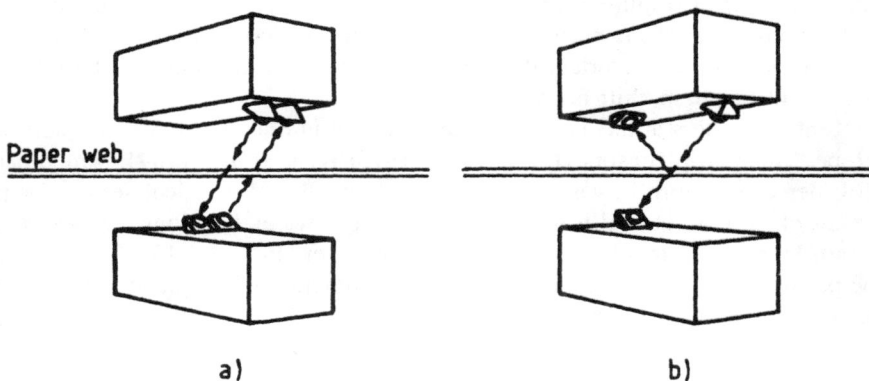

Figure 4.13 Free-space transmission sensors for measurement of thin sheets of paper and pulp. The sensors use oblique incidence and frequency modulation to eliminate the external multiple reflections: (a) double transmission is used to increase the sensitivity with thin samples; (b) the reflection is measured with thicker samples.

A versatile instrument has been developed in Poland. It is manufactured by Wilmer Instruments, and is capable of measuring attenuation, phase shift, and reflection simultaneously by using different kinds of free-space (primarily) or guided wave transmission sensors [Kraszewski *et al.*, 1980]. The device uses a complicated chopped subcarrier method to cope with the mismatch and multiple reflections in the sensors and sample [Kalinski, 1981].

Nippon Steel Corporation in Japan has studied the use of microwave transmission through a blast furnace [Ohno *et al.*, 1986]. Using a frequency of 10 GHz, researchers were able to distinguish between the coke and ore layers when measuring at different places through the chamotte brick wall. This capability is useful for measuring the descent velocity. By measuring the reflected and scattered waves the group was also able to measure the particle size in the burden.

An example of a specialized application, developed at the Oak Ridge National Laboratory (US), is the measurement of the electron density in the plasma of a Tokamak [Ma *et al.*, 1981]. The instrument is a submillimeter-wave polarimeter-interferometer, which measures (among other things) the Faraday rotation of the linearly polarized wave. The rotation of the polarization depends on the electron density and magnetic field strength along the path.

4.5.2 Guided-Wave Systems

Microstrip Transmission Sensors

At Plessey (UK), a microstrip sensor for measurement of the moisture in tobacco has been developed (Figure 4.14). The sensor is designed to be installed at the bottom of a trough measuring a continuous stream of tobacco, but it can, of course, be used with many other materials [Ozamiz *et al.*, 1979]. The sensor is based on the measurement of phase shift only.

Kent and others at the Torry Research Station in Scotland have for many years developed microstrip sensors [Kent, 1972, 1973; Kent *et al.*, 1984], which can be used to detect moisture in various materials. A small and practical sensor for measurement of liquids like oil-water emulsions, sugar-water solutions, *et cetera,* uses a 48 mm long microstrip line with a protective cover (Figure 4.15). The sensitivity of the phase shift and attenuation is studied as a function of frequency and thickness of the protective cover in [Kent *et al.*, 1984].

Waveguide Sensors

The use of a dielectric pipe across a waveguide (Figure 4.7c), for the measurement of moisture in powders and fine-granular materials, has been reported by Praxmarer from the Technical University in Leuna-Merseburg (German Democratic Republic) [Praxmarer, 1980]. He has tested the sensor successfully with PVC powder at a measurement frequency of 22 GHz.

Wyslouzil and Van Koughnett from the National Research Council of Canada have studied the use of slotted waveguide sensors (Figure 4.7a) for measurements of photographic film, paper, and cellophane [Wyslouzil *et al.*, 1974]. They measured only the attenuation at 10 GHz and found that the biggest problem was the low attenuation with almost dry materials. The position of the sample in the slot also influenced the results. The difficulties were not very significant, however, and commercial sensors were developed in cooperation with Mega System Design in Canada.

Rank Industrial Controls (UK) manufactures a table microwave moisture meter for measurement of paper samples. On top of the instrument is a lid under which the sample is inserted. A slotted waveguide is formed by a spiraling groove cut into the lid and the table beneath it. The attenuation is measured at 10.68 GHz with an accuracy of 0.01 dB when temperature compensation is used. The instrument is said to be insensitive to color, surface finish, beating, bleaching, calendering, salt content, pH, and moderate sizing [Anderson, 1980]. The sensor is equipped with isolators and wave traps at the ends to eliminate the effects of mismatch.

Valmet-Sentrol (Canada) manufactures a single-sided contacting sensor for measurement of moisture in linear board or box board at the dry end of the paper machine. The measurement frequency is 22.2 GHz and the sensor resembles the single-slot waveguide device in Figure 4.7(b). It is called a *surface wave absorption cell* (SWAC). The measurement range is 0.5 to 20% moisture for dry weights in the range of 100 to 1000 g/m^2.

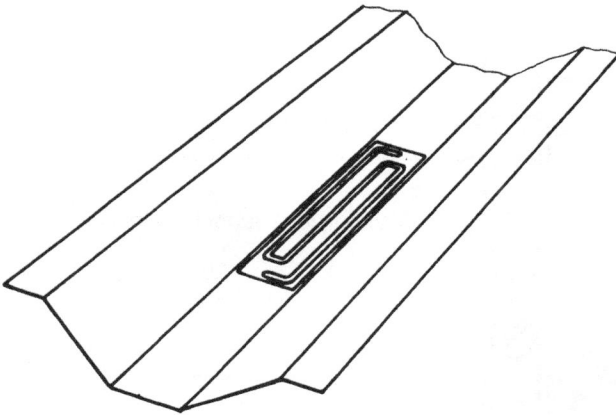

Figure 4.14 Microstrip sensor mounted in a trough for measurement of moisture in tobacco. (After [Ozamiz *et al.*, 1979].)

Other Reported Systems

A team at the Carleton University in Ottawa (Canada) have studied the use of two-conductor lines made of rods or parallel plates for the measurement of sample thickness or permittivity (e.g., moisture) [Chudobiak *et al.*, 1979]. The sample is suspended midway between the two conductors and the phase shift caused by the sample is measured with a continuous wave (CW) signal in the UHF range.

A soil moisture sensor has been developed at the University of Nebraska (US) [Bahar *et al.*, 1984]. This sensor is a leaky coaxial cable with a slotted outer conductor and a protective jacket. Part of the wave in the cable fringes out through the slots into the soil. A length of cable can be buried at a certain depth for continuously measuring the average soil moisture in a desired area. Both the phase shift and attenuation are affected by the moisture and can be used, for example, to control an automatic irrigation system. A frequency of 0.9 GHz has been used with a cable buried at a depth of 0.5 m.

While working for Kemira (Finland), Jakkula developed a sensor for measurement of chemical liquids and water suspension in a metal pipe. The sensor consists of a dielectric waveguide ring immersed in a groove cut on the inside of a metal pipe (Figure 4.16). The sensor is used to measure liquids with high dielectric losses or a moisture content over 50%. In such cases, the wave normally propagating on the surface of a dielectric waveguide does not exist, and attenuation in the waveguide is caused by incomplete reflections from the waveguide-liquid interface. Unlike sensors where attenuation is caused by dielectric losses in the sample, the attenuation

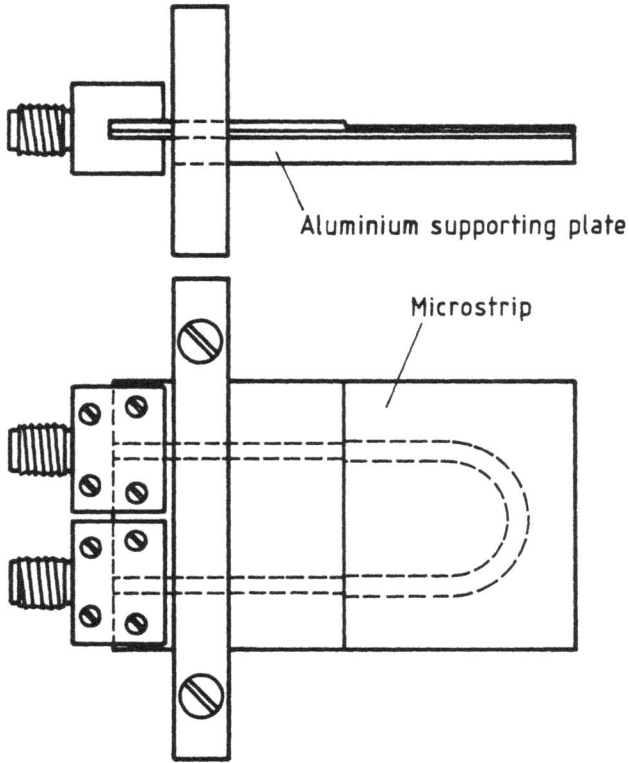

Figure 4.15 Microstrip sensor for measurement of moisture in liquids. (After [Kent *et al.*, 1984].)

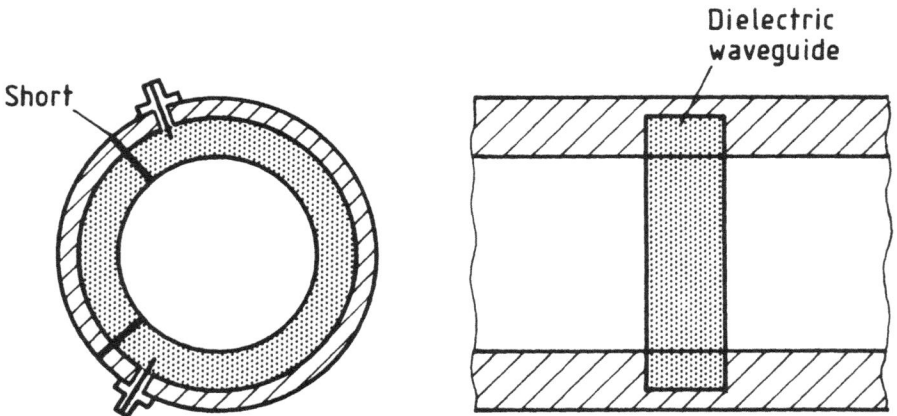

Figure 4.16 Dielectric waveguide sensor for measurement of moisture in liquids in a process pipe. (After [Jakkula, 1988].)

of the wave in this sensor is lower for higher water content of the sample. This is caused by the high ε'_r and reflection coefficient of water. The sensor can be used to measure mixture ratios (e.g., water content) in liquids for the chemical industry.

REFERENCES

Altman, J.L., *Microwave Circuits,* New York: D. Van Nostrand, 1964, 462 p.

Anderson, J.G., "Paper/Board Moisture Measurement by Microwave Loss," *Proc. 4th IFAC Conf.,* Ghent, Belgium, June 1980, pp. 75–84.

Bahar, E., and J.D. Saylor, "A Feasibility Study to Monitor Soil Moisture Content Using Microwave Signals," *IEEE MTT-S Int. Microwave Symp. Digest,* San Francisco, June 1984, pp. 362–364.

Brodwin, M., and J. Benway, "Experimental Evaluation of a Microwave Transmission Moisture Sensor," *J. Microwave Power,* Vol. 15, No. 4, December 1980, pp. 261–265.

Chudobiak, W.J., M.R. Beshir, and J.S. Wight, "An Open Transmission Line UHF CW Phase Technique for Thickness/Dielectric Constant Measurement," *IEEE Trans. Instr. Meas.,* Vol. IM-28, No. 1, March 1979, pp. 18–25.

Jacobsen, R., W. Meyer, and B. Schrage, "Density Independent Moisture Meter at X-Band," *Proc. 10th European Microwave Conf.,* Warsaw, September 1980, pp. 216–220.

Jakkula, P., "Method and Apparatus for Measuring the Moisture Content or Dry-Matter Content of Materials Using a Microwave Dielectric Waveguide," U.S. Patent No. 4,755,743, July 5, 1988.

Kaliński, J., "A Chopped Subcarrier Method of Simultaneous Attenuation and Phase-Shift Measurement under Industrial Conditions," *IEEE Trans. Ind. Electron. Contr. Instrum.,* Vol. IECI-28, No. 3, August 1981, pp. 201–209.

Khalid, K.B., T.S.M. Maclean, M. Razaz, and P.W. Webb, "Analysis and Optimal Design of Microstrip Sensors," *IEE Proc.,* Vol. 135, Pt.H, No. 3, June 1988, pp. 187–195.

Kent, M., "The Use of Strip-Line Configurations in Microwave Moisture Measurements," *J. Microwave Power,* Vol. 7, No. 3, September 1972, pp. 185–193.

Kent, M., "The Use of Strip-Line Configurations in Microwave Moisture Measurements—II," *J. Microwave Power,* Vol. 8, No. 2, September 1973, pp. 190–194.

Kent, M., J. Köhler, "Broadband Measurement of Stripline Moisture Sensors," *J. Microwave Power,* Vol. 19, No. 3, September 1984, pp. 173–179.

Kent, M., and W. Meyer, "Density Independent Moisture Metering in Fish Meal Industry," *Proc. 11th European Microwave Conf.,* Amsterdam, September 1981, pp. 448–453.

Klein, A., "Microwave Determination of Moisture in Coal: Comparison of Attenuation and Phase Measurement," *J. Microwave Power,* Vol. 16, No. 3–4, 1981, pp. 289–304.

Klein, A., "Microwave Determination of Moisture Compared with Capacitive, Infrared and Conductive Measurement Methods. Comparison of On-Line Measurements at Coal Preparation Plants," *Proc. 14th European Microwave Conf.,* Liége, September 1984, pp. 661–666.

Kraszewski, A., S. Kuliński, J. Madziar, and K. Zielkowski, "Microwave On-Line Moisture Content Monitoring in Low-Hydrated Organic Materials," *J. Microwave Power,* Vol. 15, No. 4, December 1980, pp. 267–275.

Ma, C.H., D.P. Hutchinson, P.A. Staats, and K.L. Vander Sluis, "Measurements of Electron Density and Plasma Current Distributions in Tokamak Plasma," *IEEE 6th Int. Conf. on Infrared and Millimeter Waves,* Miami Beach, December 1981, 1 p.

Metaxas, A.C., and R.J. Meredith, *Industrial Microwave Heating,* IEE Power Engineering Series 4, London: Peter Peregrinus, 1983, 357 p.

Ohno, J., H. Yashiro, Y. Shirakawa, A. Tsuda, S. Watanabe, T. Hirata, M. Higuchi, and M. Nakagome, "Microwave Burden Sensor for Blast Furnaces," *IFAC Automation in Mining, Mineral and Metal Processing,* Tokyo, 1986, pp. 353–358.

Okamura, S., "High-Moisture Content Measurement of Grain by Microwaves," *J. Microwave Power,* Vol. 16, No. 3-4, 1981, pp. 253-256.

Ozamiz, J.M., and S.J. Hewitt, "Microwave Moisture Measurement System," *Proc. 9th European Microwave Conf.,* Brighton, England, September 1979, pp. 340-344.

Praxmarer, W., "Zur Feuchtebestimmung von pulverförmigen Hochpolymeren mittels Mikrowellen," *Plaste und Kautschuk,* Vol. 27, No. 5, May 1980, pp. 252-253.

Schafer, G.E., "A Modulated Subcarrier Technique of Measuring Microwave Phase Shift," *IRE Trans. Instrum.,* Vol. I-9, 1960, pp. 217-219.

Schafer, G.E., and R.R. Bowman, "A Modulated Subcarrier Technique of Measuring Microwave Attenuation," *PIEE Supl. 23B,* Vol. 109, 1962, pp. 783-786.

Schilz, W., and B. Schiek, "Microwave Systems for Industrial Measurements," *Advances in Electronics and Electron Physics,* Vol. 55, New York: Academic Press, 1981, pp. 309-381.

Shiraiwa, T., S. Kobayashi, A. Koyama, M. Tokuda, and S. Koizumi, "Microwave Moisture Gauge for Limestone," *J. Microwave Power,* Vol. 15, No. 4, December 1980, pp. 255-260.

Skolnik, M.I., *Introduction to Radar Systems,* 2nd Ed., New York: McGraw-Hill, 1981, pp. 286-298.

Wyslouzil, W., and A.L. VanKoughnett, "An Attenuation Based Microwave Moisture Gauge for Sheet Materials," *J. Microwave Power,* Vol. 9, No. 2, June 1974, pp. 91-98.

Wyslouzil, W., and S.C. Kashyap, "Microwave Moisture Profile Gauges for Sheet Materials," *14th Microwave Power Symp. Digest,* Monaco, June 1979, pp. 153-155.

Chapter 5
Special and Hybrid Sensors

5.1 INTRODUCTION

In some cases, we can base the measurement principle on a specific feature of the object to be measured (the sample). This feature may be the shape or a certain characteristic in the electrical properties that offers an opportunity. This chapter consists of only such examples and meters combining several measurement principles in an original way. It is a collection of odd examples without much in common. Some of the meters use resonators or transmission sensors, but are presented in this chapter because of some special quality. One of the objectives of the chapter is to give an idea of the diversity of microwave techniques and the possibilities provided by the samples themselves.

5.2 SELECTED EXAMPLES

5.2.1. Measurement of Thickness and Permittivity of a Dielectric Layer on a Conducting Plane Using Surface Waves

A dielectric layer on a conducting plane will support both TE_n and TM_n surface wave modes [Adams, 1981]. Based on the propagation characteristics of the modes, Ou and others have developed a technique to measure the thickness and permittivity of such a layer [Ou et al., 1983]. The sensor consists of a horn antenna (vertical polarization for TM_n waves or horizontal for TE_n waves) and a prism to launch the wave and an identical structure for measurement of the amplitude (Figure 5.1). The prism launcher couples best to a wave mode when the x-component of the propagation vector in the prism is equal to that of the surface wave (k_{0x}), which means that the optimum coupling at a certain frequency ($f = ck_0/2\pi$) occurs for a certain angle θ:

$$k_0\sqrt{\varepsilon_{rp}} \sin\theta = k_{0x} \tag{5.1}$$

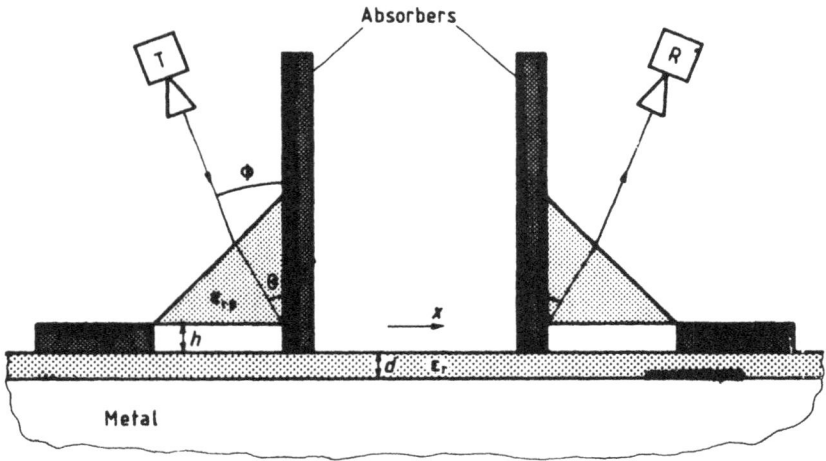

Figure 5.1 A surface wave sensor with prism launchers. The optimum air gap between the prisms and the surface is $h \approx \lambda/2$. The absorbers prevent direct coupling between the horns. (After [Ou *et al.*, 1983].)

where ε_{rp} is the permittivity (assumed to be real) of the prism. The corresponding angle ϕ in the air is easily calculated from Snell's law (1.43). If the angle (i.e., k_{0x}) is measured at a minimum of two frequencies, ε_r and d can be calculated from the propagation characteristics. The following expression can be derived for TE_n modes:

$$\left[\left(\frac{k_{0x}}{k_0} \right)^2 - 1 \right]^{1/2} = (\varepsilon_r - 1)d \left(\frac{k_c}{k_0} \right)^{1/2} (k_0 - k_c) \tag{5.2}$$

where k_c is the propagation vector at the cut-off frequency:

$$k_c = \frac{(2n + 1)\pi}{2d(\varepsilon_r - 1)^{1/2}}; \quad n = 0, 1, 2, \ldots \tag{5.3}$$

Close to cut-off, where $(k_c/k_0)^{1/2} \approx 1$, we obtain a linear relationship if we plot $[(k_{0x}/k_0)^2 - 1]^{1/2}$ *versus* k_0. The intercept is k_c and the slope is

$$S = (\varepsilon_r - 1)d \tag{5.4}$$

From (5.3) and (5.4), we can calculate ε_r and d:

$$\varepsilon_r = \left[\frac{2k_c S}{(2n + 1)\pi} \right]^2 + 1 \tag{5.5}$$

$$d = \frac{S}{\varepsilon_r - 1} \tag{5.6}$$

There are two possible procedures to perform the measurement. One is to measure k_{0x} at a few different frequencies, and then to determine S and k_c. The other, usually more practical, is to measure k_c directly. From (5.1), when $k_0 = k_c = k_{0x}$, we obtain a fixed angle θ_c for cut-off. We can start with the sensor at θ_c and measure the cut-off frequency (maximum transmission), then increase the frequency (or angle) slightly and measure the new optimum angle (or frequency). For exact results, ε_r and d can be solved from (5.1), (5.2), and (5.3).

Similar results can be derived for the TM_n modes. For the intercept and slope, we then have

$$k_c = \frac{n\pi}{d(\varepsilon_r - 1)^{1/2}}; \quad n = 0, 1, 2, \ldots \tag{5.7}$$

$$S = \frac{(\varepsilon_r - 1)d}{\varepsilon_r} \tag{5.8}$$

In contrast to the lowest TE mode, TM_0 has zero cut-off frequency and can therefore be used to measure only one variable. We can solve either ε_r or d if we know the other. An advantage with TM_0 is that we can use lower frequencies with thinner layers than with other modes. The higher TM modes have two disadvantages compared to TE modes: the slope becomes saturated for high ε_r values, and the solution of ε_r is bivalued, which requires some prior knowledge of the approximate value. In addition, the cut-off frequency of TM_1 is twice as high as that of TE_0.

The surface wave method has the advantages of being nondestructive and fairly accurate. It provides a means of measuring something that is difficult to measure with other methods. The disadvantages are that the angles ϕ must be changed during the measurements (or at least two pairs of horns are needed), the method assumes ε_r to be real (but the losses will not affect the measured results as long as $\varepsilon_r'' \ll \varepsilon_r'$), the possible confusion between modes, high frequency, and large size. It is possible to include the measurement of ε_r'' by varying the distance between the launchers. For the calculation of radiation and dielectric losses, see [Adams, 1981, pp. 64–65]. Ou and others have measured layers in the centimetre range at 8–12 GHz using prisms $20 \times 20 \times 20$ cm [Ou et al., 1983]. The distance between the horns (gain of 20 dB) and the prisms was 80 cm, and the absorbers were 50 cm high with a gap of 9 cm to the dielectric layer. The achieved accuracy for d and $\varepsilon_r'-1$ was typically 1–2%.

The most useful applications of the method may be the checking of dielectric coating materials on engine, turbine, and metal parts. Because the layers are thin and the measurement area should be small, millimeter waves are to be used. For example, if $\varepsilon_r' = 10$ and $d = 0.5$ mm, (5.3) predicts a cut-off frequency of 50 GHz for TE_0.

5.2.2 Detection of Knots in Sawed Timber

An important factor for the yield of a sawmill is the effective use of the raw material. For that purpose, Innotec (Finland) has developed a computerized system for the automatic edging of boards (raw-edged slice from a log) [Heikkilä et al., 1982]. For the steering of the process, the quality and visual appearance of the board must be evaluated. It is primarily made optically by using light sources, video cameras and image processing. The optical means cannot, however, distinguish between knots and pieces of bark or surface blemishes. Neither can they detect light-colored knots, knots covered by sawdust or frost, or knots in vane areas partly covered by bark. The knots are therefore detected with microwaves.

The microwave knot detecting system [Jakkula, 1986] is shown in Figure 5.2. A plane wave is transmitted with a horn antenna from above, and 32 separate detectors immediately below the board detect the location of the knots. The detectors are made of circular waveguide coupled at the shorted end in such a way that they detect only modes with an axial electric field component (i.e., TM modes). The frequency and diameter of the waveguide are chosen so that only the TE_{11} and TM_{01} modes can propagate, but the TE_{11} mode is not detected in the TM mode detector. Because the electric field of the TM_{01} mode is circularly symmetric (Figure 1.7), the field will not be excited by the plane wave transmitted from the horn. However, the knot forms a dielectric waveguide through the board supporting the mode HE_{11}, which is generated in the knot from the plane wave [Adams, 1981]. The field pattern of the HE_{11} is shown in Figure 5.3. The electric field has an axial component, which is largest at the borders of the knot. When the knot comes in front of a detector, this field component excites the TM_{01} mode and thus couples power from the plane wave to the detector, and so reveals the knot. Because of the distribution of the field, each knot will give a characteristic two-peak signal when it moves past a detector. HE_{11} is the dominant mode and it has zero cut-off frequency, which means that even small knots are detected.

5.2.3 Strength Grading of Timber by Combining Several Measurement Methods

The quality of sawed timber, especially its strength, is important for many customers. The sawmills should therefore be able to sort the boards into strength classes. Innotec (Finland), in cooperation with the Helsinki University of Technology, has developed a measurement system for this purpose [Heikkilä et al., 1982]. The determination

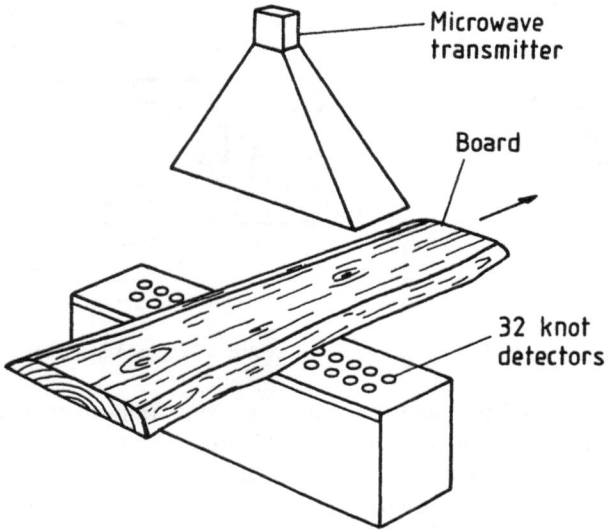

(a) the sensor consisting of a common transmitter and 32 detectors

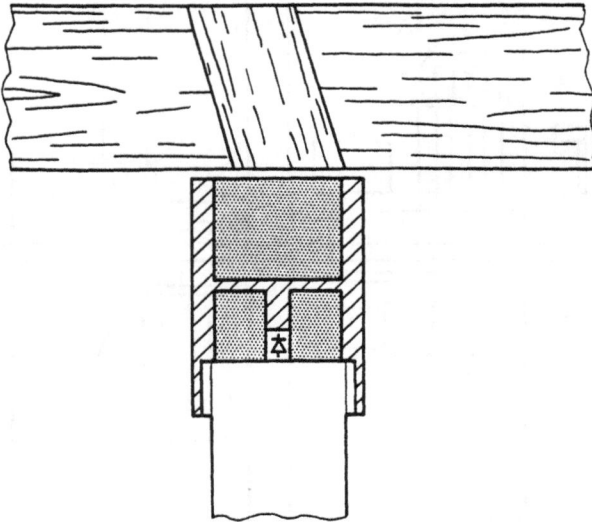

(b) an individual detector

Figure 5.2 Sensor for measurement of the position of knots in boards (After [Heikkilä *et al.*, 1982].)

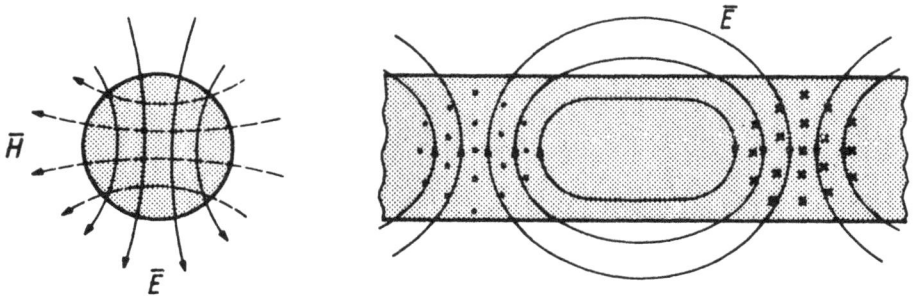

Figure 5.3 The (qualitative) field pattern of the dielectric waveguide mode HE_{11}.

of the strength is based on measurement of the most important parameters affecting the strength and then calculating the strength from the empirically known relationship. The parameters measured are the *dry density, knottiness,* and *slope of grain*. The sensor arrangement is shown in Figure 5.4. The sensors utilize three kinds of electromagnetic radiation: microwaves, infrared (IR) radiation, and γ-rays.

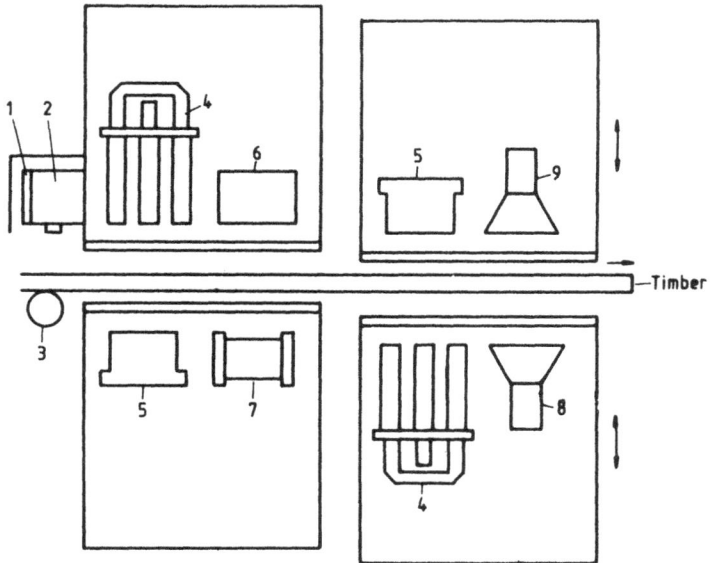

Figure 5.4 The sensor arrangement in the strength grading equipment: (1) board detector; (2) surface temperature sensor; (3) conveyor speed sensor; (4) microwave transmitter for knot detector; (5) microwave receiver (15 channels); (6) γ-ray source; (7) γ-ray detector; (8) microwave transmitter (moisture, angle of grain); (9) microwave receiver. (After [Heikkilä *et al.*, 1982].)

The dry density is determined by measuring the microwave attenuation, γ-ray attenuation, and surface temperature. The microwave attenuation is measured at 10 GHz with a free-space transmission sensor equipped with two lens-corrected horn antennas, and the surface temperature is measured with an IR thermometer. The moisture and dry density are then calculated by using the same model as in the moisture meter also developed at Innotec [Tiuri *et al.*, 1980].

The knot sensor differs from that described in Section 5.2.2. It is a microwave bridge based on free-space transmission through the board in both channels (Figure 5.5). The transmitted signal is split in two equal parts with opposite phase in a T-junction and fed to two separate horn antennas. On the other side of the board, the signals are combined in a U-section of waveguide. The diode detector is mounted in the symmetry point, where the received signals cancel when the bridge is balanced. When a knot is situated in front of either end of the U-section, a signal is detected. To determine the location of the knots, 15 U-shaped receivers are used to cover a width of 300 mm.

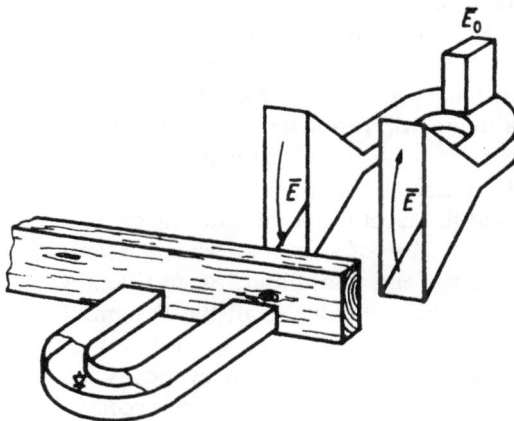

Figure 5.5 The knot detector in the strength grading equipment. 15 identical adjacent receivers are used (only one is shown). (After [Heikkilä *et al.*, 1982].)

The slope of grain measurement, performed with the same free-space transmission sensor as is the moisture measurement, is based on the anisotropy of the permittivity of wood. The permittivity is larger when measured with an electric field parallel to the grains than if it is perpendicular. For a field applied at an intermediate angle, the internal polarization vector will therefore not be parallel to the field. This has the effect of turning the polarization of a propagating wave. The sensor transmits a linearly polarized wave with the field parallel to the length axis of the board. If the angle of grain deviates from the ideal, it causes cross polarization (a field component perpendicular to the transmitted field) at the receiver, which measures both polarizations with a dual-polarized horn.

The measurement system has been approved by the Finnish authorities for strength grading of timber for supporting structures in buildings.

5.2.4 Detection of Fatigue Cracks Based on Third-Order Nonlinearities

Metal junctions are known to act like nonlinear semiconductor components producing harmonic frequency components of an incident frequency, and also to produce mixer products of two simultaneously incident frequencies. This has caused interference problems with frequencies generated in the neighborhood of transmitting radar and communication antennas (or in the actual antenna structures) on ships. The phenomenon has also been used for detection of metal objects hidden by vegetation, for example. A possible explanation for the phenomenon is the tunneling of electrons through an oxide layer between two metal surfaces.

Experiments have been performed at the Remote Measurements Laboratory of SRI International (US) to demonstrate the possiblity of using the phenomenon to detect fatigue cracks in metals [Bahr, 1980]. The sensor was a $\lambda/2$ dipole on a printed circuit board, fed by a coaxial cable. When pressed against a metal surface, the sensor became a microstrip resonator. The transmitted signal contained two frequencies ($f_1 = 10.6$ GHz, $f_2 = 9.8$ GHz) generated by a balanced modulator from a common 10.6 GHz source. Both components were amplified to the level of 2 W. The receiver was tuned to detect the third-order mixer product $f_{rec} = 2f_2 - f_1 = 9.0$ GHz. The most critical feature of such a meter is the very high requirement of isolation between the receiver and transmitter. Steep filters are needed to separate 9.0 from 9.8 GHz and all other frequencies in the system. Leakage through external paths must also be hindered. The achieved sensitivity was about 150 dB below the transmitted power. Samples were prepared by cyclically loading notched metal plates in air at room temperature. Nothing was done to accelerate the formation of oxide in the cracks. The metals used were aluminum and stainless steel.

The experiments showed that the third-order signal at 9.0 GHz was clearly detectable with the meter. The method was demonstrated to be a powerful technique for nondestructive testing of metal objects. For different situations, when the typical length and depth of the cracks are different, other frequencies may give optimal results.

5.2.5 Gas Analysis by Microwave Spectrometry

The absorption spectral lines associated with rotational transitions in gases are narrow at low pressure (Section 2.2.5). By measuring the absorption spectrum in a frequency band, the concentrations of the different constituents in a gas sample can be determined. This applies, of course, only to molecules, which have spectral lines in the measurement bandwidth. Gas analysis with microwave spectrometry is a well known laboratory method, but it can be used also for industrial applications. The team at Philips Research Laboratories in Hamburg has developed two different spectrometers for industrial use. One is a guided-wave transmission sensor [Schiek et al., 1977] and the other is a resonator [Reinschlüssel et al., 1985]. In both cases the gas is kept at a pressure of about 10 Pa (0.1 mbar).

The greatest difficulty of microwave gas spectrometry is its limited sensitivity, which follows from the low pressure (small amount of molecules in the sample). The sensitivity can, however, be improved considerably by using so-called *Stark modulation*. Under the influence of a static electric field, most absorption lines split. When the electric field is turned on and off with a typical frequency 50 kHz, this will modulate the amplitude of the lines. The receiver is tuned to detect the 50 kHz modulation. In addition, special care must be taken to reduce the noise of the signal source and receiver as much as possible.

The transmission sensor developed at Philips Research Laboratories is an oversized waveguide with the flat Stark electrode in the middle. The sensor can be tuned over the range 10 to 26 GHz, and is 88 cm long. The Stark voltage is typically 100 to 200 V. The sensor actually consists of two absorption cells modulated with a phase shift of 180° to produce a balanced system for suppression of the noise. The reported detection limit for the freon gas ($C_2H_4F_2$) was 20 parts per million (ppm) and for ammonia (NH_3) it was 1 ppm, using an integration time of 10 s.

The resonator sensor is an 88 cm long section of cylindrical helix waveguide with a rod-shaped Stark electrode in the middle (Figure 5.6). The sensor is designed to be used in the frequency range of 26 to 40 GHz, where it has TE_{01n} resonances with an even spacing of 200 MHz. The helix waveguide suppresses all other modes. The resonator sensor is more complicated to use than the transmission sensor because the device must be tuned with a sliding plunger to cover the whole spectrum. The meter automatically measures the transmission amplitude at every resonance peak once, then moves the plunger (motor drive controlled by the computer) to shift slightly the resonance peaks, and measures again until each peak has been shifted 200 MHz and the whole spectrum has been covered. The advantage of the resonator is its high sensitivity, which is better by a factor of 100 as compared to the transmission sensor.

5.2.6 Automatic Moisture Meter Based on Microwave Drying

Measurement of moisture in discrete samples is an important part of quality control throughout industry. Many different methods have been developed, of which the measurement of loss of weight during oven drying is one of the most widely used.

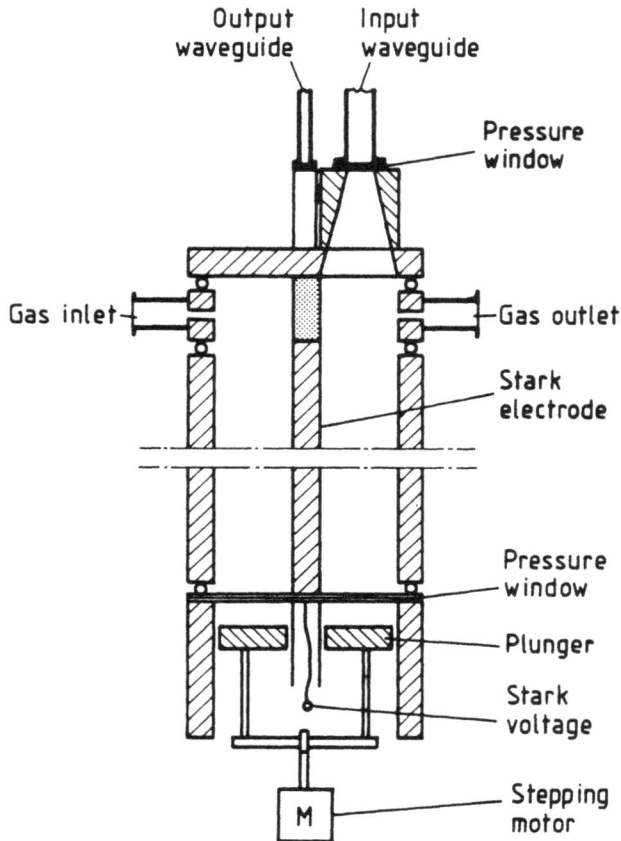

Figure 5.6 Microwave resonator for gas analysis by measurement of molecular rotational spectral lines. After [Reinschlüssel *et al.*, 1985].

At the Philip Morris Research Center (US) a method involving microwave drying has been developed for the measurement of tobacco samples [Thomas *et al.*, 1979]. The method involves drying of samples in the range 4.5 to 13.0 g in a microwave oven, with a power of 480 W, at a frequency of 2.45 GHz, typically for a time of 100 to 200 s. The sample holder rests on a shaft extending through the oven floor and acting on a scale. Both the oven and scale interface with a microcomputer. By continuously monitoring the rate of weight loss, the weight curve has been shown asymptotically to approach the final value in a predictable way. By proper modeling, achieving very accurate results has therefore been possible with the short drying times mentioned above. Tobacco is a material containing both thermally stable and volatile organic compounds. During normal oven drying, evaporation of compounds other than water is therefore inevitable. With the microwave method, this has been reduced to a minimum because of the short drying time, low ambient temperature, and proper modeling, which removes the need to evaporate all of the water in the sample. The method can, of course, be used with almost any material. For different materials and sample sizes, the optimum power and drying time are different. The model must also be calibrated for different materials.

5.2.7 Laboratory Meter for Fast and Simultaneous Determination of Permittivity and Density

A measurement system for determinating the weight and complex permittivity of samples, with fixed shape and volume, has been developed at the Radio Laboratory of the Helsinki University of Technology (Figure 5.7). The system consists of a coaxial resonator, with a flat center conductor, and a scale (Figure 5.8). The sample is placed in a plastic sample holder on a plastic plate and rod, extending through a hole in the bottom wall of the resonator and rests on the plate of the scale. The resonance meter and scale can interface with a computer, which calculates the permittivity (and possibly, for example, the moisture) and density from a previously calibrated model.

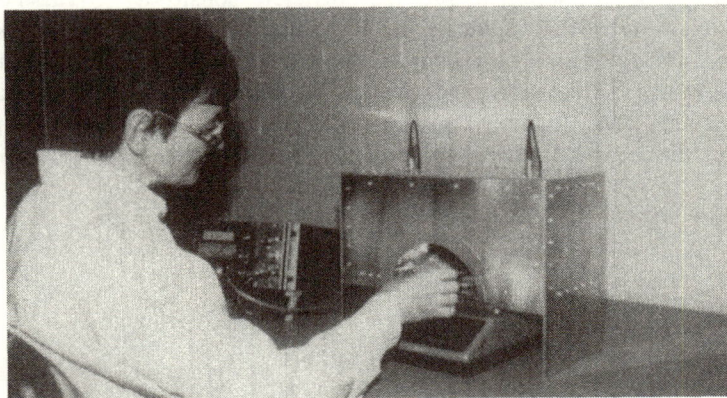

Figure 5.7 Coaxial resonator and laboratory scale for fast determination of complex permittivity and weight (density) of samples.

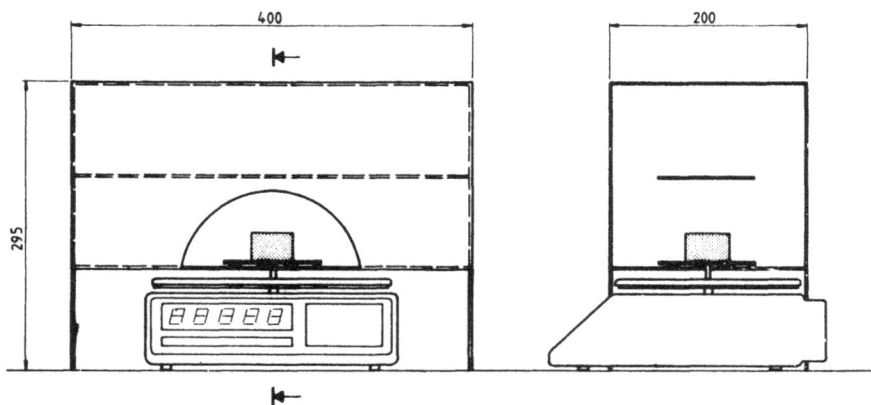

Figure 5.8 Front view and cross section of the measurement system in Figure 5.7.

The resonator has been designed to have a uniform vertical electric field at the sample, but the sample must still always be put in exactly the same location. This is achieved by using guiding pins or a shallow depression for the sample holder in the plastic plate.

The sample size should be small enough so as not to change the field pattern very much, but large enough to provide sufficient sensitivity. With the meter in Figure 5.7 (f_r = 480 MHz), samples in the range of 10 to 100 cm^3, depending on the mean ε'_r, are optimal. The sensitivity as a function of ε'_r is determined by the shape of the sample. The perturbation theory (Section 3.2.2) gives the limiting functions $\Delta f_r/f_r \propto (\varepsilon'_r-1)/\varepsilon'_r$ for very flat samples, and $\Delta f_r/f_r \propto \varepsilon'_r-1$ for high and narrow samples. For practical cylindrical samples, the function can be empirically approximated by

$$
\frac{\Delta f_r}{f_r} \approx \frac{\varepsilon'_r - 1}{2(\varepsilon'_r)^\alpha} \cdot S; \quad 1 \leq \alpha \leq 2
$$

$$
\alpha \approx 0.89 - 0.284\,\frac{h}{d} + 0.036\left(\frac{h}{d}\right)^2 ; \quad \begin{cases} \varepsilon'_r \leq 25 \\[2mm] 0.2 \leq \dfrac{h}{d} \leq 1.7 \end{cases} \tag{5.9}
$$

where h is the height and d is the diameter of the sample.

Equation (5.9) is based on the calibration curves in Figure 5.9. In the same way, we obtain for the losses:

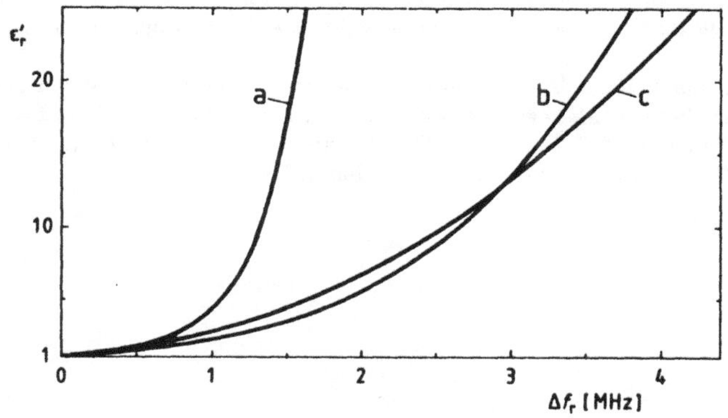

Figure 5.9 Calibration curves for the resonator in Figures 5.7 and 5.8 measured with cylindrical samples: (a) $h/d = 0.20$, $V = 57.2$ cm^3; (b) $h/d = 0.84$, $V = 52.4$ cm^3; (c) $h/d = 1.63$, $V = 34.0$ cm^3.

$$\Delta\left(\frac{1}{Q}\right) \approx \frac{\varepsilon_r''}{(\varepsilon_r')^\beta} \cdot S; \quad 0 \leq \beta \leq 2 \tag{5.10}$$

The measurement system is well suited for measurement of liquids, powders, and fine grained materials, or solid samples cut to a fixed shape. Coarse grained materials produce scatter in the results because of the varying shape of the sample. The advantage of this system is the possibility of measuring both weight (density) and complex permittivity of a large number of samples with a minimum effort. A disadvantage is the fixed measurement frequency, which for practical reasons is more or less restricted to the VHF and UHF bands.

REFERENCES

Adams, M.J., *An Introduction to Optical Waveguides,* Chichester: John Wiley and Sons, 1981, 401 p.

Bahr, A.J., "Microwave Detection of Third-Order Nonlinearities in Fatigue Cracks," *Electronics Letters,* Vol. 16, No. 4, February 1980, pp. 150–152.

Heikkilä, S., P. Jakkula, and M. Tiuri, "Microwave Methods for Strength Grading of Timber and for Automatic Edging of Boards," *Proc. 12th European Microwave Conf.,* Helsinki, September 1982, pp. 599–603.

Jakkula, P., "Method and Apparatus for Knot Detection in Sawn Timber," U.S. Patent No. 4,607,212, 1986.

Ou, W., C.G. Gardner, and S.A. Long, "Nondestructive Measurement of a Dielectric Layer Using Surface Electromagnetic Waves," *IEEE Trans. Microwave Theory Tech.,* Vol. MTT-31, No. 3, March 1983, pp. 255–261.

Reinschlüssel R., and B. Schiek, "A Sensitive Cavity-Based Gas-Spectrometer for Broadband Operation at 26–40 GHz," *Proc. 15th European Microwave Conf.,* Paris, September 1985, pp. 895–900.

Schiek, B., T. Paukner, and W. Schilz, "A Microwave Spectrometer—Suitable for Gas Analysis in Industrial Applications," *Proc. 7th European Microwave Conf.*, Copenhagen, September 1977, pp. 251–255.

Tiuri, M., K. Jokela, and S. Heikkilä, "Microwave Instrument for Accurate Moisture and Density Measurement of Timber," *J. Microwave Power*, Vol. 15, No. 4, 1980, pp. 251–254.

Thomas, C.E., M.C. Bourlas, T.S. Laszlo, and D.F. Magin, "Automatic Microwave Moisture Meter," *Proc. 14th Microwave Power Symp.*, Monaco, June 1979, pp. 150–152.

Chapter 6
Reflection and Radar Sensors

6.1 INTRODUCTION

In this chapter we will discuss sensors that are based on measurement of the reflection of microwaves from an object (sample). The topic is not a homogeneous one, but comprises both contacting probes and noncontacting free-space sensors for measurement of dielectric properties of materials as well as radar sensors for measurement of distance and movement. The reflection sensors for measurement of materials often have features in common with resonators or transmission sensors. Reflection sensors form a group of their own, mainly from the point of view of the electronics. The sensors have predominantly been used for scientific and medical purposes, and in such cases an automatic network analyzer has usually been employed. The radar sensors use more or less standard radar technology, but the range rarely exceeds a few tens of meters and the transmitted power is only a few milliwatts. Radar sensors are often used for measurement of a single target instead of imaging. Only the impulse radar forms an exception, in which case the time dependence of the echo is recorded. A two-dimensional picture may be produced by moving the impulse radar linearly along the target.

6.2 REFLECTION SENSORS

6.2.1 Open-Ended Transmission Line Sensors

The open end of a transmission line causes a reflection coefficient Γ with an amplitude close to unity (except for an open waveguide). The deviation from unity is caused only by radiation. When the end is held against or immersed into a dielectric medium, the reflection coefficient changes. The ε_r' of the medium influences the phase of Γ, and the dielectric losses together with the altered radiation influence the amplitude. This offers a simple and fast way to measure nondestructively the permittivity of dielectric materials, over a broad frequency range by using a simple sensor structure and a commercial ANA. The sensor structures most frequently used are the coaxial line and waveguide.

Coaxial Sensors

Open-ended coaxial reflection sensors have been studied by many authors. Numerous modifications to the basic structure have also been reported (e.g., ground-plane flange, extended center conductor, capacitively loaded center conductor), but, because the basic structure (Figure 6.1) seems to be preferred by many authors, it will be discussed in more detail. A good review article about coaxial reflection measurements of dielectric materials has been compiled by the Stuchlys [Stuchly *et al.*, 1980].

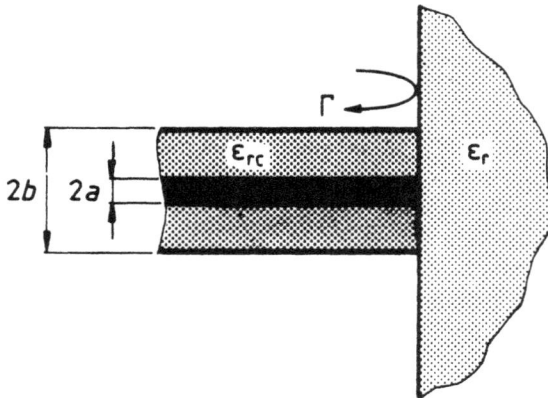

Figure 6.1 An open-ended coaxial reflection sensor.

An open end of a transmission line, like the coaxial sensor in Figure 6.1, can be characterized by the equivalent circuit of Figure 6.2 ($a,b \ll \lambda_0/\sqrt{\varepsilon_r}$), where G is the radiation conductance, C_f is the capacitance of the fringing field inside the sensor, and C is the capacitance of the field outside the sensor. Both C and G depend on the permittivity of the sample, and C_f, C, and G depend on the dimensions of the transmission line and permittivity of the dielectric filling the line. The values in air (C_f, C_0, G_0) can be calculated analytically [Marcuvitz, 1951]. Simply measuring the total end capacitance ($C_T = C_f + C_0$) and C_f is also possible [Stuchly *et al.*, 1982a]. As a rule of thumb C_f is at least an order of magnitude lower than C_0. Both $C_T = B_0/\omega$ and G_0 can be estimated from the diagram in Figure 6.3. As a first approximation, C_f and the radiation conductance G_0 can be neglected. When the

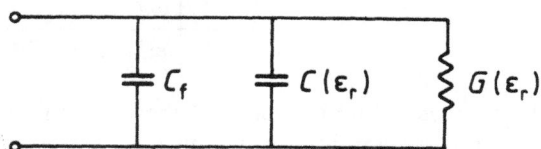

Figure 6.2 Equivalent circuit for the open-ended coaxial sensor.

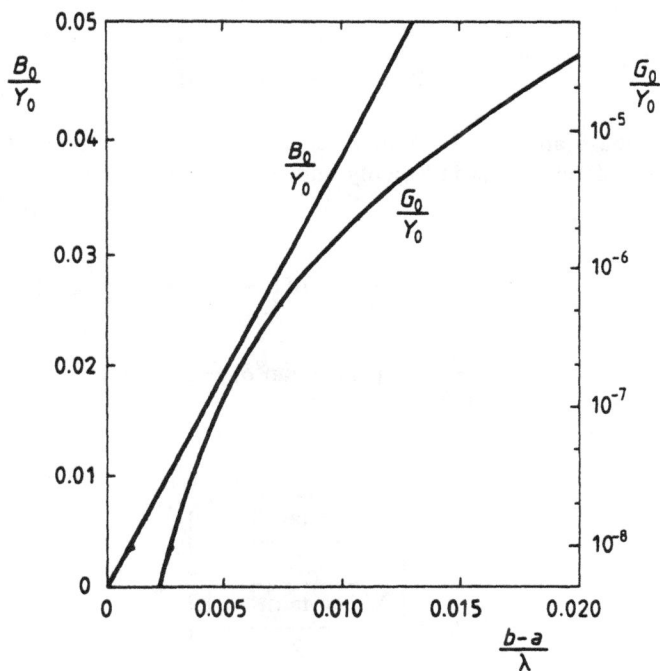

Figure 6.3 The end susceptance (B_0) and conductance (G_0) of an open-ended coaxial line sensor in air. The characteristic admittance of the line is Y_0 ($Z^{-1} = Y = G + jB$). (After [Stuchly *et al.*, 1982b].)

sensor is held against the sample, the end capacitance changes according to the permittivity of the sample, giving the load admittance:

$$Y = G + jB = j\omega C_0(\varepsilon'_r - j\varepsilon''_r) = \omega C_0\varepsilon''_r + j\omega C_0\varepsilon'_r$$

The reflection coefficient is, then,

$$\Gamma = \frac{-Z_0 + 1/[j\omega C_0(\varepsilon_r' - j\varepsilon_r'')]}{Z_0 + 1/[j\omega C_0(\varepsilon_r' - j\varepsilon_r'')]} \tag{6.1}$$

where Z_0 is the characteristic impedance of the transmission line.

If we solve for the permittivity, we have

$$\varepsilon_r' = \frac{-2|\Gamma|\sin\phi}{\omega C_0 Z_0(1 + 2|\Gamma|\cos\phi + |\Gamma|^2)} = A$$

$$\varepsilon_r'' = \frac{1 - |\Gamma|^2}{\omega C_0 Z_0(1 + 2|\Gamma|\cos\phi + |\Gamma|^2)} = B \tag{6.2}$$

where ϕ is the phase angle of Γ. A more accurate analysis (taking into account C_f and G_0) has been done by Maria Stuchly and others [Stuchly et al., 1982b] resulting in

$$\varepsilon_r' = A - \frac{C_f}{C_0} + \frac{G_0 \varepsilon_r'^{5/2}}{\omega C_0}[\beta(1 - \tan^2\delta) + 2\alpha\tan\delta]$$

$$\varepsilon_r'' = B - \frac{G_0 \varepsilon_r'^{5/2}}{\omega C_0}[\alpha(1 - \tan^2\delta) - 2\beta\tan\delta] \tag{6.3}$$

where

$$\alpha = \left[\frac{\sqrt{1 + \tan^2\delta} + 1}{2}\right]^{1/2}$$

$$\beta = \left[\frac{\sqrt{1 + \tan^2\delta} - 1}{2}\right]^{1/2} \tag{6.4}$$

and $\tan\delta = \varepsilon_r''/\varepsilon_r'$. Equation (6.3) contains ε_r on both sides and can only be solved with an iterative process. The topic is further studied in [Gajda et al., 1983; Athey et al., 1982; Epstein et al., 1987; and Moschüring et al., 1987].

At a single frequency the reflection coefficient is conveniently measured with high accuracy by using the coaxial line as a resonator (see Figures 3.10, 3.14, and 3.39), in which case the relations between the resonant frequency, quality factor, and Γ are given by (3.7) and (3.19). One of the advantages of the open-ended coaxial line, however, is the broad bandwidth (typically, two decades). A long resonator exhibits a series of evenly spaced resonances, but the useful bandwidth is, in practice, limited to one decade. To take full advantage of the wide frequency range, an ANA (for scientific and medical use) or a specially designed reflection measurement device (for field use) must be used to measure Γ directly. The greatest difficulties in practice are calibration and evaluation of accuracy. These matters have been studied extensively in [Marsland et al., 1987; Stuchly et al., 1982a; Stuchly et al., 1987; Burdette et al., 1980]. Several authors have shown for a given reflection coefficient measurement accuracy that the greatest accuracy in determining the permittivity is obtained for an optimum value of the capacitance:

$$C_{0,opt} = \frac{Y_0}{\omega} (\varepsilon_r'^2 + \varepsilon_r''^2)^{-1/2} \qquad (6.5)$$

When measuring at solid surfaces, care must be taken to achieve good contact. Any air gap between the sensor and sample will also reduce the measurement accuracy. The thickness of the sample here has been assumed to be "infinite." In practice, the field penetrates into the sample to a depth approximately equal to the radius of the sensor, which is the minimum thickness required for measurement with the sensor [Brunfeldt, 1987]. A method of increasing the measurement accuracy at low frequencies and extending the measurement range down to 10 kHz has been reported in [Esselle et al., 1988]. The idea is to increase C_0 with the narrow-gap arrangement shown in Figure 6.4. The sensor has been developed for measurement of biological tissues. It is not well suited for solid surfaces because the field penetrates a very short distance into the sample due to the narrow opening in the sensor.

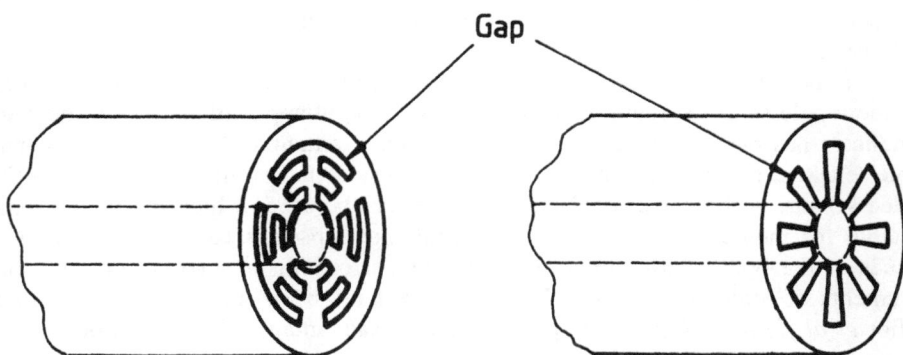

Figure 6.4 Open-ended coaxial sensors with increased C_0 for measurement at low frequencies (down to 10 kHz). (After [Esselle et al., 1988].)

The fields of application of coaxial reflection sensors are many, ranging from medicine to measurement of liquids and soil moisture. Marsland and Evans from Cambridge University (UK) have developed a laboratory measurement system for the study of cryopreserved tissue during the rewarming process [Marsland et al.,

1987]. They report the advantages that measuring a wide range of tanδ is possible, the sensor is convenient for insertion into a temperature-controlled environment, a given size of probe can cover a two-decade frequency range, and little or no sample preparation is needed. Marsland and Evans use a 300 mm long sensor made of 6.4 mm semirigid coaxial cable ($2a$ = 1.63 mm, $2b$ = 5.28 mm) for measurements in the range 50 MHz to 2.6 GHz. An ANA is used to perform the measurements and to calculate the results. To overcome the problems with the calibration, they have developed a method based on the measurement of three materials of known permittivity (e.g., methanol, ethanediol, saline water, or air). The achieved accuracy is generally good, but in the low end of the frequency range the results are particularly sensitive to measurement errors, and in the upper end the radiation increases, which causes errors if (6.2) is used. However, the calibration procedure corrects for the radiation.

Applied Microwave Corporation (US) manufactures a field portable meter for measurement of the complex permittivity of almost any material [Brunfeldt, 1987]. The most important application is the measurement of soil moisture. The meter is a two-channel reflectometer. One channel measures the sensor and the other is used for reference. The length of both cables is 7.6 m. The FMCW radar measurement technique (see below) is used. The transmitted and reflected signals are mixed in both channels to produce two low-frequency signals, which are compared. Using narrowband filters, the reflections from the sensor and reference termination are separated from other reflections (e.g., from connectors). Different sensor heads are used for measurement at 1.25, 4.8, and 18 GHz. The measurement accuracy for the 1.25 GHz head is reported to be ε_r' = ±0.05 ± 2.5%, ε_r'' = ±0.05 ± 7.5%.

In addition to the basic structure of open-ended coaxial sensors (Figure 6.1), the monopole sensor (Figure 6.5) has especially useful applications. Here, the center conductor has been extended to form a monopole antenna. This structure is much more sensitive to the permittivity of the medium surrounding the sensor than the flat-ended sensor. A thinner coaxial line can therefore be used, which makes the sensor suitable for medical applications. The needlelike sensor is easily pushed into the tissues of the body, without any damage, for performing *in vivo* permittivity measurements of biological tissues. These matters have been studied in, for example, [Bliot *et al.*, 1980; Misra, 1987]. The major disadvantage of the monopole sensor as compared to the flat-ended sensor is the tendency to radiate, which is strongly dependent on the length of the monopole. Especially in low-loss media, the radiation causes reflections from surfaces and objects in the neighborhood of the sensor, which will disturb the measurement. The monopole should therefore be much shorter than a quarter-wavelength (in the medium).

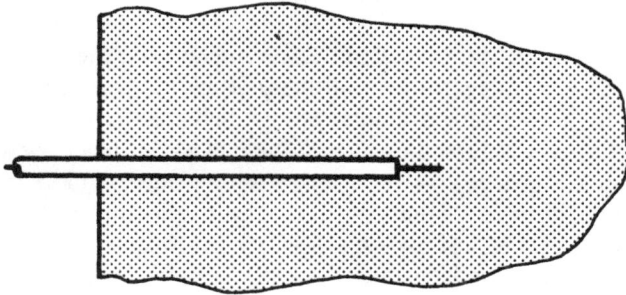

Figure 6.5 The monopole coaxial reflection sensor is suitable for *in vivo* permittivity measurements of biological tissues.

Waveguide Sensors

Open-ended waveguides can be used as reflection sensors in much the same way as coaxial sensors. However, the waveguides radiate much more effectively. This can cause problems if reflecting surfaces or objects are in the vicinity of the sensor, and the sample must be much thicker (in wavelengths) than is the case with coaxial sensors. Waveguide sensors are therefore best suited for measurements at millimeter waves, where coaxial sensors become extremely thin (see Section 2.7.4) and the penetration depth, as a consequence, becomes very small. The waveguide can also be terminated in a metal plate with a hole to reduce the radiation and penetration depth [Decréton *et al.*, 1975]. Sphicopoulos and others have studied the reflection coefficient in rectangular and cylindrical waveguide sensors as a function of permittivity, both theoretically and experimentally. They report simple polynomial expressions for the permittivity as a function of Γ, which have been fitted to the computed results [Sphicopoulos *et al.*, 1985].

Spalla (from the Center for Electronic Research in Sicily C.R.E.S.) and others from the University of Palermo (Italy) have measured the reflection from samples of bacteria in water in the range of 66–73 GHz [Spalla *et al.*, 1976]. They measured through the bottom of a small glass sample holder, to which the bacteria had slowly descended. The reflection was found to vary with time and frequency in such a way that could be explained as arising from multiple reflections in the layer of sediment. Specific maxima due to absorption in protein, DNA, and RNA have been reported by other authors, but no such effects could be detected, probably due to the high attenuation in water.

6.2.2 Shielded and Other Transmission Line Sensors

In some cases, performing the reflection measurements with the transmission line as a sample holder may be more convenient than putting the open end against the surface of the sample. Bussey, for example, has studied coaxial sensors with the outer conductor extending as a sample holder past the end of the inner conductor [Bussey 1980]. The dielectric insulator is retracted to allow part of the sample to be set between the inner and outer conductors. The sensor is more sensitive than the conventional flat-ended sensor and radiates less, but care must be taken so that the sample holder is packed homogeneously.

Bastida and others have used the corresponding waveguide structure [Bastida et al., 1979]. The method is equivalent to the infinite sample technique discussed as a laboratory method in Section 2.7.2. They use a cylindrical waveguide sample holder for measurement of soil moisture. By using a suitable model to describe the dependence of the permittivity of soils on the moisture, they have developed a method based only on measurement of the amplitude of the reflection coefficient. According to the report, the results are independent of temperature, type of soil, density, salinity, and grain size, without compensation, which is remarkable for a one-parameter measurement. This is possible by making proper use of "inherent" compensation provided by optimization of measurement frequency and the mixing formula. Two versions of the electronic unit have been built, one for the laboratory and the other for operational field use.

Martinson and others have reported a waveguide reflection sensor [Martinson et al., 1985], which is complementary to the open-ended sensors. The measured reflection coefficient is low in cases for which it is high with the open-ended sensors, and vice versa. The sensor consists of a section of slotted rectangular waveguide terminated in a matched load. The slots are machined in the broad wall of the waveguide perpendicular to the length axis. The separation between the slots is $\lambda_g/2$ so that the reflections from the slots add constructively. When the slotted wall is held against the sample surface, the permittivity affects the mismatch caused by the slots. A disadvantage is the large measurement area, which must be plane in the case of solids.

6.2.3 Free-Space Sensors

By measuring at a distance the amplitude and phase of the wave reflected from a flat service, we can calculate the complex reflection coefficient and permittivity of the sample. If the measurement is repeated at different frequencies, more information can be deduced.

Laminated Samples

Scientists at the Dnepropetrovsk State University (USSR) have studied the problem of calculating the thicknesses and permittivities of the layers in a laminated dielectric material from the measured frequency dependence of the reflection coefficient [Akhmetshin, 1986; Akhmetshin *et al.*, 1986; Bartashevskii *et al.*, 1986]. They have reported several methods to perform the calculations. The most straightforward technique is the method of optimization of parameters, which means that the frequency dependence of the reflection coefficient is calculated for different values of the parameters (e.g., using the signal-flow graph technique, see Section 9.5). The values are then optimized until the calculated result agrees with the measured result. In practice, to avoid ambiguous results, the method requires some prior knowledge of the number of layers, approximate thicknesses, and permittivities.

Another method is that of identification of natural resonances, detected as minima in the reflection coefficient. Each layer has a series of resonances, which occur according to the condition in (3.7). The quality factor of each resonance is determined by (3.19).

A third method is the calculation of the envelope of the pulse response by Fourier transforming the frequency response. The interfaces between the layers can then be identified from the delays of the reflections. The most accurate results are obtained with the method of optimization of parameters, where the results given by identification of resonances or calculation of the pulse response are used as the first approximation.

The calculation methods discussed above are fairly effective for lossless materials, but when losses are included, the calculation of ε_r'' usually only gives approximate results. In the literature [Bartashevskii *et al.*, 1986], results are reported for measurements in the range of 7.0–12.0 GHz for a two-layer structure, where the layers were 8.5 cm of $\varepsilon_r' = 2.1$ and 2.5 cm of $\varepsilon_r' = 3.5$. The deduced values agree with the true values to within 6% for both thickness and permittivity when the pulse response plus optimization of parameters technique was used.

Dielectric Coating

The methods described above are applicable when the thicknesses of the layers are at least of the same order of magnitude as the wavelength. Konev and others from the Institute of Applied Physics at the Academy of Sciences in Minsk (USSR) have studied the measurement of a thin layer of dielectric coating using microwave ellipsometry at millimeter-wave frequencies (i.e., measuring the change of polarization at reflection for oblique incidence [Konev *et al.*, 1985; Konev *et al.*, 1986]). The former reference presents a detailed analysis of the method (in Russian). The latter is a short summary in English, dealing mainly with error analysis. The method is based on the fact that the reflection coefficient is different for vertically polarized and horizontally polarized waves when the incidence angle θ_1 is $0° < \theta_1 < 90°$. When two interfaces are present, the interference further changes the reflection coefficients. The basic equation for ellipsometry is the ratio of the complex reflection coefficients for the two polarizations in the case of two interfaces. From Figure 4.3, we have

$$\frac{\Gamma_v}{\Gamma_h} = \frac{\Gamma_{v12} + \Gamma_{v23}\exp(-jg)}{1 + \Gamma_{v12}\Gamma_{v23}\exp(-jg)} \cdot \frac{1 + \Gamma_{h12}\Gamma_{h23}\exp(-jg)}{\Gamma_{h12} + \Gamma_{h23}\exp(-jg)} \tag{6.6}$$

$$g = \frac{2\omega d}{c}(\varepsilon_r - \sin^2\theta_1)^{1/2}$$

where d is the thickness and ε_r is the permittivity. The subscript v denotes vertical polarization, h represents horizontal polarization, 1 is the upper medium (usually air), 2 is the dielectric coating, and 3 is the medium below the coating layer. Because of the difficulty in measuring phase at millimeter-wave frequencies, the ration Γ_v/Γ_h is measured as a change in ellipticity of the polarization. The ellipticity can be measured, for example, by turning the receiving antenna (horn) around its axis. If the permittivity (or conductivity, if metal) of medium 3 is known, both d and ε_r can be solved from (6.6). Solving the complex equation, however, is not quite so simple. Different methods have been described in the references cited above. In [Konev et al., 1986], the effect of measurement errors on the results have been evaluated. When a measurement frequency of 150 GHz and a coating thickness of 150 μm were assumed, a measurement accuracy of $\pm1\%$ of the polarization parameters led to a maximum error of $\pm6\%$ for the permittivity and $\pm8\%$ for the thickness. Obtaining the above-mentioned measurement accuracy requires effective shielding to prevent direct coupling between the transmitting and receiving antennas. Neither the shielding nor complexity of calculation, however, will pose any major difficulties today. Because of the ability to measure very thin dielectric layers by using frequencies within the capacity of modern technology, the method may find many useful applications.

Corona and others have proposed a somewhat similar method for measurement of lossy dielectric materials [Corona et al., 1987]. They measure the back-reflected (scattered) wave from a dihedral corner reflector (a flat plane that is bent to 90° along the line) covered by the unknown material. Both polarizations are measured and compared to the case where the reflector lacks the dielectric coating. If the unknown material is lossy, the reflector surface can be depicted with a surface impedance, in which case no phase measurement is necessary for calculating the permittivity. This greatly simplifies the measurement as compared to other methods. Because of the shape of the reflector, however, the method as such is suitable only for laboratory measurements.

6.3 RADAR SENSORS

6.3.1 Introduction

So far, we have been concerned mainly with the measurement of material properties, but radar (i.e., *radio detection and ranging*) sensors are used for the measurement of distance, movement, vibration, interfaces, *et cetera*. The measurement signal is primarily the time of flight, which depends on the distance to the object (often called a *target*), and secondarily the strength of the echo, which contains information on the shape, size, and ε_r of the reflecting object. The main applications of the radar, however, are not industrial, but rather to be found in the fields of aviation, maritime, space, and defense technology. In the growing field of remote sensing, airborne and satellite-borne radars are also becoming important. Several books have been written about the technology and applications [Barton, 1988; Skolnik, 1980; Ulaby *et al.*, 1982]. Here, we will limit the discussion to a presentation of the basic principles of operation, a list of reported industrial applications, and some typical examples.

6.3.2 Basic Principles of Operation

Pulse Radar

In the conventional *pulse radar,* a short section of the carrier frequency is transmitted (Figure 6.6). If the distance to the target is l and the speed of propagation is c, the echo returns after a time interval:

$$\tau = \frac{2l}{c} \tag{6.7}$$

which is thus a measure of the distance to the target. The range resolution is usually considered to be determined by the length (Δt) of the pulse. If several targets are present, they can be separated (resolved) by the radar only if they are far enough apart:

$$\Delta l \geqslant \frac{c\Delta t}{2} \tag{6.8}$$

The shorter is the pulse, the better is the resolution. However, transmitting shorter pulses requires a broader bandwidth ($\Delta f \approx 1/\Delta t$) and the received energy is smaller. Therefore, other methods, such as pulse compression (chirp radar) [Skolnik, 1980] are conventionally used. For a single target, which is often the case in industrial

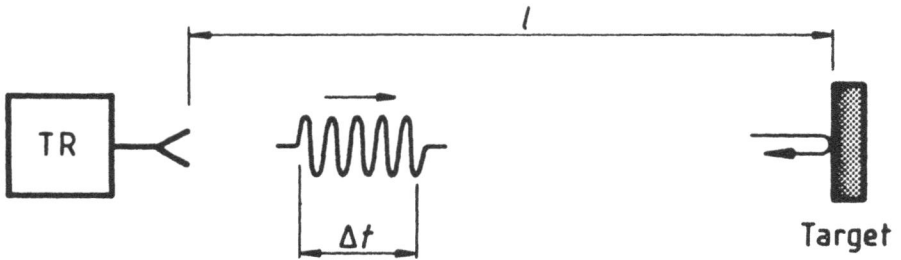

Figure 6.6 Operation of a pulse radar.

applications, the distance measurement accuracy is not limited by (6.8). If the signal-to-noise ratio is sufficiently high, we are able to detect the envelope of the pulse and to determine the time of arrival of its leading edge with high precision.

Impulse Radar and Time Domain Reflectometry

If high distance resolution is needed and the maximum distance is small, we can use *impulse radar*. It transmits a very short pulse, called an *impulse* (Figure 6.7). The center frequency is $f_c \approx 1/\Delta t$, and the bandwidth is very broad because of the shortness of the impulse. Indeed, the breadth of the spectrum roughly equals the center frequency. Therefore, there is also a risk of interference problems. The average power, however, is usually low and the radiation is directed into the target (e.g., the ground, a tree, a wall). The resolution is roughly given by (6.8), where c is the speed of propagation in the medium, which means that a better resolution is achieved in a medium with high permittivity. Impulse radars require special receiver technology. Electronic circuitry is not fast enough to record the complete time dependence of the echo from one transmitted impulse. Therefore, the received signal is sampled only once for each transmitted impulse. By changing the time interval between the transmission and sampling, the whole received signal is gradually reconstructed. The antennas also differ from those of other radars. First, the antennas must be broadband because of the broad spectrum of the impulse. They must also be absolutely free of resonances, which will otherwise lengthen the "tail" of the impulse. Triangular dipoles loaded with resistors or exponential TEM horns (two broadening and diverging strips) are most often used.

Impulse radars are used for the detection of interfaces (e.g., cables and pipes in the ground and walls of houses, rot in trees, bottom of the peat layer in marshes, thickness of snow cover, interface between ice and water), but the same technique can be used with guided waves. Such devices are called *cable radars* and are mainly used for locating cable failures. If the waves are guided in a TEM transmission line, a step can be used instead of an impulse. Measurement of the step response in a

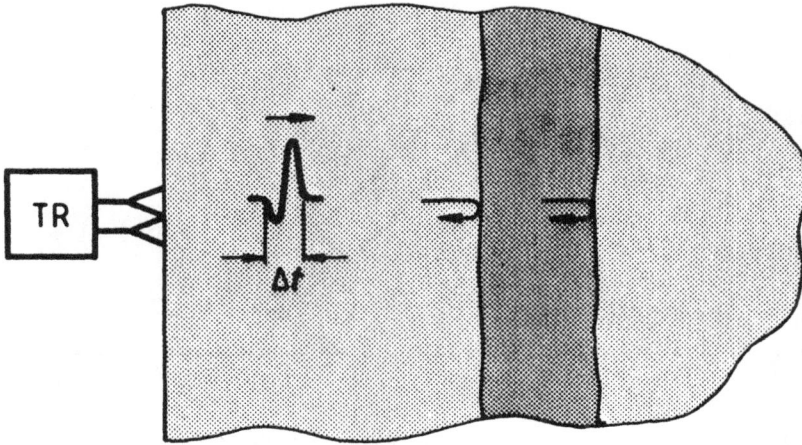

Figure 6.7 Operation of an impulse radar.

TEM transmission line is called *time-domain reflectometry* (TDR). The TDR technique is sometimes used for performing exact measurements of the impedance variations along a transmission line. If the variations are caused by a sample, the TDR sensor can be used as a profiling device. If the reflections are strong, the interpretation of the signal becomes difficult because of multiple reflections and attenuation (by reflections) of distant echoes.

Frequency Modulated Continuous Wave Radar (FMCW)

A frequently used type of radar for short ranges is the *frequency modulated continuous wave radar* (FMCW). The frequency of the transmitted wave is continuously swept up and down between values f_1 and f_2 (Figure 6.8). The signal reflected from a target when received will therefore have a different frequency from the signal currently being transmitted. The difference (f_{IF}) is detected by feeding the received signal and a part of the transmitted signal into a mixer. The difference frequency is directly proportional to the distance to the target:

$$f_{IF} = \tau \frac{f_2 - f_1}{T} = l \frac{2\Delta f}{cT} \tag{6.9}$$

When the sweep is linear (df/dt = constant), the detected low-frequency spectrum is therefore a picture of the reflection amplitude as a function of distance. If we are interested only in targets at a certain distance interval, all other echoes can be filtered

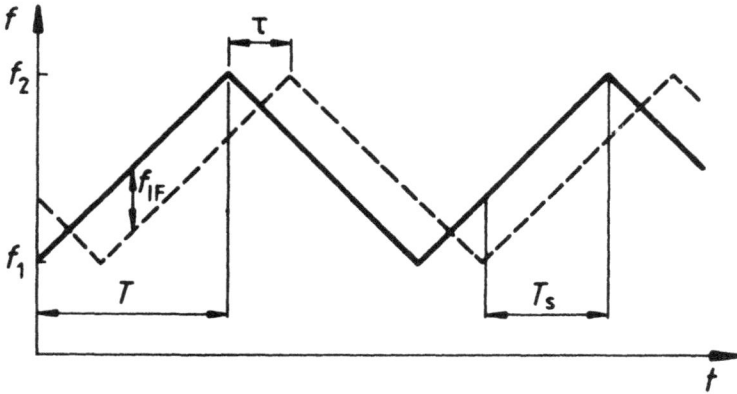

Figure 6.8 Operation of an FMCW radar. The solid line marks the transmitted signal and the broken line delineates the echo from a single target.

with a bandpass filter. The turning points in Figure 6.8, however, will disturb the spectrum. This can be overcome by using computer-controlled sampling. The samples are taken during the period T_s (see Figure 6.8) and then Fourier transformed to yield the spectrum.

Because of the finite length of the difference frequency waveform (approximately T), f_{IF} can be determined only approximately. Because the number of periods of f_{IF} is n ($= f_{IF} \cdot T$), which can be determined with the accuracy ± 1 period, we have for the resolution:

$$\Delta l = \frac{1}{n} \cdot l = \frac{l}{f_{IF}T} = \frac{c}{2\Delta f} \tag{6.10}$$

which is the same ratio between bandwidth and resolution as (6.8). The reader may note that the resolution depends only on the sweep bandwidth. To obtain a high resolution, therefore, we need to use a broad bandwidth. In the case of a single target, detecting f_{IF} coherently (i.e., to measure the total phase shift during the sweep from f_1 to f_2) is possible. In that case, the distance measurement accuracy also depends on the phase measurement accuracy. The sweep need not even be linear, which means that much less expensive electronics can be used [Schilz et al., 1981]. Because of the broad bandwidth ($f_2 - f_1$) of the transmitted signal in a standard FMCW radar, the risk for interference problems is the same as with impulse radars. The spectrum, however, is more clearly defined, but the average power is usually higher.

Interferometer and Doppler Radar

A radar that transmits a continuous sine wave at a constant frequency and compares the phase of the wave reflected from a single target to the phase of the transmitted wave can measure relative movements with great precision. Such a radar is called an *interferometer*. It is used for measurement of small displacements and vibrations. If the target moves with a constant speed relative to the radar, the phase difference changes continuously. This is equivalent to a constant frequency difference between the transmitted and received waves. The difference frequency is called the *doppler frequency,* and is given by

$$f_d = \frac{1}{2\pi} \cdot \frac{d\phi}{dt} = \frac{1}{2\pi} \cdot \frac{4\pi v}{\lambda} = \frac{2v}{c} \cdot f \qquad (6.11)$$

where v is the radial speed of the target. Radars measuring f_d are called *doppler radars* and used mainly for measurement of speed. For example, the radars used by the police for speed control are doppler radars. In the case of only one target, the phase of f_d can be tracked continuously, as in an interferometer, for precise measurement of large displacements.

Practical Systems

The short description given above of the different types of radar gives a complete summary of the working principles of the industrial radar sensors. If we look at the actual systems in more detail, however, they are often more complicated because of the various kinds of modulation and signal processing methods used. Some desired feature for a special application may be achieved by them. Describing the radar technology in further depth, however, is beyond the scope of this book. More details are found in the references cited in Table 6.1.

6.3.3 Applications of Radar Sensors

Measurement of Level

Both liquids and granular, solid materials are stored and processed in tanks and silos in almost every factory. Measurement of the level of the material surface is therefore a common and important type of measurement. Especially where the material surface is covered by foam (which does not reflect ultrasonic waves) or the meter must be electrically separated from the atmosphere in the tank because of explosive gases, microwave radar sensors provide a good choice. For example, Autronica (Norway)

Table 6.1

Applications of Radar Sensors

(Key to status at time of publication of reference: C–commercial, O–operational, T—prototype tested, S—scientific paper, R—review article)

Type of Radar and Application	Status	Reference
FMCW		
Level of molten steel in mold of continuous casting machine	C	[Shiraiwa *et al.*, 1974]
Slag level in a basic oxygen furnace for preventing slopping	O	[Kobayashi *et al.*, 1983]
Distance (nonlinear sweep, spatial filtering technique)	S	[Ostwald *et al.*, 1982]
Level of liquid on-board ships	C	[Edvardsson, 1979]
Detection of buried objects	T	[Clarricoats *et al.*, 1977]
Furnace wall thickness	C	[Hobson *et al.*, 1987]
Thickness of ice	T	[Jakkula *et al.*, 1980]
Miscellaneous	R	[Nowogrodzki, 1983]
	R	[Bailey, 1980]
	R	[Schilz *et al.*, 1981]
	R	[Baker *et al.*, 1988]
Doppler-Interferometer		
Casting speed of a bottom cast ingot	O	[Shiraiwa *et al.*, 1974]
		[Shiraiwa *et al.*, 1981]
		[Kobayashi *et al.*, 1985]
Flow rate of solid particulars	T	[Atek *et al.*, 1984]
Flow rate of solid particulars	S	[Hamid *et al.*, 1975]
		[Stuchly *et al.*, 1977]
	R	[Thorn *et al.*, 1982]
Velocity meter for vehicles	T	[Stuchly *et al.*, 1978]
Proximity detector for piece goods	R	[Morris, 1980]
Camber of forged axles in hot forming	T	[Shiraiwa *et al.*, 1974]
Eccentricity of tubing in a threading machine	O	[Shiraiwa *et al.*, 1974]
Displacement, vibration of steel wire (close range, 5-port)	T	[Kobayashi *et al.*, 1984]
Displacement, vibration of power line	O	[Tiuri *et al.*, 1983]
Displacement of biological subject	S	[Arai *et al.*, 1984]
Displacement of biological subject	S	[Thansandote *et al.*, 1983]
Small displacements	S	[Thansandote *et al.*, 1982]
Thickness profile of snowpack	T	[Aarholt, 1984]
Pulse Impulse		
Thickness of peat layer in marshes	O	[Tiuri *et al.*, 1983]
Distribution of frazil under river ice	T	[Toikka, 1987]
Rot in trees	T	unpublished
Buried pipes	T	[Michiguchi *et al.*, 1988]
Numerous applications	R	[Baker *et al.*, 1988]

Table 6.1 (continued)

Type of Radar and Application	Status	Reference
Time-Domain Reflectometry		
Permittivity of biological substances (two-conductor sensor)	S	[Bose *et al.*, 1986]
Permittivity of biological substances (open-ended coaxial sensor)	O	[Gabriel *et al.*, 1986]
Permittivity of soil samples, field instrument	S	[Delaney *et al.*, 1984]

markets a level-measuring radar system, which is primarily intended to measure the level of liquids in tanks on-board ships, but also finds use in industry. The device is a single-sweep FM radar, which means that the frequency is swept from 9.5 to 10.5 GHz during the transmission, which lasts 0.1 s. The measurement range is 0–40 m and the accuracy is ±2 mm, if the atmospheric conditions are known. The diameter of the tank-mounting hole is 215 mm. Several radars can interface to the same electronic unit, which may be programmed to display the level, cargo volume, and weight.

Measurement of Vibration

Measurement of vibration is an important means of predicting bearing failures so that due measures can be taken during the next service break. Unforeseen interruptions in production are very expensive. A microwave radar sensor can be used in places that are difficult to reach by contacting acceleration transducers or hostile or dangerous because of, for example, high temperature or high voltage.

At the Radio Laboratory of the Helsinki University of Technology a radar was developed in cooperation with the company Imatran Voima for measurement of wind-induced vibrations in power lines [Tiuri *et al.*, 1983]. Such vibrations cause failures of the conductors and transmission towers, unless damped by properly located weights. The purpose of the radar, which is placed on the ground under the line, is to collect statistical data on the duration, amplitude, and frequency of the vibrations. The radar interfaces with a data acquisition unit, a wind speed and direction meter, and a thermometer. The radar is a 16 GHz interferometer, where the transmission frequency is continuously changed so that the phase difference between the transmitted and received waves is kept constant. The dc voltage controlling the output frequency of the VCO therefore changes in accordance with the distance to the power line.

Keltronics (Sweden) manufactures a radar for measurement of vibration in industry. The radar is an interferometer operating at about 15 GHz. It is connected to a portable computer, which performs the signal analysis. Special software has been developed for making diagnoses and predicting failures. The maximum measurement distance is 4 m, and the radar is equipped with a laser, which points at the target being measured.

Applications of Impulse Radar

Impulse radars have been used mainly as subsurface interface radars for locating cables and pipes in the ground. For example, Geophysical Survey Systems (US) manufactures radars for such purposes. The same electronics can be used with different transceivers working at different frequencies. The transceiver box contains the impulse transmitter, sampler, and one or two triangular dipole antennas. The center frequencies of the different radars are about 80, 120, 300, 500, and 900 MHz, and 1 GHz. At the Radio Laboratory of the Helsinki University of Technology, such radars have been tested and adapted for different applications. The 120 MHz unit has been successfully used for surveying the peat resources of marshlands (Figure 6.9), and for detecting frazil under the ice on rivers in the winter. The frazil is formed at open rapids in cold weather, and it may block the river downstream causing floods. The 900 MHz unit has been found useful in detecting rot in park trees. The resolution, however, is not good. At present (1989), a radar with a center frequency of 2.5 GHz is being developed at the Radio Laboratory in cooperation with the University's Laboratory of Applied Electronics. The radar will be a lightweight, battery-operated, field portable instrument.

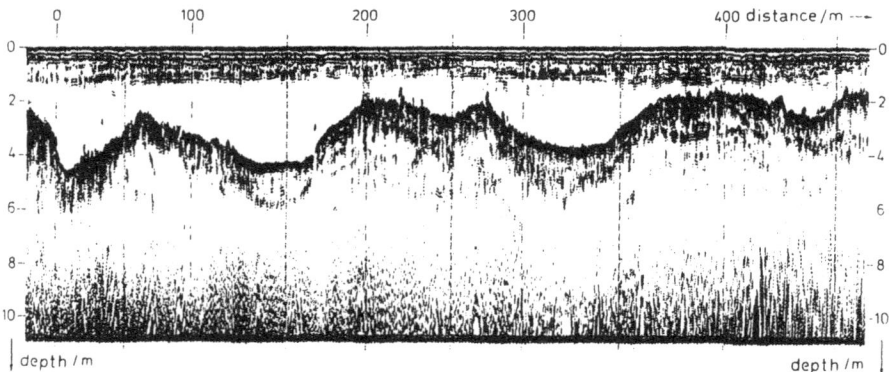

Figure 6.9 A profile of a marshland measured with the 120 MHz impulse radar and plotted with a graphic recorder. The bottom of the peat layer can be clearly seen. Larger stones or fallen tree trunks can also be distinguished. Between 125 and 175 m, the valley is probably filled with clay.

Reported Applications

Above we mentioned only a few of the more important industrial applications of radar sensors. Table 6.1 lists some reported applications and indicates the status of the project if mentioned in the reference or known to the authors.

REFERENCES

Aarholt, E., "Variation in Microwave Transmissions and Snowpack Densities," *CNES CESR URSI Intern. Symp. on Microwave Signatures in Remote Sensing,* Toulouse, January 1984, 8 p.

Akhmetshin, A.M., "Determining Parameters of Laminated Dielectrics Near Minima of Absolute Reflection Coefficients," *Sov. J. Nondest. Test.* (US), Vol. 22, No. 3, March 1986, pp. 213–219.

Akhmetshin, A.M., and A.Y. Kurin, "Determining Parameters of Laminated Materials Having Dielectric Losses by Measurements in a Range of Frequencies," *Sov. J. Nondest. Test.* (US), Vol. 22, No. 3, March 1986, pp. 205–213.

Arai, I., and T. Suzuki, "A Microwave Interferometer System for the Displacement Measurements of Biological Subject," *Electronics and Communications in Japan,* Vol. 67-C, No. 10, 1984, pp. 108–118.

Atek, K.A., S.A. Mawjoud, and A.S. Jabber, "The Effect of Humidity and Bulk Density on Flow Rate Measurements of Rice, Salt and Sugar," *J. Inst. Electron. and Telecommun. Eng.* (India), Vol. 30, No. 5, September 1984, pp. 124–128.

Athey, T.W., M.A. Stuchly, and S.S. Stuchly, "Measurement of Radio Frequency Permittivity of Biological Tissues with an Open-Ended Coaxial Line: Part I," *IEEE Trans. Microwave Theory Tech.,* Vol. MTT-30, No. 1, January 1982, pp. 82–86.

Bailey, S.J., "Level Sensor '80: A Key Partner in Productivity," *Control Eng.* (US), Vol. 27, No. 10, October 1980, pp. 75–79.

Baker, J.M., J. Clarke, and P.M. Grant (eds.), *IEE Proc.,* Vol. 135, Pt.F., No. 4, August 1988 (Special Issue on Subsurface Radar), pp. 277–392.

Barton, D.K., *Modern Radar System Analysis,* Norwood: Artech House, MA, 1988, 590 p.

Bartashevskii, E.L., V.F. Borul'ko, O.O. Drobakhin, and I.V. Slavin, "Effectiveness of Data Processing in Determining Parameters of Dielectrics by the Multifrequency Method," *Sov. J. Nondest. Test.* (US), Vol. 22, No. 6, June 1986, pp. 372–374.

Bastida, E.M., N. Fanelli, and E. Marelli, "Microwave Instruments for Moisture Measurements in Soils, Sands and Cements," *14th Microwave Power Symp.,* Monaco, June 1979, pp. 147–149.

Bliot, F., A. Castelain, and B. Dujardin, "Numerical Simulation and Models for Microcoaxial Probes. Application to 'In Vivo' Measurements of Dielectric Parameters of Biological Media in the Microwave Band (1–12 GHz)," *Proc. 10th European Microwave Conf.,* Warsaw, September 1980, pp. 531–535.

Bose, T.K., A.M. Bottreau, and R. Chahine, "Development of a Dipole Probe for the Study of Dielectric Properties of Biological Substances in Radiofrequency and Microwave Region with Time-Domain Reflectometry," *IEEE Trans. Instr. Meas.,* Vol. IM-35, No. 1, March 1986, pp. 56–60.

Brunfeldt, D.R., "Theory and Design of a Field-Portable Dielectric Measurement System," *Proc. IGARSS '87 Symp.,* Ann Arbor, MI, May 1987, pp. 559–563.

Burdette, E.C., F.L. Cain, and J. Seals, "*In Vivo* Probe Measurement Technique for Determining Dielectric Properties at VHF Through Microwave Frequencies," *IEEE Trans. Microwave Theory Tech.,* Vol. MTT-28, No. 4, April 1980, pp. 414–427.

Bussey, H.E., "Dielectric Measurements in a Shielded Open Circuit Coaxial Line," *IEEE Trans. Instr. Meas.,* Vol. IM-29, No. 2, June 1980, pp. 120–124; correction in Vol. IM-30, No. 3, September 1981.

Clarricoats, P.J.B., R. Kularajah, R.R. Lentz, and G.T. Poulton, "Detection of Buried Objects by Microwave Means," *Proc. 7th European Microwave Conf.,* Copenhagen, September 1977, pp. 409–413.

Corona, P., G. Ferrara, and C. Gennarelli, "A New Technique for Free-Space Permittivity Measurements of Lossy Dielectrics," *IEEE Trans. Instr. Meas.,* Vol. IM-36, No. 2, June 1987, pp. 560–563.

Decréton, M.C., and M.S. Ramachandraiah, "Nondestructive Measurement of Complex Permittivity for Dielectric Slabs," *IEEE Trans. Microwave Theory Tech.,* Vol. MTT-23, No. 12, December 1975, pp. 1077–1080.

Delaney, A.J., and S.A. Arcone, "A Large-Size Coaxial Waveguide Time Domain Reflectometry Unit for Field Use," *IEEE Trans. Geosci. and Remote Sensing,* Vol. GE-22, No. 5, September 1984, pp. 428–431.

Edvardsson, O., "An FMCW Radar for Accurate Level Measurements," *Proc. 9th European Microwave Conf.,* Brighton, UK, September 1979, pp. 712–715.

Epstein, B.R., M.A. Gealt, and K.R. Foster, "The Use of Coaxial Probes for Precise Dielectric Measurements: A Reevaluation," *IEEE MTT-S Int. Microwave Symp. Digest,* Las Vegas, June 1987, pp. 255–258.

Esselle, K.P.A.P., and S.S. Stuchly, "Capacitive Sensors for In-Vivo Measurements of the Dielectric Properties of Biological Materials," *IEEE Trans. Instr. Meas.,* Vol. 37, No. 1 , March 1988, pp. 101–105.

Gabriel, C., E.H. Grant, and I.R. Young, "Use of Time Domain Spectroscopy for Measuring Dielectric Properties with a Coaxial Probe," *J. Phys. E: Sci. Instrum.,* Vol. 19, No. 10, October 1986, pp. 843–846.

Gajda, G., and S.S. Stuchly, "Numerical Analysis of Open-Ended Coaxial Lines," *IEEE Trans. Microwave Theory Tech.,* Vol. MTT-31, No. 5, May 1983, pp. 380–384.

Hamid, A., and S.S. Stuchly, "Microwave Doppler-Effect Flow Monitor," *IEEE Trans. Industr. Electron. Control Instr.,* Vol. IECI-22, No. 2, May 1975, pp. 224–228.

Hobson, G.S., R.C. Tozer, J.M. Rees, P.L. Judd, and R. Devayya, "Microprobe—Microwave Measurement of Furnace Wall Thickness," *Proc. 17th European Microwave Conf.,* Rome, September 1987, pp. 881–886.

Jakkula P., P. Ylinen, and M. Tiuri, "Measurement of Ice and Frost Thickness with an FM-CW Radar," *Proc. 10th European Microwave Conf.,* Warsaw, September 1980, pp. 584–587.

Kobayashi, S., A. Hatono, K. Katohgi, A. Kuriyama, and K. Ichihara, "Prediction and Control of Slag Slopping in BOF Using Microwave Gauge," *Proc. 4th IFAC Symp.,* Helsinki, August 1983, pp. 297–301.

Kobayashi, S., and S. Miyahara, "Applications of Microwave Interferometry for Displacement Measurement," *Sumitomo Search* (Japan), No. 31, November 1985, pp. 169–180.

Kobayashi, S., and S. Miyahara, "A Microwave Displacement Meter Using a Five-Port Device," *Proc. IECON '84,* Tokyo, October 1984, pp. 633–636.

Konev, V.A., E.M. Kuletshov, and N.N. Punkho, *Radiowave Ellipsometry* (In Russian), Minsk: Science and Technology, 1985, 104 p.

Konev, V.A., S.A. Tikhanovich, and M.I. Zhigalko, "Error in the Determination of the Dielectric Properties and Thickness of Surface Layers by the Method of Radiometric Ellipsometry," *Sov. J. Nondestr. Test.* (US), Vol. 22, No. 6, June 1986, pp. 363–366.

Marcuvitz, N., *Waveguide Handbook,* New York: McGraw-Hill, 1951, 428 p.

Marsland, T.P., and S. Evans, "Dielectric Measurements with an Open-Ended Coaxial Probe," *IEE Proc.,* Vol. 134, Pt.H, No. 4, August 1987, pp. 341–349.

Martinson, T., T. Sphicopoulos, and F.E. Gardiol, "Nondestructive Measurement of Materials Using a Waveguide-Fed Series Slot Array," *IEEE Trans. Instr. Meas.,* Vol. IM-34, No. 3, September 1985, pp. 422–426.

Michiguchi, Y., K. Hiramoto, M. Nishi, T. Ootaka, and M. Okada, "Advanced Subsurface Radar System for Imaging Buried Pipes," *IEEE Trans. Geosci. Remote Sensing*, Vol. GE-26, No. 6, November 1988, pp. 733–740.

Misra, D.K., "A Study on Coaxial Line Excited Monopole Probes for *In Situ* Permittivity Measurements," *IEEE Trans. Instr. Meas.*, Vol. IM-36, No. 4, December 1987, pp.1015–1019.

Morris, H.M., "Object Detection Techniques Range from Limit Switches to Lasers," *Control Eng.*, (US), Vol. 27, No. 11, November 1980, pp. 65–70.

Moschüring, H., and I. Wolff, "The Measurement of Inhomogeneities and of the Permittivity Distribution in MIC-Substrates," *IEEE Instr. Meas. Tech. Conf.*, Boston, April 1987, pp. 154–159.

Nowogrodzki, M., "Microwave CW Radars in Industrial Applications," *Electro '83. Electronics Show and Convention*, New York, April 1983, pp. 2/5/1–7.

Ostwald, O., and B. Schiek, "A Spatial Filtering Technique for Microwave Distance Measurements," *Proc. 12th European Microwave Conf.*, Helsinki, September 1982, pp. 604–609.

Schilz, W., and B. Schiek, "Microwave Systems for Industrial Measurements," *Advances in Electronics and Electron Physics*, New York: Academic Press, Vol. 55, 1981, pp. 309–381.

Shiraiwa, T., and S. Kobayashi, "The Application of Microwave Techniques for Measurement and Control in the Steel Industry," *Proc. SIMAC '74 Conf.*, Sheffield, UK, October 1974, pp. P16/ 1–12.

Shiraiwa, T., Y. Sakamoto, S. Kobayashi, S. Anezaki, H. Kato, and A. Kuwabara, "Automatic Control of Casting Speed in Ingot Casting," *Automatica* (GB), Vol. 17, No. 4, 1981, pp. 613–618.

Skolnik, M.I., *Introduction to Radar Systems*, 2nd Ed., New York: McGraw-Hill, 1980, 587 p.

Spalla, M., C. Tamburello, and L. Zanforlin, "A Method to Measure the Characteristics of Biological Materials at Millimeter Waves," *Proc. 6th European Microwave Conf.*, Rome, 1976, pp. 122–125.

Sphicopoulos, T., V. Teodoridis, and F.E. Gardiol, "Simple Nondestructive Method for the Measurement of Material Permittivity," *J. Microwave Power*, Vol. 20, No. 3, September 1985, pp. 165–172.

Stuchly, M.A., T.W. Athey, G.M. Samaras, and G.E. Taylor, "Measurement of Radio Frequency Permittivity of Biological Tissues with an Open-Ended Coaxial Line: Part II—Experimental Results," *IEEE Trans. Microwave Theory Tech.*, Vol. MTT-30, No. 1, January 1982a, pp. 87–92.

Stuchly, M.A., M.M. Brady, S.S. Stuchly, and G. Gajda, "Equivalent Circuit of an Open-Ended Coaxial Line in a Lossy Dielectric," *IEEE Trans. Instr. Meas.*, Vol. IM-31, No. 2, June 1982b, pp. 116–119.

Stuchly, M.A., and S.S. Stuchly, "Coaxial Line Reflection Methods for Measuring Dielectric Properties of Biological Substances at Radio and Microwave Frequencies—A Review," *IEEE Trans. Instr. Meas.*, Vol. IM-29, No. 3, September 1980, pp. 176–183.

Stuchly, S.S., A. Kraszewski, and M.A. Stuchly, "Uncertainties in Radio Frequency Dielectric Measurements of Biological Substances," *IEEE Trans. Instr. Meas.*, Vol. IM-36, No. 1, March 1987, pp. 67–70.

Stuchly, S.S., M.S. Sabir, and A. Hamid, "Advances in Monitoring of Velocities and Densities of Particulates Using Microwave Doppler Effect," *IEEE Trans. Instr. Meas.*, Vol. IM-26, No. 1, March 1977, pp. 21–24.

Stuchly, S.S., A. Thansandote, J. Mladek, and J.S. Townsend, "A Doppler Radar Velocity Meter for Agricultural Tractors," *IEEE Trans. Vehicul. Tech.*, Vol. VT-27, No. 1, February 1978, pp. 24–30.

Thansandote, A., S.S. Stuchly, and A.M. Smith, "Monitoring Variations of Biological Impedances Using Microwave Doppler Radar," *Phys. Med. Biol.* (GB), Vol. 28, No. 8, 1983, pp. 983–990.

Thansandote, A., S.S. Stuchly, and J.S. Wight, "Microwave Interferometer for Measurements of Small Displacements," *IEEE Trans. on Instr. Meas.*, Vol. IM-31, No. 4, December 1982, pp. 227–232.

Thorn, R., M.S. Beck, and R.G. Green, "Non-intrusive Methods of Velocity Measurement in Pneumatic Conveying," *J. Phys. E: Sci. Instrum.* (GB), Vol. 15, No. 11, 1982, pp. 1131–1139.

Tiuri, M.E., and J. Hyyryläinen, "Microwave Radar for Measuring Vibrations in Power Transmission Lines," *Proc. 13th European Microwave Conf.*, Nürnberg, September 1983, pp. 663–666.

Tiuri, M., M. Toikka, I. Marttila, and K. Tolonen, "The Use of Radio Wave Probe and Subsurface Interface Radar in Peat Resource Inventory," *Proc. Symp. IPS Comm. I*, Aberdeen, Scotland, 1983, pp. 131–143.

Toikka, M., "The Use of Radars to Measure the Distribution of Ice and Frazil in Rivers," *Proc. IGARSS '87*, Ann Arbor, MI, May 1987, pp. 1407–1408.

Ulaby, F.T., R.K. Moore, and A.K. Fung, *Microwave Remote Sensing, Active and Passive, Vol. II, Radar Remote Sensing and Surface Scattering and Emission Theory*, Norwood, MA: Artech House, 1982, pp. 457–1064.

Chapter 7
Radiometer Sensors

7.1 INTRODUCTION

All the other chapters in this book describe sensors that create an electromagnetic field and measure the interaction of the field with the object to be measured (here called the *sample*). Such devices are called *active sensors*. Electromagnetic waves, however, are also created by the sample. This radiation is broadband electromagnetic noise. The strength of the radiation leaving the surface at a specific frequency depends on the temperature, permittivity, and some other features of the sample, and can therefore be used for measurement purposes. The receivers used for measurement of the noise power emitted by a sample in a frequency band are called *radiometers*. They are *passive sensors* because they merely "listen" to the sample.

7.2 BASICS OF RADIOMETRY

7.2.1 Black-Body Radiation

All matter both radiates and absorbs electromagnetic radiation. The mechanism is the same. An incident wave puts the electric charges into motion and charges in motion radiate (see Section 1.3.5). For thermodynamical reasons, the tendency to radiate exactly equals the tendency to absorb radiation at any location in space. Otherwise, a more radiative region would be colder and a more absorptive region would be warmer than the environment, even at thermodynamic equilibrium.

If we assume that the power transmission coefficient of the surface of a sample is unity, this also means that the surface does not reflect any radiation. The sample is therefore called a *black body,* and the radiation leaving the surface is called *black-body radiation*. The radiation is noise radiated by charges in thermal motion. Radiation can be calculated with the aid of quantum physics, which states that the energy of the charges obey Boltzmann's statistics. The result is *Planck's law* for the *brightness* (B) of the surface of a black body:

$$B_{bb} = \frac{2hf^3}{c^2[\exp(hf/kT_p) - 1]} \tag{7.1}$$

where T_p is the (physical) temperature, h is Planck's constant ($h = 6.626 \times 10^{-34}$ Js), k is Boltzmann's constant ($k = 1.38 \times 10^{-23}$ J/K), c is the speed of light ($c = 2.998 \times 10^8$ m/s), and f is the frequency. The brightness is the power per unit frequency bandwidth radiated by a surface unit area into a unit solid angle [given in $W/(m^2 \cdot Hz \cdot sr)$]. The black-body radiation is equal in all directions and depends only on the temperature of the black body (Figure 7.1). At normally encountered temperatures, the brightness increases with frequency through the whole microwave band. In the infrared band, it reaches a maximum and falls steeply thereafter. The frequency of the maximum brightness increases with temperature. At oven temperatures, the upper end of the spectrum enters the optical band from the red end, which is observed as glowing.

The brightness concept applies both to transmitting and receiving radiation, that is, to the radiating surface and to an observer measuring the brightness of the surface. The observer measures the power per unit bandwidth per unit area received from a unit solid angle. Note that the observed brightness is independent of the distance to the radiating surface. The *intensity* per bandwidth (power per unit bandwidth per unit area) radiated by a surface element (point source) depends inversely on the square of the distance (see Section 1.3.3), but the surface area seen in a given solid angle depends on the square of the distance. Hence, the brightness of a surface is independent of the distance.

At microwave frequencies, we can simplify (7.1) because the energy of the quantum (hf) is much smaller than the thermal energy of a particle (kT_p), which means that the turning point of the spectrum is, in practice, always at higher frequencies. We therefore have

$$\exp(hf/kT_p) - 1 \approx hf/kT_p$$

which leads to

$$B_{bb} = \frac{2f^2kT_p}{c^2} = \frac{2kT_p}{\lambda^2} \tag{7.2}$$

Equation (7.2) is known as the *Rayleigh-Jeans law*. We can see that the brightness in the microwave range is directly proportional to the temperature.

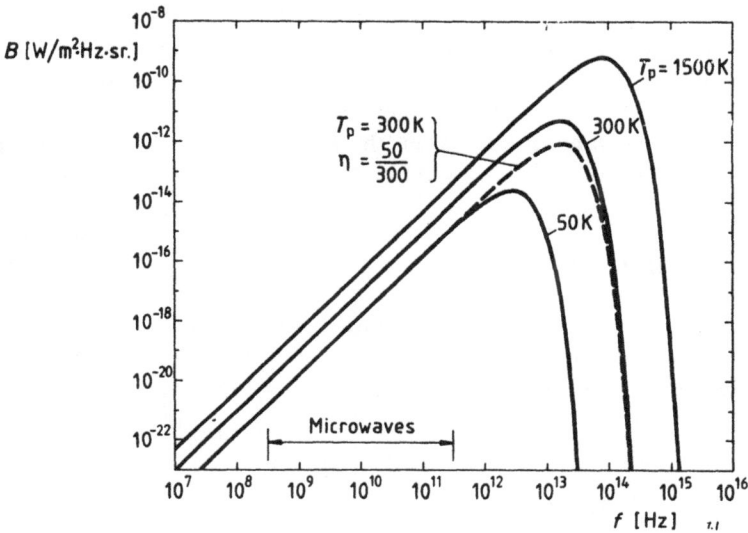

Figure 7.1 The spectrum of the black-body radiation. The spectra of the physical temperature 50 K and the same brightness temperature, when the physical temperature is 300 K, differ only in the infrared and optical bands.

7.2.2 Brightness Temperature

In practice, the surface of a sample always reflects back some of the black-body radiation, making the brightness of a practical body lower than that of a black body at the same temperature. We must therefore multiply the right-hand side of (7.2) by the power transmission coefficient, called the *emissivity* (η) of the surface:

$$B = \frac{2k\eta T_p}{\lambda^2} \tag{7.3}$$

However, common practice is to define instead another temperature, called the *brightness temperature* T_b:

$$B = \frac{2kT_b}{\lambda^2} \tag{7.4}$$

Here, we therefore have

$$T_b = \eta T_p \tag{7.5}$$

Equation (7.4) is the general definition of the brightness temperature, also taking into account other radiation transfer phenomena discussed in Section 7.3.

Hence, a black body at the physical temperature T_b radiates the same brightness as a practical body with the brightness temperature T_b at the physical temperature T_p. The definition is meaningful only when the Rayleigh-Jeans law is valid. In the IR and optical bands, a low emissivity and low temperature result in different spectra (Figure 7.1).

7.2.3 Antenna Temperature

When the brightness in different directions from an antenna is $B(\theta, \phi)$, the received power P_r depends on the normalized directional pattern $P_n(\theta, \phi)$ (see Section 1.3.5) and the effective area A_e of the antenna. For a lossless antenna, we have

$$\frac{P_r}{\Delta f} = \frac{A_e}{2} \int_{4\pi} P_n B \, d\Omega \tag{7.6}$$

where the division by 2 derives from the fact that an antenna always receives only one polarization, whereas the noise radiation normally is unpolarized (i.e., includes equal amounts of two orthogonal polarizations). The effective area depends on the maximum gain G_m of the antenna and the wavelength:

$$A_e = \frac{\lambda^2 G_m}{4\pi} \tag{7.7}$$

Substituting into (7.6), we have

$$\frac{P_r}{\Delta f} = \frac{\lambda^2 G_m}{8\pi} \int_{4\pi} P_n B \, d\Omega \tag{7.8}$$

The maximum gain, conversely, depends on P_n:

$$G_m = \frac{4\pi}{\displaystyle\int_{4\pi} P_n d\Omega} = \frac{4\pi}{\Omega_A} \tag{7.9}$$

where Ω_A is the so-called *solid angle* of the antenna. For a high-gain antenna, Ω_A is roughly equal to the solid angle bounded within the half-power limits of the main lobe. By using (7.4) and (7.9) in (7.8), we have

$$\frac{P_r}{\Delta f} = k \frac{\displaystyle\int_{4\pi} P_n T_b \mathrm{d}\Omega}{\displaystyle\int_{4\pi} P_n \mathrm{d}\Omega} \qquad (7.10)$$

If we compare the power available at the antenna terminals to the power P_{ml} emitted by a matched load at the temperature T_{ml}:

$$\frac{P_{ml}}{\Delta f} = k T_{ml} \qquad (7.11)$$

we may by analogy define the *antenna temperature* T_A:

$$\frac{P_r}{\Delta f} = k T_A \qquad (7.12)$$

We now have

$$T_A = \frac{\displaystyle\int_{4\pi} P_n T_b \mathrm{d}\Omega}{\displaystyle\int_{4\pi} P_n \mathrm{d}\Omega} = \frac{\displaystyle\int_{4\pi} P_n T_b \mathrm{d}\Omega}{\Omega_A} = \frac{G_m}{4\pi} \int_{4\pi} P_n T_b \mathrm{d}\Omega \qquad (7.13)$$

If the sample is sufficiently large enough to fill the entire directional pattern of the antenna and the brightness temperature is constant, T_b can be moved to the other side of the integral sign in (7.13). In this case $T_A = T_b$, independent of the distance to the sample. If, however, the sample is so small that $P_n \approx 1$ over the whole sample, we have

$$T_A \approx \frac{\Omega_s}{\Omega_A} T_b \qquad (7.14)$$

where Ω_s is the solid angle of the sample. In other cases, we must use (7.13).

From the discussion above, the brightness temperature is clearly the most convenient quantity for describing the radiation emitted by a sample. T_b is closely related to both the physical temperature of the sample and the antenna temperature of the radiometer.

However, note that if the emissivity of a surface is low, it will reflect a considerable part of the radiation incident on the surface from the outside. Thus, the measured antenna temperature contains contributions from both the sample and the background (Figure 7.2):

$$T_A = \eta T_p + (1 - \eta)T_{bg} \tag{7.15}$$

In remote sensing of the ground, the background will be the sky, which is very cold in the range from 500 MHz to 10 GHz, where the cosmic 2.7 K radiation dominates [Ulaby *et al.*, 1981]. Above 10 GHz the atmospheric emission increases, and below 500 MHz the galactic radiation is significant. In microwave remote sensing, the background therefore has little influence on the brightness temperature, but in other cases the background must be taken into account.

7.2.4 Radiometer

A radiometer is essentially a sensitive receiver that only measures the received noise power. It can be realized in many different ways, as described by Ulaby, Moore, and Fung [Ulaby *et al.*, 1981], Tiuri and Räisänen in [Kraus, 1986] and Skou [Skou, 1989], but here we will limit the discussion to the basic configurations. Because the received power is always very low, the total gain in the radiometer must be high. To avoid problems of stability, thermal drift, and gain variation, a mixer for changing the frequency and some system for continuous calibration are usually used.

The most common type of radiometer is the *Dicke radiometer,* shown in Figure 7.3. Part of the time (usually 50%), it measures the noise emitted by a matched load (see (7.11)) for reference, and part of the time the radiometer measures the signal coming through the antenna. The temperature of the matched load is kept constant or is measured. Because the measurement of the unknown antenna temperature is based on comparison to a known reference temperature, the effect of gain variations is decreased, or eliminated if the temperatures are equal. The switching frequency is usually in the range of 10–1000 Hz, depending on the application.

In the Dicke radiometer, the Dicke switch may be followed by a low-noise amplifier or directly by the mixer. The intermediate frequency (IF) branch contains a bandpass filter, which determines the bandwidth Δf of the radiometer, and some amplifiers. The radiometer circuit is terminated by a synchronous detector and an integrator. The synchronous detector produces a dc signal proportional to the difference $T_{ml} - T_A$. The detector is controlled by the same circuit as the Dicke switch. The integrator takes the time average of the detected signal. The time of averaging for one result is called the *integration time*. The integrator may be a digitally implemented true integrator or simply a low-pass filter. The integration time determines the fastest variations that can be detected as well as the rms noise level at the output.

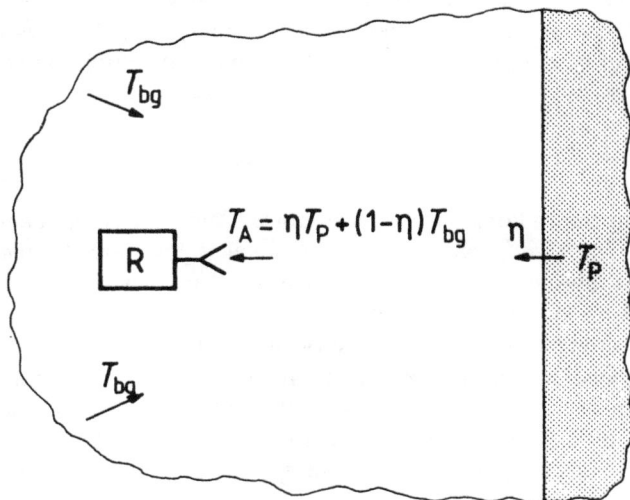

Figure 7.2 The radiometer receives both radiation emitted by the sample and reflected from the background.

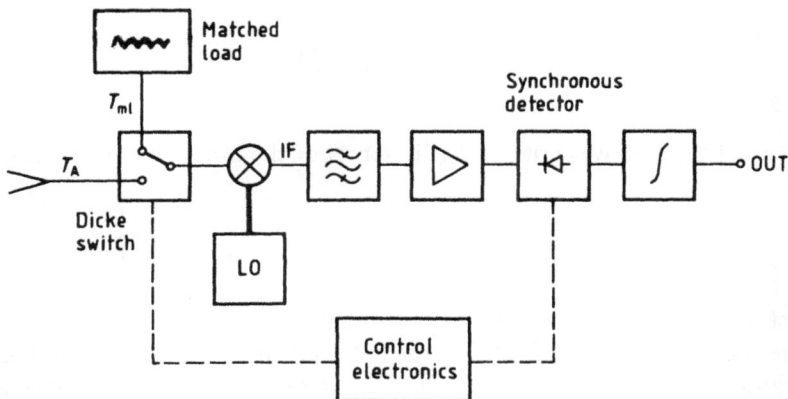

Figure 7.3 The Dicke radiometer.

The longer is the integration time, the smoother is the output signal and the more slowly it adapts to a change in antenna temperature.

If the difference between T_{ml} and T_A is large, the null-balancing technique can be used. Additional noise from a source is injected through a probe or a directional coupler in the branch with the lower temperature. The injected noise power is controlled so as to balance exactly the difference between T_{ml} and T_A, thus completely eliminating the effect of the gain variations. The signal controlling the power of the injected noise is now the output signal of the radiometer.

7.2.5 Receiver Noise Temperature and Sensitivity

Every component in the radiometer produces noise. The first components after the antenna, called the *front end*, where the power level is low, are the most critical. Therefore, adding a low-noise amplifier between the Dicke switch and the mixer is preferable (Figure 7.3). An amplifier reduces the effect of the noise of the subsequent components by the inverse of the gain because the signal level is correspondingly higher. The noise generated in a component can be characterized by a *noise temperature*, T_N, which is the equivalent noise at the input produces the same output signal as the noise of the component. If the noise temperature of each component in the receiver is T_{Ni} (Figure 7.4), the noise temperature of the receiver is

$$T_N = T_{N1} + \frac{T_{N2}}{G_1} + \frac{T_{N3}}{G_1 G_2} + \dots + \frac{T_{Nn}}{G_1 G_2 \dots G_{n-1}} \tag{7.16}$$

Instead of the noise temperature, the *noise figure, F,* is often used:

$$F = 1 + \frac{T_N}{T_0} \tag{7.17}$$

where $T_0 = 290$ K. If we write (7.16) in terms of the noise figure, we have

$$F = F_1 + \frac{F_2 - 1}{G_1} + \frac{F_3 - 1}{G_1 G_2} + \dots + \frac{F_n - 1}{G_1 G_2 \dots G_{n-1}} \tag{7.18}$$

The noise figure is often given in decibels relative to 1.

The noise temperature T_N of the receiver, given by (7.16), can be considered as an additional antenna temperature in a noiseless receiver. For a standard radiometer in room temperature, T_N is normally between 200 and 2000 K. The total noise is called the *system noise temperature*:

$$T_s = T_A + T_N \tag{7.19}$$

When the output is calibrated to show antenna temperature, the rms variation depends on the total noise, bandwidth, and integration time. The situation can be compared to estimating the mean value of a normally distributed random variable from a certain

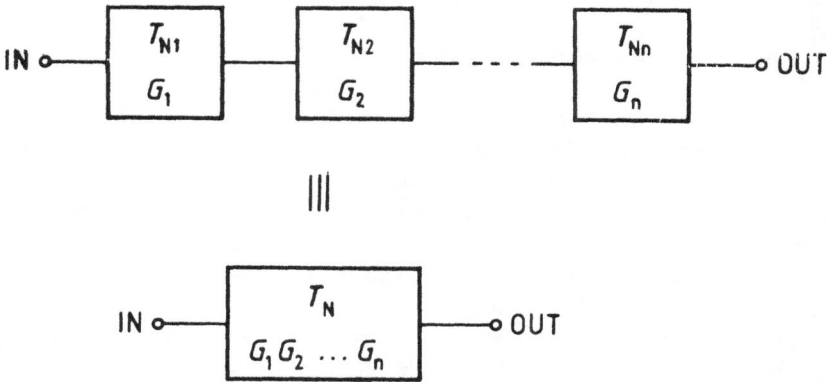

Figure 7.4 The noise temperature and gain of a cascaded system of components. T_N is given by (7.16).

number of samples. The standard deviation of the estimated mean value depends on the inverse of the square root of the number of the samples. In the radiometer, the number of samples is proportional to the highest frequency present (i.e., IF bandwidth) multiplied by the integration time. We can show [Tiuri, 1964] that the rms variation of the output is

$$\Delta T_A = \frac{K T_s}{\sqrt{\Delta f \tau}} \tag{7.20}$$

where τ is the integration time and K is a constant, which depends on the type of radiometer. For example, a traditional Dicke radiometer measures the signal coming from the antenna only half of the time, resulting in $K = 2$. The rms noise ΔT_A is considered to be the smallest detectable change in T_A, and is therefore a measure of the sensitivity of the radiometer. Usually, ΔT_A is reported given the assumption that $\tau = 1$ s.

Because ΔT_A depends on T_s, T_N is to be minimized. One way is to add a low-noise amplifier in front of the mixer, as mentioned above. At millimeter-wave frequencies, this is not usually possible because such amplifiers are extremely expensive if available. Another possibility is to cool the front end as in many radio astronomy receivers, but this complicates the radiometer. The third way is provided by the mixer. If no filter is used in front of the mixer, both image frequencies will be used, thereby doubling the effective bandwidth. If, for example, the IF bandwidth is 100–500 MHz ($\Delta f = 400$ MHz) and the local oscillator frequency is 10 GHz, the radiometer will receive the noise in the bands 9.5–9.9 GHz and 10.1–10.5 GHz, giving an effective bandwidth of $\Delta f = 800$ MHz. Such a radiometer is called a *double-sideband* (DSB) receiver.

7.3 SUBSURFACE RADIOMETRY AND THERMOGRAPHY

7.3.1 Equation of Radiation Transfer

In the previous section, we assumed the radiation to be emitted from a surface. In reality, the radiation originates from a layer of material below the surface. The thickness of the layer contributing to the measured brightness temperature depends on the attenuation. Only if the temperature is uniform in the layer will the brightness temperature be directly given by (7.5) as we have assumed above. If the temperature varies with the depth, the resulting brightness temperature is determined by the *equation of radiation transfer* (see below). This provides the possibility of measuring the temperature distribution inside the sample by using *subsurface radiometry*. Let us study what happens to the radiation inside the material in a thin layer. The radiation entering the layer from the right in Figure 7.5 corresponds to the brightness temperature T_{b0}. It is attenuated in the layer so that the portion:

$$T_{b1} = aT_{b0} \tag{7.21}$$

leaves the layer to the left. According to thermodynamics, a material that absorbs radiation also emits radiation, according to the equation:

$$T_{b2} = (1 - a)T_p \tag{7.22}$$

where T_p is the physical temperature of the material in the layer (e.g., [Ulaby *et al.*, 1981]). The resulting brightness temperature is, therefore,

$$T_b = aT_{b0} + (1 - a)T_p$$
$$= \exp(-2k''d)T_{b0} + [1 - \exp(-2k''d)]T_p \tag{7.23}$$

If $T_p = T_{b0}$, the brightness temperature is unchanged, but if $T_p \neq T_{b0}$, it is not. The higher is the attenuation, the closer to T_p is the observed brightness temperature. If the temperature changes continuously with depth, the brightness temperature observed at the surface can be calculated from (7.23) by adding the effect of many consecutive thin layers. If we express the brightness temperature as an integration, we have it as a function of x, when the brightness temperature at $x = 0$ is $T_b(0)$:

$$T_b(x) = T_b(0) \exp\left[-2 \int_0^x k''(x') \, dx' \right]$$
$$+ 2 \int_0^x k''(x')T_p(x') \exp\left[-2 \int_{x'}^x k''(x'') \, dx'' \right] dx' \tag{7.24}$$

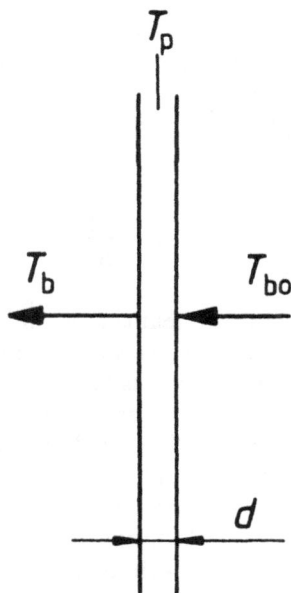

Figure 7.5 The brightness temperature changes as the radiation propagates in a lossy medium. In a thin layer of the medium, the background brightness temperature T_{bo} is attenuated and new radiation is emitted, resulting in the brightness temperature T_b.

where x' and x'' are variables for integration with respect to the x-coordinate. Relation (7.24) is the equation of radiation transfer for a lossy medium where no scattering occurs [Ulaby *et al.*, 1981]. To obtain the real observable brightness temperature of a sample, we must multiply the calculated result by the emissivity. If the sample contains other interfaces that cause internal reflections or it is so thin or lossless that the back surface influences the results, a combination of the equation of radiation transfer and the signal flow graph technique will yield the accurate result. However, in most practical cases, the simple relation (7.23) corrected for the emissivity gives a good estimate of the brightness temperature.

7.3.2 Visibility

Many of the different applications of subsurface radiometry that have been proposed in the literature are related to the detection of so-called "hot spots" inside the sample by using contacting antenna probes. Such mapping of temperature is called *thermography*. An example is detection of cancerous tumors in human tissues by using small antennas pressed against the skin. The temperature of a tumor is usually higher than that of the environment because of accelerated metabolism. In such cases, the hot spot is often smaller than the antenna lobe, and lossy material between the anomalous region and the surface further decreases the observed antenna temperature difference. For evaluation of the probability to detect a hot spot, *visibility* is often used. It is defined as

$$N = \frac{\Delta T_A}{\Delta T_p} \tag{7.25}$$

where ΔT_p is the difference in physical temperature and ΔT_A is the observed antenna temperature difference when the antenna is moved past the hot spot.

Different methods for calculation of visibility have been reported. One method [Myers *et al.*, 1979] is based on the equation of radiation transfer, but the situation is slightly complicated by the fact that the sample is located in the near field of the antenna. If the directional pattern of the antenna is known for the measurement situation, however, reasonable accuracy can be obtained by using (7.13) or (7.14) in conjunction with the equation of radiation transfer.

Another method is based on the reciprocity principle [Nguyen *et al.*, 1980]. The electric field of the antenna, when used for transmitting microwaves into the sample, must be known. The electric field can, for example, be measured with a small probe in a liquid with the same dielectric properties as the sample. In the case of the transmitting antenna, the local power absorption is proportional to $\varepsilon_r''|E|^2$. According to the reciprocity principle, the portion of the total radiation absorbed in a certain volume, V_{hs}, is equal to the portion of the total antenna temperature which originates in that volume. Hence, we have

$$N = \eta \, \frac{\displaystyle\int_{Vhs} \varepsilon_r''|E|^2 \, \mathrm{d}V}{\displaystyle\int_\infty \varepsilon_r''|E|^2 \, \mathrm{d}V} \tag{7.26}$$

where η is the mismatch loss in the sample-to-antenna interface. By use of proper matching, η can be made closer to unity than is the emissivity of the surface for radiation into air. With the method of reciprocity, the dielectric inhomogeneities that may be present in the sample are difficult to take into account.

Both methods give only approximate results, but they can still help evaluate the usefulness of contacting radiometer sensors.

7.3.3 Contacting Antennas

The antenna, used as a contacting probe, should be shielded so that it only received radiation from inside the sample. A further advantage is if the operator can move the antenna by holding it in his or her hand. These requirements are fulfilled by the open-ended waveguide antenna (Figure 7.6). It should be filled with a dielectric material of such a permittivity that the antenna is matched to the sample under normal conditions. Local variations in the permittivity of the sample cause sample-to-antenna interface mismatch loss, which influences the antenna temperature. However, the reflection coefficient caused by the mismatch, for example, can be measured by injecting noise toward the antenna through a directional coupler (Figure 7.7). The noise is injected in short intervals, which modulates the output of the receiver. The modulation gives the reflection coefficient, used for correcting the antenna temperature [Lüdeke *et al.*, 1978], [Osterrieder *et al.*, 1982]. Different balancing techniques can also be used to cancel the effect of mismatch loss [Schilz *et al.*, 1981]. In addition, the front-end components of the receiver radiate noise toward the antenna. In a contacting antenna, part of this noise is reflected back and must be taken into account.

The spatial resolution, achieved with subsurface radiometry by using a single contacting antenna, depends primarily on the shape and size of the antenna aperture. The attenuation in practical situations is usually so high that most of the radiation originates from within the near field of the antenna. A better resolution is therefore achieved with a smaller antenna (i.e., by increasing the permittivity of the dielectric filling the antenna), or at a higher frequency. However, the attenuation is usually stronger at higher frequencies, leading to a smaller *penetration depth*. The choice of frequency is therefore a compromise between the lateral resolution, mismatch loss, and penetration depth. Figure 7.8 shows the field of an open-ended waveguide antenna measured in glycerol, which closely mimics the permittivity of fatty tissue.

For imaging purposes, the contacting antennas can be built in arrays [Mamouni *et al.*, 1987], thus eliminating the need to move the antenna along the sample.

7.3.4 Correlation Radiometer

Subsurface radiometry is used primarily for mapping of local temperature differences inside the sample (i.e., for thermography). The spatial resolution achieved by a system is therefore an important feature. Poor resolution not only smears the details and makes difficult the exact location of anomalies, but also makes the temperature deviation in a small hot spot appear smaller.

In the previous section, we saw that in the near field, where most of the radiation originated, the resolution was equal to the size of the antenna. A better resolution, however, can be achieved with a *correlation radiometer*.

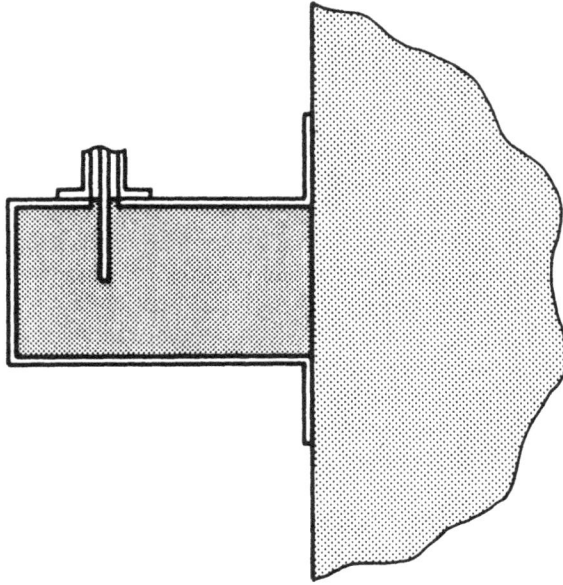

Figure 7.6 Open-ended waveguide antenna for subsurface radiometry.

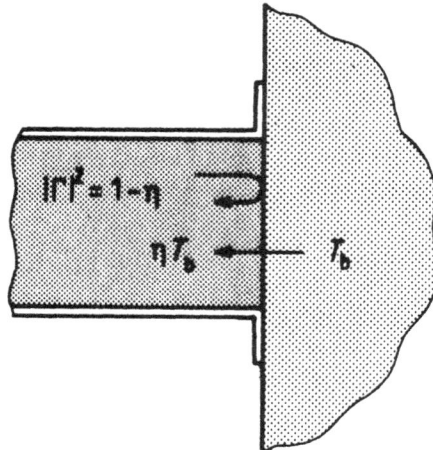

Figure 7.7 The mismatch loss can be measured by injecting noise through a directional coupler toward the antenna.

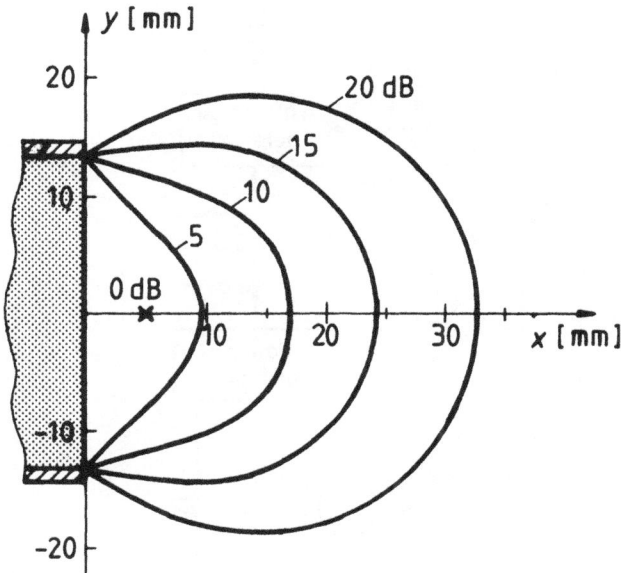

Figure 7.8 Sensitivity pattern of a cylindrical antenna measured in glycerol. The diameter of the aperture was 27 mm and the frequency was 4.75 GHz.

The principle of the correlation radiometer (as suggested in [Mamouni *et al.*, 1983]) is shown in Figure 7.9. It is a two-channel instrument, where the signals, received by two closely spaced antennas, are processed coherently. The other channel is equipped with a circulator, the third port of which is terminated in a short circuit or open circuit, depending on the position of the switch. The phase of the reflection coefficient in that port is therefore either $+\pi/2$ or $-\pi/2$. When the switch is turned, the phase of the signal fed to the power combiner (called a *hybrid*) is reversed. For the uncorrelated noise in the two channels, the position of the switch has no effect on the output of the combiner. The correlated noise, however, is added either constructively or destructively. When the switch is turned frequently the correlated part of the noise modulates the power level at the output of the combiner. The correlated noise can thus be measured separately.

The correlated noise is that which is received by both antennas from the same source. The noise must therefore originate in the overlapping volume of the antenna patterns, shown by heavy shading in Figure 7.9. This volume is narrower in one direction than the pattern of a single antenna, thus providing better spatial resolution (Figure 7.10). In low-loss samples, the noise originating near the surface of the sample can be eliminated by choosing the optimal spacing of the antennas, which

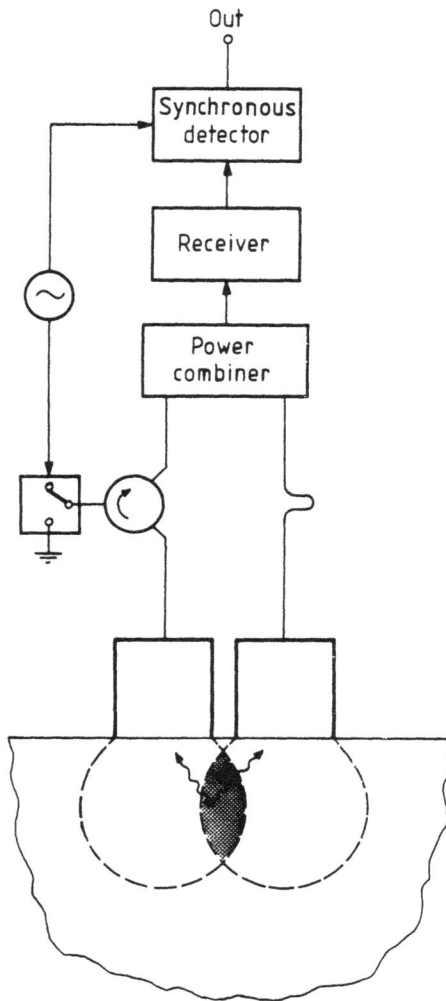

Figure 7.9 The correlation radiometer detects the noise originating in the volume covered by both antennas.

is a great advantage in many cases. Performing limited scanning of the measurement volume in one direction is also possible by changing the phase difference between the channels.

The major problem with the correlation radiometer is its low sensitivity. In high-loss samples (such as human tissues), only very little correlated noise is received (see Figure 7.8). However, use of a lower frequency is now possible, which

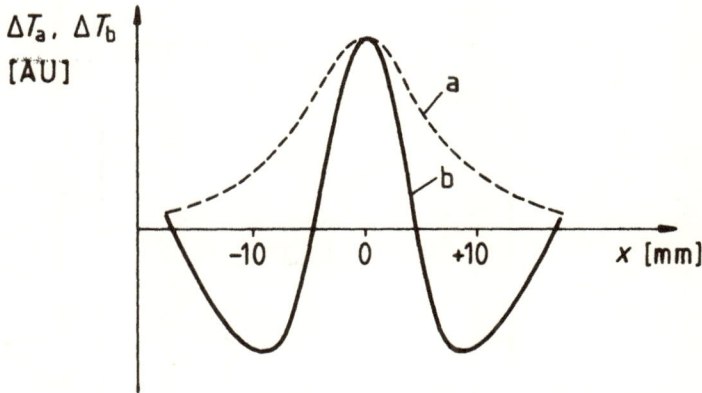

Figure 7.10 Measured resolution with (a) normal radiometer and (b) correlation radiometer. The thermal structure was a cylinder (diameter, 11 mm; closest point from surface, 9.5 mm) of hot water embedded in cold water. Frequency: 2.2–3.8 GHz. (After [Mamouni *et al.*, 1983].)

usually means a higher penetration depth. The antennas can also be tilted toward each other to increase the overlapping volume. The correlation radiometer technique has been studied extensively by Yves Leroy and others at the University of Lille (France) [Mamouni *et al.*, 1983; Bellarbi *et al.*, 1984].

7.3.5 Aperture Synthesis Thermography

The idea of receiving noise coherently with several antennas can be further extended from the correlation radiometer described above. By using one antenna as reference and one or more movable antennas (Figure 7.11) to measure the correlation in a certain area relative to the reference antenna, we are able to synthesize (simulate) an antenna with an aperture of the size of the covered area. The method is well known from radio astronomy. For close-range applications, the antenna is computationally focused to produce a thermographic image of an area of the sample of about the size of the synthesized aperture [Haslam *et al.*, 1984]. The distance to the focal plane can be chosen freely and varied for the retrieval of the temperature depth profile. The attenuation, however, limits the visibility of all subsurface features. If the sample were lossy, the antennas should be used at some distance from the sample to reduce the path of the radiation through the lossy medium. The interspace should be filled with a lossless material with the same ε_r' as the sample to reduce the reflection and refraction of the radiation.

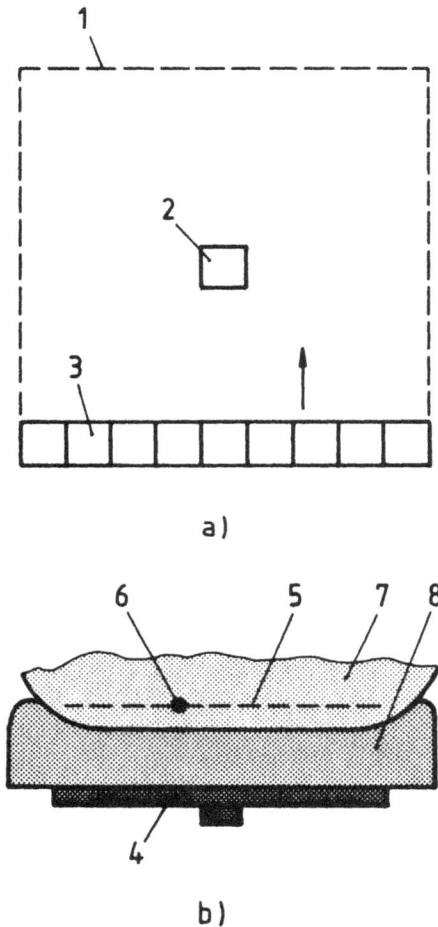

a)

b)

Figure 7.11 Possible future application of aperture synthesis thermography in medicine: (a) 1—Synthesized square antenna aperture, 2—stationary reference antenna, 3—movable antenna array; (b) 4—antenna plane, 5—focal plane, 6—hot spot, 7—patient, 8—matching lossless dielectric. (After [Haslam *et al.*, 1984].)

The technique can be used with other than planar geometries. Schaller, for example, has used a cylindrical water tank with one reference and another movable antenna around the periphery [Schaller, 1987]. The synthesized antenna is focused in the cross plane of the tank (Figure 7.12).

The mathematics involved in aperture synthesis techniques is somewhat complicated and beyond the scope of this book. The resolution achievable with aperture synthesis thermography is on the order of $\lambda/4$–$\lambda/2$ in the material. The resolution

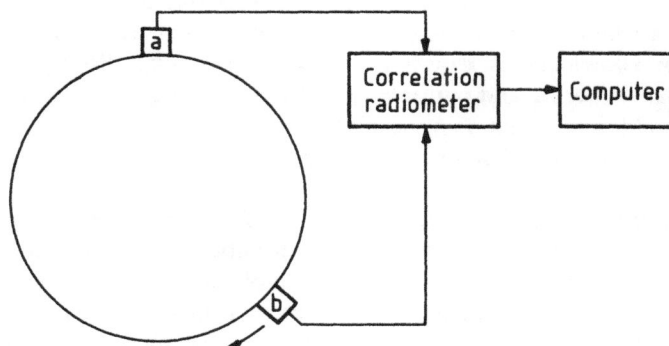

Figure 7.12 Aperture synthesis thermography in a cylindrical sample holder. Antenna (a) is stationary and antenna (b) is moved around the cylinder. The focal plane is inside the circle. Schaller has tested such a system on noise sources in a water tank. (After [Schaller, 1987].)

depends on the measurement arrangement and the reconstruction algorithm. In cases of strongly inhomogeneous samples or those with high losses, major difficulties are encountered. To the knowledge of the authors, aperture synthesis techniques in microwave thermography are still at the laboratory stage.

7.3.6 Multifrequency Radiometry

Local variations in the brightness temperature reveal the presence of hot spots or other anomalies below the surface of the sample. More information could in many cases be deduced if the depth profile of the temperature were also known. The correlation radiometer provides a means to obtain some depth resolution, but the temperature profile is out of reach. Synthetic aperture techniques give better depth resolution, but are very complicated. The best way is to use the *multifrequency technique,* namely, to measure the brightness temperature as a function of frequency.

The multifrequency technique is based on the assumption that the attenuation in the sample depends on the frequency, which normally is the case. From (7.23) or (7.24), we can see that, for a certain temperature profile, we have a different brightness temperature for different values of k''. The attenuation acts like a weighing function. At a higher frequency, where the attenuation is usually higher (see (4.26)), the brightness temperature is primarily determined by that of the layers close to the surface. At a lower frequency, the deeper layers also influence the measurement. By measuring the brightness temperature as a function of frequency in a sufficiently broad band, in principle, solving the inverse problem (i.e., to calculate the temperature profile down to a depth determined by the smallest attenuation) is possible.

The attenuation or antenna sensitivity pattern as a function of both frequency and depth must be known. In inhomogeneous media, such as biological tissues, substantial prior knowledge is required for proper modeling. Strong inhomogeneities cause internal scattering and multiple reflections, making modeling even more complicated. In addition, when using a multichannel radiometer with a single antenna, the total measuring time becomes longer because some kind of time sharing must be used.

The multifrequency technique can be combined with thermography for mapping of the three-dimensional temperature distribution. In the complete system, however, this leads to calculations at least as complicated as for the aperture synthesis techniques. The sensitivity pattern of the antenna must be known in the sample as a function of frequency. The calculations involve the retrieval of the depth profile in combination with the two-dimensional deconvolution of the sensitivity pattern. Bardati and others have performed tests in the two-dimensional case with a four-channel (1.1–5.5 GHz) radiometer in a tank with saline water [Bardati *et al.*, 1987]. The hot spot was simulated with a tube (2 cm diameter) of hot water at 1 cm from the surface. The results were promising.

7.4 APPLICATIONS OF RADIOMETER SENSORS

7.4.1 Medicine

Microwave thermography has several potentially useful medical applications, and research has been active in this field since the mid-1970s. The applications are related to the detection of inflammated areas (e.g., after receiving acute radioactive radiation) or cancerous tumors, which, as mentioned earlier, are usually warmer than the environment due to accelerated metabolism. Another application is the monitoring of local temperature during hyperthermia treatment of cancer. (*Microwave hyperthermia* is the treatment of cancer by repeatedly raising the temperature using microwave power.) The best results seem to have been achieved with radiometers working at about 3 GHz using contacting antennas. Such an instrument is capable of detecting local temperature deviations down to a depth of a few centimeters.

Some equipment is already commercially available. Odam (France) markets radiometers equipped with single contacting antennas. The frequency is 2.5 GHz or 9 GHz and the results are displayed on a TV monitor. Odam also has equipment for simultaneous treatment of cancer by hyperthermia and monitoring of temperature with the same antenna. The frequencies are different for treatment and monitoring.

The greatest advantage with microwave thermography is the noninvasive nature of the measurement. With the more complicated systems described above, the imaging and temperature depth profile measurement capabilities further increase the usefulness. The greatest difficulties are related to the high attenuation in the skin, which has a high water content, and the inhomogeneous nature of the tissues. Mainly because of these difficulties microwave thermographical methods are not yet operational. More sophisticated systems, such as aperture synthesis, correlation, multifrequency, and multiprobe radiometers, still need to be developed together with the data interpretation methods. More information can be found in the literature [Leroy, 1982; Leroy et al., 1987; Land et al., 1987; Bach Andersen, 1987; Giaux et al., 1988].

7.4.2 Industrial Applications

Microwave radiometers, in some cases, may be employed as thermometers where no other sensors can be used. For example, in high temperature rotating ovens, contacting temperature sensors are difficult to use, and infrared thermometers are impossible if the oven is filled with smoke and dust. At the Radio Laboratory of the Helsinki University of Technology, a radiometer (10 GHz) was developed for measurement of the temperature inside a cement oven, where the level is above 1000° C. The radiometer was successfully tested in such a manner that the antenna looked into the oven through a ceramic window.

Instead of using a normal antenna pointing toward the sample (or making contact with it), transmission sensors or even resonators can also be employed. Leroy and others from the University of Lille have proposed the use of a split waveguide sensor (see Figure 4.7a), which is shorted or left open at the other end,for the measurement of the temperature of a textile web during thermal treatment [Leroy et al., 1986]. Such a sensor provides good shielding against noise from the environment. The antenna temperature can be calculated with the equation of radiation transfer, (7.23) or (7.24). If the attenuation in the sensor is low, the signal depends on the attenuation, which in turn, depends on the moisture. If the attenuation can be controlled, for example, by measuring the reflection coefficient at the input of the sensor, the temperature profile can be measured with a set of sensors of different length.

7.4.3 Remote Sensing

Microwave radiometers have been used for remote sensing of the environment from helicopters, airplanes, satellites, and the ground [Ulaby et al., 1981]. The most important applications are the measurement of soil moisture, ice cover, water equivalent of the snow cover, thickness of oil slicks on the sea surface, and temperature profile or ozone content of the atmosphere. The measurement of soil moisture is based on local variations in emissivity due to variations in permittivity. The method works only over open terrain because the canopy disturbs the measurement.

The brightness temperature of the sea surface is very low, about 100 K, because of the high permittivity of water. Therefore, ice-covered water shows a clear contrast to open water. In polar regions, where both multiyear ice and first-year ice are present, the two types can be separated because of the lower brightness temperature of multiyear ice. This is due to scattering in the coarse-grained weathered surface layer. Skou and others from the Technical University of Denmark have studied the remote sensing of sea ice in Greenland waters [Skou et al., 1979].

Measurement from satellite of the water equivalent of snow cover on land is also based on scattering. Dry snow is almost lossless, and the measured brightness temperature is therefore due to radiation emitted from the ground. On the way through, the snow-cover part of the radiation is scattered back. The more snow there is, the stronger is the scattering and the lower is the brightness temperature. The scattering is also stronger at higher frequencies. The water equivalent can be estimated by comparing the brightness temperature at two different frequencies. Hallikainen and others from the Helsinki University of Technology have used the 18 and 37 GHz data from the Nimbus-7 satellite and studied the influence of different surface types (forest, farmland, lake district, et cetera) [Hallikainen et al., 1987].

When the sea surface is covered by a layer of oil, multiple reflections occur in the oil layer. This causes the brightness temperature to increase periodically with the thickness of the layer. By using two different frequencies, the thickness of the oil slick can be calculated [Lääperi et al., 1982]. In practice, the oil mixes with sea water (and forms lumps) relatively soon after the release, depending on the sea state. The mixture is very lossy. An estimate of the amount of oil can still be made if a sufficiently low frequency is used (≈ 5 GHz) so that the mixed layer is partly transparent to microwaves [Lääperi, 1983; Skou, 1982]. An airplane or helicopter equipped with a microwave radiometer can greatly aid the successful collection of the pollution.

REFERENCES

Bach Andersen, J., "Medical Applications of Microwaves," Proc. 17th European Microwave Conf., Rome, September 1987, pp. 50–56.

Bardati, F., G. Calamai, M. Mongiardo, B. Paolone, D. Solimini, and P. Tognolatti, "Multispectral Microwave Radiometric System for Biological Temperature Retrieval: Experimental Tests," Proc. 17th European Microwave Conf., Rome, September 1987, pp. 386–391.

Bellarbi, L., A. Mamouni, J.C. Van de Velde, and Y. Leroy, "On Possibilities of Thermal Pattern Recognition by Correlation Microwave Thermography," Proc. 14th European Microwave Conf., Liége, September 1984, pp. 645–650.

Giaux, G., J. Delannoy, D. Delvalee, Y. Leroy, B. Bocquet, A. Mamouni, and J. Van de Velde, "Microwave Imaging at 3 GHz for the Exploration of Tumors of the Breast," IEEE Trans. Microwave Theory Tech., Vol. 36, No. 5, May 1988.

Hallikainen, M., and P. Jolma, "Retrieval of Snow Water Equivalent from Satellite Microwave Radiometer Data," Proc. IGARSS '87, Ann Arbor, MI, May 1987, pp. 853–858.

Haslam, N.C., A.R. Gillespie, and C.G.T. Haslam, "Aperture Synthesis Thermography—A New Approach to Passive Microwave Temperature Measurements in the Body," IEEE Trans. Microwave Theory Tech., Vol. MTT-32, No. 8, August 1984, pp. 829–835.

Kraus, J.D. (*see* Ch. 7 by M. Tiuri and A. Räisänen), *Radio Astronomy,* 2nd Ed., Powell, OH: Cygnus-Quasar Books, 1986, 690 p. (*See also* 1st Ed., New York: McGraw-Hill, 1964).

Land, D.V., and V.J. Brown, "Subcutaneous Temperature Measurement by Microwave Thermography," *Proc. 17th European Microwave Conf.,* Rome, September 1987, pp. 896–901.

Leroy, Y., "Microwave Radiometry and Thermography: Present and Prospective," *Biomedical Thermology,* 1982, pp. 485–499.

Leroy, Y., A. Mamouni, J. Van de Velde, B. Bocquet, and B. Dujardin, "Microwave Radiometry for Non-Invasive Thermometry," *Automedica,* Vol. 8, 1987, pp. 181–202.

Leroy, Y., J.C. Van de Velde, A. Mamouni, B. Meyer, and J.F. Rochas, "Contactless Thermometry of a Textile Web by Microwave Radiometry," *Proc. 16th European Microwave Conf.,* Dublin, September 1986, pp. 377–381.

Lüdeke, K.M., B. Schiek, and J. Köhler, "Radiation Balance Microwave Thermograph for Industrial and Medical Applications," *Electronics Letters,* Vol. 14, No. 6, March 1978, pp. 194–195.

Lääperi, A., "Experimental Results with the Microprocessor Controlled Microwave Radiometer System in the Combined Oil Experiment," *Specialist Meeting on Microwave Radiometry and Remote Sensing Applications,* Rome, April 1983.

Lääperi, A., and E. Nyfors, "Microprocessor Controlled Microwave Radiometer System for Measuring the Thickness of an Oil Slick," *Proc. IGARSS '82,* Munich, June 1982, pp. 4.1–4.6.

Mamouni, A., B. Bocquet, J.C. Van de Velde, and Y. Leroy, "Microwave Thermal Imaging by Radiometry," *Proc. 17th European Microwave Conf.,* Rome, September 1987, pp. 381–385.

Mamouni, A., Y. Leroy, J.C. Van de Velde, and L. Bellarbi, "Introduction to Correlation Microwave Thermography," *J. Microwave Power,* Vol. 18, No. 3, September 1983, pp. 285–293.

Myers, P.C., N.L. Sadowsky, and A.H. Barrett, "Microwave Thermography: Principles, Methods and Clinical Applications," *J. Microwave Power,* Vol. 14, No. 2, June 1979, pp. 105–115.

Nguyen, D.D., M. Robillard, M. Chive, Y. Leroy, J. Audet, C. Pichot, and J.C. Bolomey, "Modeling of Probes and Interpretation of the Thermal Patterns in Microwave Thermography," *15th Int. Microwave Power Symp.,* Iowa City, IA, May 1980.

Osterrieder, S., and G. Schaller, "An Improved Microwave Radiometer for Measurements on the Human Body," *Proc. 12th European Microwave Conf.,* Helsinki, September 1982, pp. 559–564.

Schaller, G., "Synthetic Aperture Radiometry for the Imaging of Hot Spots in Tissue," *Proc. 17th European Microwave Conf.,*Rome, September 1987, pp. 902–907.

Schilz, W., and B. Schiek, "Microwave Systems for Industrial Measurements," *Advances in Electronics and Electron Physics,* New York: Academic Press, Vol. 55, 1981, pp. 309–381.

Skou, N., "Microwave Radiometer Measurements on Oil Slicks. Results from an Experiment on Kattegat on 13 May 1982," Technical University of Denmark, Electromagnetics Institute, Report R261, July 1982.

Skou, N., *Microwave Radiometer Systems: Design and Analysis,* Norwood, MA: Artech House, 1989, 160 p.

Skou, N., and F. Sondergaard, "Radiometer Signatures and SLAR Imagery of Sea Ice in Greenland. Results from July 1977 and November 1978," Technical University of Denmark, Electromagnetics Institute, Report R216, October 1979.

Tiuri, M.E., "Radio Astronomy Receivers," *IEEE Trans. Antennas Propag.,* Vol. AP-12, December 1964, pp. 930–938.

Ulaby, F.T., R.K. Moore, and A.K. Fung, *Microwave Remote Sensing, Active and Passive, Vol. I, Microwave Remote Sensing Fundamentals and Radiometry,* Norwood, MA: Artech House, 1981, 456 p.

Chapter 8
Active Imaging

8.1 INTRODUCTION

Microwaves can be used for producing images of samples in several different ways. We have already discussed passive imaging (i.e., radiometric methods of producing thermographic images) in Chapter 7. In Chapter 6, we discussed radars, some of which are used for active imaging. The spatial resolution achieved with radars is determined by the directive properties of the antennas. In this chapter, we will discuss sensors with higher resolution. These are based on the coherent recording (amplitude and phase) of the electric fields scattered in different directions, thus making possible the focusing of the image. The discussion will be limited to close-range applications, with a maximum range of a few meters. The sensors are divided into two groups: *holographic* sensors to measure reflected radiation and *tomographic* sensors to measure transmitted radiation. Actually, holography means producing "whole" (*holos* in Greek) images, referring to the three-dimensional virtual images produced by optical holograms. Tomography means producing cross-sectional images (*tomos,* the Greek word for "section"). The greatest advantage of microwave imaging as compared to normal photography is the ability to image samples embedded in a dielectric environment, which is opaque to visible light.

8.2 HOLOGRAPHY

8.2.1 True Holography

Optical Holography

Holography is best known from optics, where it has been used since the 1950s. The principle is shown in Figure 8.1. The sample is illuminated with a coherent light source, a laser. From the same source, a reference beam is formed with a beam

---

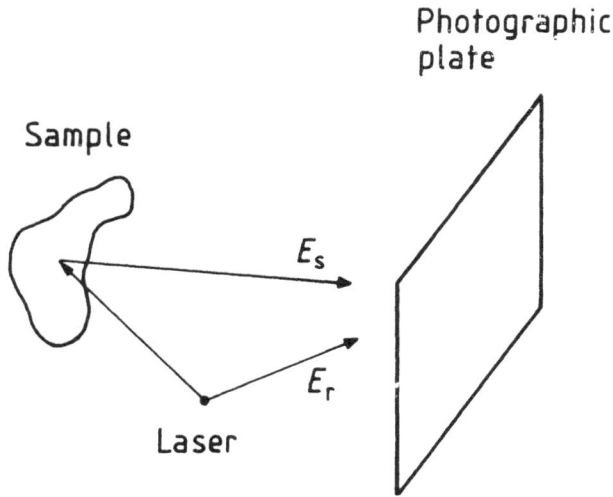

Figure 8.1 Formation of an optical hologram. The photographic plate records the intensity of the interference pattern caused by the light reflected (scattered) from the sample and reference beam.

splitter. The light scattered from the surface of the sample (E_s) and the reference beam (E_r) form an interference pattern on the photographic plate, which records the intensity of the radiation:

$$I_h \propto (E_r + E_s)(E_r + E_s)^*$$

$$\propto |E_r|^2 + |E_s|^2 + E_r^*E_s + E_rE_s^* \tag{8.1}$$

where the asterisk denotes the complex conjugate. When the developed and copied (positive) transparent film is illuminated with the reference beam, the resulting wavefront is

$$E_i \propto |E_r|^2E_r + |E_s|^2E_r + |E_r|^2E_s + E_r^2E_s^* \tag{8.2}$$

The third term on the right-hand side of (8.2) is a reproduction of the scattered wave, and therefore produces a virtual image (Figure 8.2) of the sample. The fourth term produces a displaced and reversed (complex conjugate) virtual image. The images appear three-dimensional to an observer because the hologram accurately reconstructs the original wavefront. The appearance of the reversed image was due to the *rectification* of the interference pattern when it was recorded by the film. The original interference pattern resembles a standing wave, where the phase difference is π radians between adjacent lobes, but the film records only the intensity. The first and

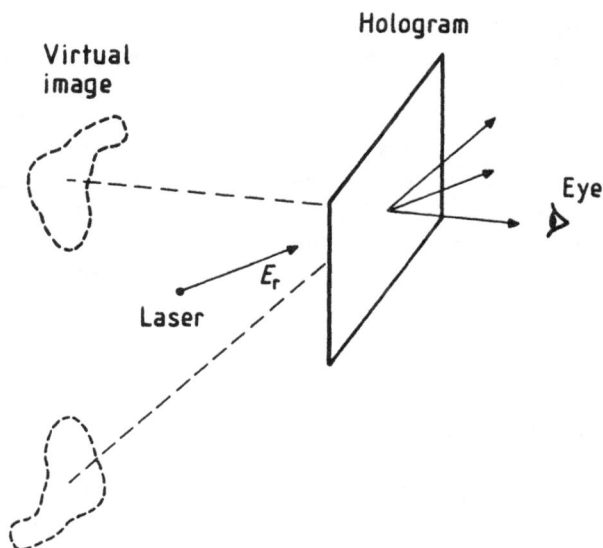

Figure 8.2 When the hologram is illuminated with a coherent light source, the original field intensity distribution in the plane of the hologram is reconstructed, producing a virtual image and a complex conjugate of the virtual image of the sample.

second terms are undiffracted light. A hologram can also be made to reflect light, which is perhaps the more familiar case to the public. By using properly filtering layers of coating, the holograms can even be made to work with normal noncoherent light. Optical holograms are used mainly because of the three-dimensional images, but the virtual image can always be converted into a two-dimensional real image in a plane by replacing the eye in Figure 8.2 with a convex lens.

Microwave Holography

The holographic principle can also be used for microwave imaging. In that case, the laser in Figure 8.1 is replaced by a microwave source, and the photographic plate must be replaced by some other recording device. Both liquid crystal plates and other thermally sensitive materials have been used, but these are impractical because of their low sensitivity. The best solution is to scan the desired aperture with a power-detecting probe or an array of probes (e.g., small dipole or waveguide adapter with a diode detector). In the early days of microwave holography, the detected signal was used for modulating a light bulb that exposed a photographic film, or the measured intensity pattern was displayed on an oscilloscope and photographed. After the film was developed and reduced in scale to fit the wavelength of light, the microwave image could be visualized with a laser beam, as in Figure 8.2, and a lens to produce a real image, which was photographed. Such a hologram is too small for the virtual image to be viewed directly. The hologram is about the size of the scanned aperture multiplied by the ratio of the wavelengths of light and microwaves (10^{-4}–10^{-5}).

The theoretical resolution of the microwave image is limited by diffraction (i.e., the size of the aperture in wavelengths compared to the range). We will return to this condition in Section 8.2.2. The resolution of optically processed images is usually much worse than the theoretical limit because of the difficulty of producing a photographic image of the interference pattern. The optical processing is also very cumbersome because of the many phases from measurement to image. Therefore, numerical processing is used at present. More details about the optical processing can be found in the references [Orme *et al.*, 1973; Anderson, 1977; Tricoles *et al.*, 1977; Anderson, 1979].

Numerical processing of the recorded interference pattern is based on the well known dependence of the far-field directional pattern of an antenna on the two-dimensional Fourier transform of the aperture fields (see (1.30)). In a similar manner, the aperture fields of an antenna can be calculated from the inverse Fourier transform of the far-field pattern. Therefore, the microwave reflection image of the sample is calculated from the inverse Fourier transform of the recorded intensity, computationally accounting for the reference wave. The reconstructed image can then be processed numerically (e.g., enhanced, smoothed, or filtered to enhance or eliminate certain characteristics) by using commercially available digital image processing programs [Anderson *et al.*, 1976]. The disadvantage of this method is that it produces two images separated by a bright spot, exactly as in optical holography. For some directions, the images will overlap. The arrangement with the reference source in front of the scanning aperture is also troublesome in practice. This procedure is therefore preferable only at millimeter-wave frequencies, where direct recording of both the phase and amplitude of the scattered wave (without the reference wave) is difficult.

8.2.2 Quasiholography

General

The most straightforward way to produce a microwave image of the sample from the scattered wave is to record the amplitude and phase of the scattered field in a certain aperture and to compute the (inverse) Fourier transform. Because holography traditionally means recording the intensity of the interference pattern produced with the aid of a reference wave, this is sometimes called *quasiholography*. For the computation of the Fourier transform, there are two completely different approaches, the *quasioptical* and *numerical* method.

Quasioptical Reconstruction

The quasioptical method is very simple and obvious. The recording aperture is formed by a dielectric lens, which produces a real image in a plane behind the lens (Figure 8.3). The microwave detector (intensity is enough) scans the image plane and directly yields the image. Numerical calculations are necessary only if further processing of the image is needed. A disadvantage of the method is the fixed focal length of the lens. For focusing the system to different ranges, the distance between the lens and scanning detector must be changed. However, we can use the lens for only compression of the field distribution and calculate the image from the Fourier transform [Anderson, 1977], in which case, the above-mentioned distance can be fixed. Another disadvantage is the weight of the lens, which will be considerable if the diameter is large. However, a larger diameter gives a better resolution.

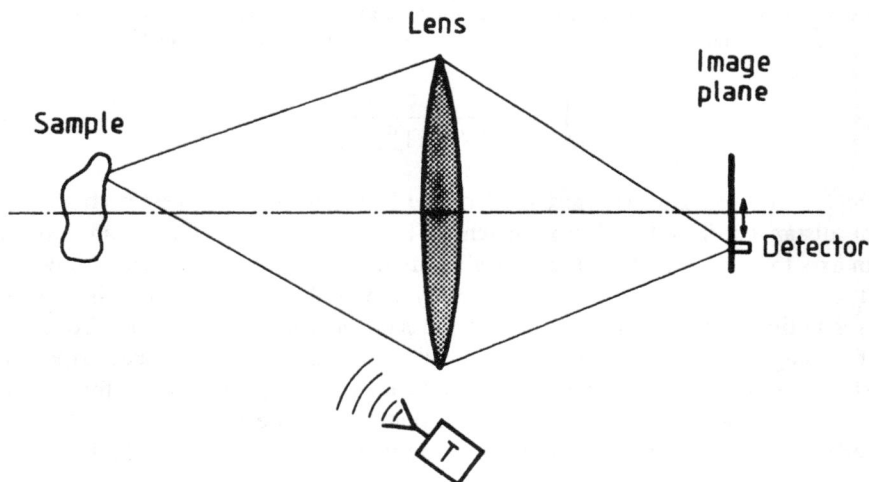

Figure 8.3 Microwave imaging with a lens-focused system.

For calculation of the design parameters and resolution of the system, the propagation of microwaves can be approximated with *Gaussian beams*. This is the fundamental analytic technique used in quasioptics, applicable when the diffraction of the rays cannot be omitted because the diameter of the beam is not much greater than the wavelength. For a lens (Figure 8.4), with the focal length f, we have

$$\frac{d_2}{f} = 1 + \frac{(d_1/f) - 1}{[(d_1/f) - 1]^2 + (\pi w_{01}^2 / \lambda f)^2} \tag{8.3a}$$

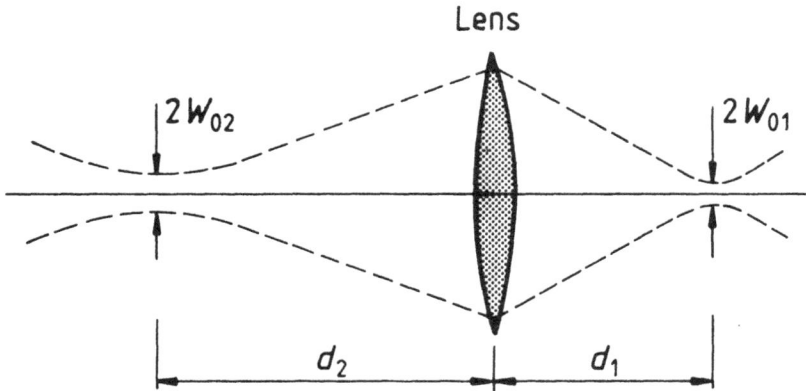

Figure 8.4 In quasioptics, Gaussian beams are used instead of rays. The Gaussian beam accounts for the diffraction effects, which cause the focal points to grow in diameter.

$$\left(\frac{w_{02}}{w_{01}}\right)^2 = \frac{1}{[(d_1/f)-1]^2 + (\pi w_{01}^2/\lambda f)^2} \tag{8.3b}$$

where w_{01} and w_{02} are the waist radii of the beam at the narrowest points ("foci"), at the distances d_1 and d_2 from the lens. The waist radius is the distance from the beam axis to the point where the Gaussian distributed electric field strength decreases by $1/e$. The first radius is determined by the probe and the other is the resulting resolution, when d_1 and d_2 are of the same order of magnitude. To calculate the focusing capacity of such a microwave imaging system, we need to know the waist radius of the probe. The Gaussian beam model is most valid for a detector probe made of a conical corrugated horn that launches a beam, where 98% of the power is contained in a Gaussian beam. The waist is located inside the horn (Figure 8.5), at

$$\frac{L}{R} = \frac{1}{1 + 6.769(M/2\pi)^2} \tag{8.4}$$

where

$$\frac{M}{2\pi} = \frac{a}{\lambda} \tan \frac{\theta_0}{2}$$

Where, a is the radius of the aperture and θ_0 is the angle in Figure 8.5. The waist radius is

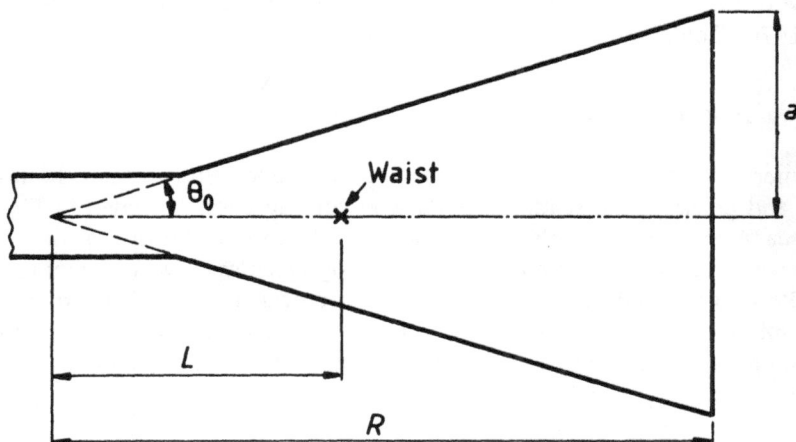

Figure 8.5 Geometry of a corrugated horn.

$$w_{01} = \frac{0.6435a}{\sqrt{1 + 0.1715M^2}} \tag{8.5}$$

For other probes, the focus is more difficult to calculate. In practice, the difference is small compared to the difficulties arising when the system is used for imaging samples embedded in a dielectric medium.

As an alternative to the configuration in Figure 8.3, a common antenna can be used for transmitting and receiving. In this case, only the point currently being measured is illuminated, which saves power. Using a common antenna, however, increases the isolation requirements between transmitter and receiver. An advantage of joint scanning is the higher resolution that is achieved [Anderson, 1976].

A focusing reflector can be used instead of the lens. Such a reflector is made of a section of an ellipsoid. The detector is located at one focal point and the sample at the other. Edrich has used a similar arrangement for remotely focused millimeter-wave thermography [Edrich, 1979].

The resolution of all microwave imaging systems is effected by *speckle,* which means that the image has a spotty appearance. It is caused by the interference of several scattering points within the area represented by one picture element, called a *pixel.* The reason is the same as that which causes the projected spot of a laser beam to appear as if composed of many tiny dots. The speckle in microwave images can be reduced by performing the measurements over a broad range of frequencies. Even noncoherent noise can be used with recording in the real-image plane because only the intensity is measured. More detailed information on the shape of the beams and size of the focal points in quasioptical systems is given, for example, in [Corn-bleet, 1976; Goldsmith, 1982; Arnaud, 1976].

Numerical Reconstruction

For numerical reconstruction of the images, the probe must measure both the amplitude and phase of the scattered field. The configuration is shown in Figure 8.6. If the scanning probe is replaced by an array of probes, almost real-time imaging may be achieved. This system is one of the holographic microwave imaging methods most likely to be used in the future. In the case of samples embedded in a dielectrically inhomogeneous environment with a rough surface, however, the situation is more complicated, and much development work still needs to be done.

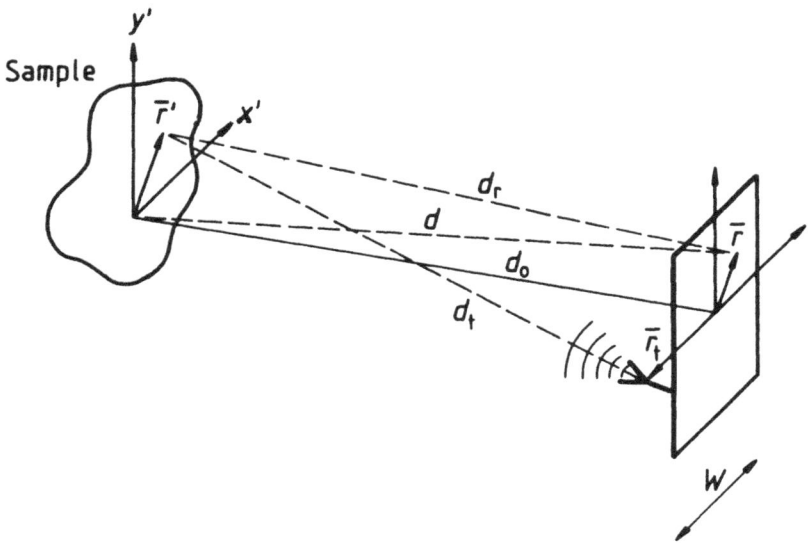

Figure 8.6 Measurement configuration for numerical reconstruction of images.

For the case of a sample in air or embedded in a homogeneous dielectric environment, the field at the probe is given by (1.30a), when the radiating source is described by $\bar{\mathbf{J}}_s$. Let us here consider $\bar{\mathbf{J}}_s$ as current density instead of a surface current and change the surface integral in (1.30a) into a volume integral. We should also recall that (1.30) was the far-field approximation, but we could show [Anderson, 1979] that (1.30) could be used when

$$W^4 < \lambda d_0^3 \tag{8.6}$$

where W is the maximum width of the scanning aperture and d_0 is the range. The current density is caused by the radiating transmitter at $\bar{r}_t$. If we denote the *sample reflectivity* by D (three-dimensional), which depends on the location ($\bar{r}'$) and the measurement frequency, and consider only the component of the current that is perpendicular to d_r (the current directed along d_r does not radiate in the direction of the receiver), we have

$$\bar{J}_\perp = \bar{a}_1 D \exp(-jkd_t) \tag{8.7}$$

where $\bar{a}_1$ is a constant that also contains the polarization vector, which is defined by the incident radiation, assumed to be uniform over the sample. Substituting into (1.30a) (slightly modified), we have approximately

$$\bar{E}(\bar{r}) = \bar{a}_2 \exp\left[-jkd_0\left(2 + \frac{r_t^2 + r^2}{2d_0^2}\right)\right]$$
$$\int_{V_s} D \exp(j2kr_z') \exp\left(jk\frac{\bar{r} + \bar{r}_t}{d_0} \cdot \bar{r}_\perp'\right) dV' \tag{8.8}$$

where the position vector in the sample is $\bar{r}' = \bar{r}_\perp' + \bar{r}_z'$. The integral has the form of a three-dimensional Fourier transform. If the sample is two-dimensional, the situation is simple. The sample reflectivity $D_\perp$ (in the $x'y'$ plane) can be calculated from the two-dimensional inverse Fourier transform (with respect to $\bar{r} + \bar{r}_t$) of the function:

$$\bar{E} \exp\left[jkd_0\left(2 + \frac{r_t^2 + r^2}{2d_0^2}\right)\right] \tag{8.9}$$

If the sample is three-dimensional, the image can be focused to different ranges by choosing different values for d_0, but, for good resolution in the z direction, the measurement must be repeated over a range of frequencies. The three-dimensional sample reflectivity is then obtained from the one-dimensional inverse Fourier transform with respect to k:

$$D = \int_{k_1}^{k_2} D_\perp \exp(-jkr_z') \, dk \tag{8.10}$$

The effect of the sweep is, in fact, the same as in an FMCW radar, and therefore the depth resolution is given by (6.10):

$$\Delta_z = \frac{c}{2\Delta f} \tag{8.11}$$

The maximum sampling interval (δf) to achieve this resolution is given by the sampling theorem and the maximum dimension of the sample in the z' direction (d_z):

$$\delta f < \frac{c}{4d_z} \tag{8.12}$$

The resolution in the transverse plane is limited by the finite scanning aperture, which can be shown to be

$$\Delta_\perp = \frac{\lambda d_0}{W} \tag{8.13}$$

when $d_0/W > 1$ (for smaller values, the approximations above are not valid), which means that the best achievable resolution is about one wavelength (in the medium) at the highest measurement frequency. To achieve the theoretical resolution, the field $\bar{E}$ must be sampled with a sufficiently small sampling interval. If the oscillating term in the phase ($\exp[-jkr^2/2d_0]$) is removed from the computation, the sampling interval is determined by the maximum transverse dimensions ($d_\perp$) of the sample alone. From the sampling theorem, we have

$$\delta r\perp < \frac{\lambda d_0}{2d_\perp} \tag{8.14}$$

Paoloni and others from the University of Wollongong (Australia) have tested a system based on this theory [Paoloni *et al.*, 1984]. They sampled the field amplitude and phase every 2 cm in 64 × 64 cm points (aperture 126 × 126 cm). The range was 2 m and the frequency range was from 8 to 12.4 GHz. The theoretical resolution, as given by (8.11) and (8.13), is slightly less than 4 cm in both the transverse plane and axial direction. Tests with sharp-edged metallic samples gave results which were very close to the theoretical values. Figure 8.7 shows results obtained with a metallic cross in two different cases.

8.2.3 Two-Dimensional Imaging of a Half-Space

An important form of microwave active imaging is where the closely spaced transmitting and receiving antennas are moved along the surface of the ground, or some other object, in a straight line (let us call it the x-axis). If the z-axis points into the

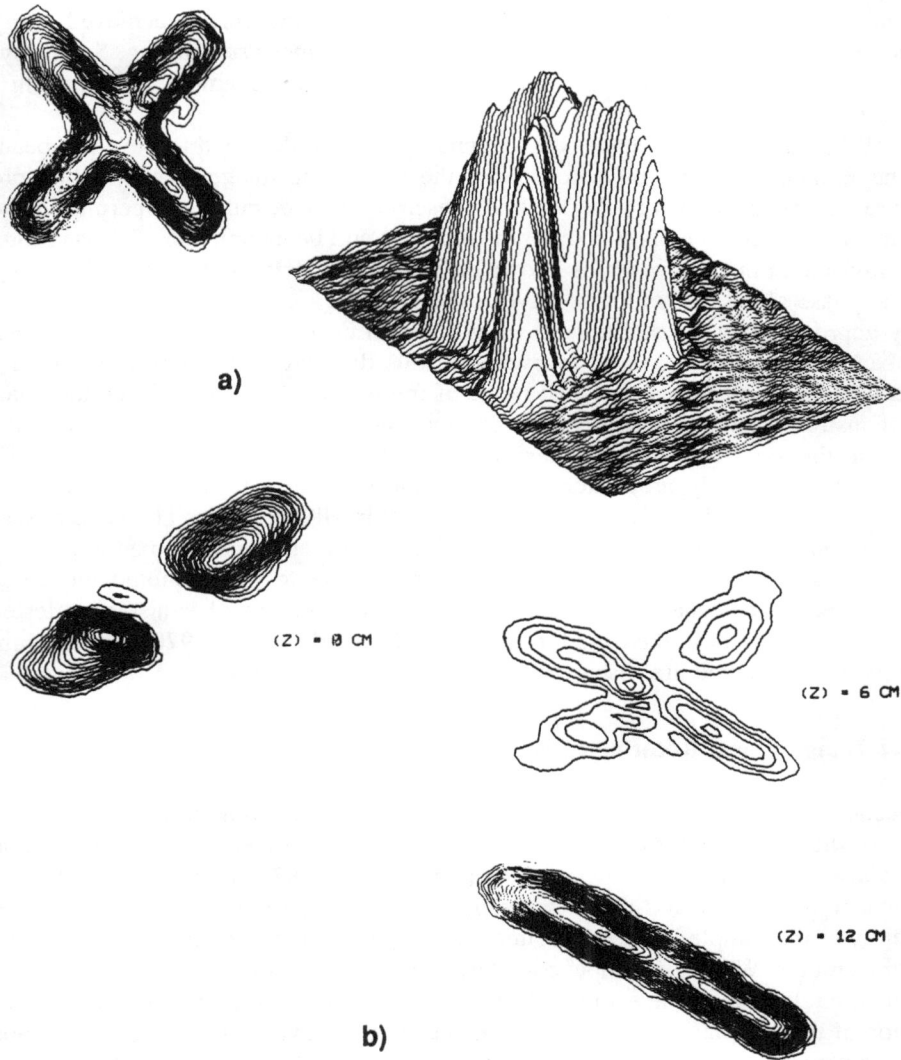

a)

b)

Figure 8.7 Reconstructed images of a metallic cross; each arm is 50 cm long and 8 cm wide. In (a) the arms are in the same plane, and in (b) the distance between the arms is 12 cm and the images are focused in three planes spaced 6 cm apart. (*Source:* Paoloni, F.J., and M.J. Duffy, "Microwave Imaging with Digital Reconstructions," *J. Electrical and Electronics Eng.* (Australia), Vol. 4, No. 1, March 1984, pp. 54–59. Reprinted with permission.)

surface, the image is a two-dimensional image in the xz-plane (like Figure 6.9). This kind of imaging is normally performed with impulse or FMCW subsurface radars, discussed briefly in Chapter 6. A better resolution, however, is achieved with a coherent, focused radar. This is often called *synthetic aperture imaging*. Such a system, based on the impulse technique, is described in, for example, [Michiguchi *et al.*, 1988].

When using CW systems, the antennas should be designed to have fan beams in the xz-plane to avoid reflections from the sides. The image reconstruction procedure can be analogous to the methods described above, but if the permittivity of the medium is not accurately known, the focusing will be incomplete. This is because the wavelength in the medium is known only approximately, and so, too, the change of the phase along the trail of a wave scattered from a certain point will be known only approximately. In addition, there will be an error in the vertical scale. Chaloupka has proposed a method to overcome this difficulty [Chaloupka, 1984]. The measurement is made over a broad range of frequencies (FMCW). When the image is reconstructed, the reflectivity of a certain point is integrated in such a way that the contributions from different directions are taken at different frequencies. The frequency is inversely proportional to the distance, relative to the vertical distance. Thus, the phase of the reflection will be constant in all directions. The vertical scale will be correct and the focus will be unspoiled. An inhomogeneous medium will, of course, blur the image, but the method is not sensitive to weak inhomogeneities.

Three-dimensional subsurface synthetic aperture imaging has also been tested. Such a system is described in, for example, [Richards *et al.*, 1978] and the limitations of the method are discussed in [Junkin *et al.*, 1988].

8.2.4 Imaging of a Rotating Sample

Especially in the case of nondestructive testing and investigation of the internal structures of dielectric samples, the use of the rotating sample method is feasible, sometimes also called *reflection tomography* [Chu *et al.*, 1987; Ermert *et al.*, 1981]. A stationary transmitter and receiver are used to measure the backscattered field coherently as the sample is rotated around an axis perpendicular to the direction toward the transmitter. When the sample completes one full revolution, the sampled aperture forms a circle around the sample. The reconstruction of the image produces a projection of the sample in the plane perpendicular to the axis of revolution. A number of different reconstruction algorithms have been developed [Mensa *et al.*, 1983; Schutz *et al.*, 1987] for slightly different measurement methods. The basic method is based on the measurement at a single frequency, and produces a relatively good theoretical resolution. The central lobe of the so-called *point-spread function* (the image of a point scatterer) has a width of $\lambda/5$, but the first sidelobe is only 8 dB below the peak. The sidelobes can be reduced by combining measurements at different frequencies or with different angles between the transmitter and receiver.

For samples of low permittivity or with a spherical shape, for example, repeating the measurement with the axis of rotation shifted to produce another projection of the sample is possible. The images can then be combined to yield a three-dimensional image.

A disadvantage of the rotating sample method is the fact that it can only be used with samples of low permittivity or having very regular and simple shapes. In the latter case, the refraction of the wavefront at the surface can be taken into account, but the calculations become more complicated. High-permittivity samples, in some cases, can be immersed in a liquid of similar permittivity to reduce the reflections from the surface. The samples should also be only weakly inhomogeneous because, in practice, only single scattering could be taken into account (the so-called *Born approximation*). High losses will, of course, also make difficult the imaging of the interior.

Ideally, this type of measurement is made in an anechoic chamber, normally used for antenna measurements, where the equipment for the controlled rotation of the sample is available and no disturbing reflections from the environment will occur.

8.2.5 Applications of Microwave Holographic Imaging

The methods described above have two properties in common: they are based on the coherent measurement in an aperture of reflections from a sample, and the image is focused computationally. Such methods have many potential applications in a wide range of fields, although few are used on a routine basis. Increased exploitation of the methods is largely due to the required computational power, which has only recently become available in portable microcomputers or in single-board computers.

The reported applications are related to nondestructive testing, detection of concealed objects (such as weapons in clothing, bugs in walls, pipes, or cables, and archeological remnants in the ground), measurement of the shape of reflector antennas, and deformation of objects under stress [Richards *et al.*, 1978; Solymar, 1977; Tricoles *et al.*, 1977; Anderson, 1977; Chaloupka, 1984].

The measurement of reflector antennas is one of the most important applications. It has been widely used since the mid-1970s. The holographic measurement of antennas is based on the relationship between the Fourier transform of the aperture fields and complex directional pattern. Because the near-field of a large parabolic antenna stretches tens of kilometers from the antenna, satellites are normally used for the measurement, as described by Goldsmith and Erickson in [Kraus, 1986]. A reference antenna with a broad beam (e.g., a horn) is mounted slightly displaced from the Cassegrain focus in such a way that it looks past the secondary reflector and receives the signal directly from the satellite, thus providing the phase reference. The complex far-field directional pattern is recorded by turning the large antenna in different directions around the satellite. The inverse Fourier transform then yields the amplitude and phase of the fields in front of the large reflector. The phase is constant over the whole aperture of a perfect antenna. For a practical antenna, local differences indicate deviations in the reflector from the ideal parabolic shape.

The methods described in Section 8.2 can be combined with pulse transmission to yield resolution at longer ranges. Such devices are called *synthetic aperture radars* (SAR), which have found many applications in remote sensing of the ground from airplanes and satellites. Even from satellites, SARs yield images with a resolution of a few meters, independent of daylight, clouds, and weather. In addition, the images reveal myriad details and properties that are invisible in optical photographs (e.g., subsurface ocean waves, pack ice, topography, *et cetera*).

8.3 TOMOGRAPHY

8.3.1 Introduction

As was pointed out in Section 8.1, tomography (x-ray, nuclear magnetic resonance) normally means producing cross-sectional images of the interior of a sample, whereas holography (light) means three-dimensional imaging of the surface of a sample. We have therefore called the imaging microwave sensors "holographic" if they are based on the measurement of reflected (back-scattered) radiation, and "tomographic," if they measure transmitted radiation. There is, however, no clear difference between the two cases because of the penetration of microwaves and the fact that small inhomogeneities scatter radiation in all directions. We have already studied methods for imaging the interior of a sample by studying backscattered radiation (Sections 8.2.3 and 8.2.4). In this section, we will concentrate on methods based on transmitted radiation.

8.3.2 General Principles

Tomographic methods are divided into *transmission tomography* and *diffraction tomography*. The former are used with very weakly inhomogeneous materials, where the wavefront propagates almost undisturbed. Either the change of phase or attenuation is measured through the sample at different locations with a pair of probes or arrays of probes. The image is constructed assuming that the waves propagate in straight lines through the sample. The result represents a projection of the sample in one plane. From data obtained by turning the sample and repeating the measurement several times, a three-dimensional image can be constructed. The method is simple, but its applicability is limited. For example, Ermert and others report experimental results obtained with a sponge phantom [Ermert *et al.*, 1981].

Samples with slightly stronger inhomogeneities, but which are still sufficiently homogeneous to fulfill the Born approximation at the measurement frequency (i.e., the assumption that the field inside the inhomogeneities is equal to the field outside), can be measured with diffraction tomography. In this case, the field after transmission is considered to contain two components, the attenuated incident field and the scattered field. The latter is radiated by an equivalent current generated from the incident field by the inhomogeneities in the sample. The reconstruction procedure is, in principle, the same as that described in Section 8.2.2, but slightly different for differing geometries. Pichot and others have given a good account of the necessary calculations in [Pichot *et al.*, 1985]. Other useful reports are, for example, [Baribaud *et al.*, 1988; Bolomey *et al.*, 1983; Broquetas *et al.*, 1987; Caorsi *et al.*, 1988; Jofre *et al.*, 1986; Paoloni, 1987; Peronnet *et al.*, 1983; Rius *et al.*, 1987].

8.3.3 Medical Applications of Diffraction Tomography

One of the most promising applications of microwave tomography is found in medicine. The resolution is worse than that achieved with x-ray or nuclear magnetic resonance tomography, but the interaction with soft tissues especially is different for microwaves and the other methods. Therefore, supplementary information can be had from microwave tomography. A human patient is, however, not an easy target to measure because of the high attenuation and relatively high contrast in permittivity among different kinds of tissue.

Pichot and others from the Laboratoire des Signaux et Systemes (France) have demonstrated the feasibility of the microwave method with measurements on isolated organs [Pichot *et al.*, 1985; Peronnet *et al.*, 1983]. To reduce the reflections at the surface, they performed the measurements with the sample immersed in water. Measuring at 3 GHz, they achieved a resolution of 6 mm, which is equal to about $\lambda/2$ in the tissue, because soft tissues (fat excluded) have a high water content and therefore high permittivity. A sketch of the measurement system is shown in Figure 8.8. The sampling of the field was realized by using an array of dipoles on a printed circuit board. Each dipole was loaded with a *pin* diode. By biasing a diode with a low frequency signal, the field scattering from the corresponding dipole became marked with the modulation, and could be separated and selected from the signal received with the collector horn antenna behind the dipole matrix. Thus, scanning the whole aperture was possible in a very short time. If interfaced with a modern computer, such a system is capable of almost real-time tomographic imaging. Figure 8.9 shows some images obtained on a horse kidney. The relatively high resolution, due to high permittivity, is evident in the striking amount of detail shown in the images.

The status of medical applications of microwave tomography is the same as that of so many other uses of microwave sensors described in this book. The feasibility of the idea has been clearly demonstrated and there is no doubt about the usefulness of the sensor, but the path to a generally accepted practical utilization is long. This is particularly apt for applications of a technology that is unfamiliar even to most engineers.

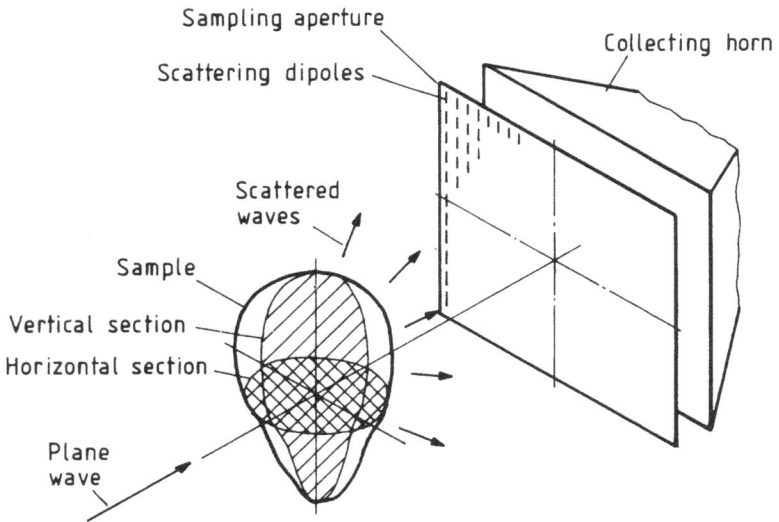

Figure 8.8 Schematic view of the 3 GHz diffraction tomography measurement setup. (After [Peronnet *et al.*, 1983].)

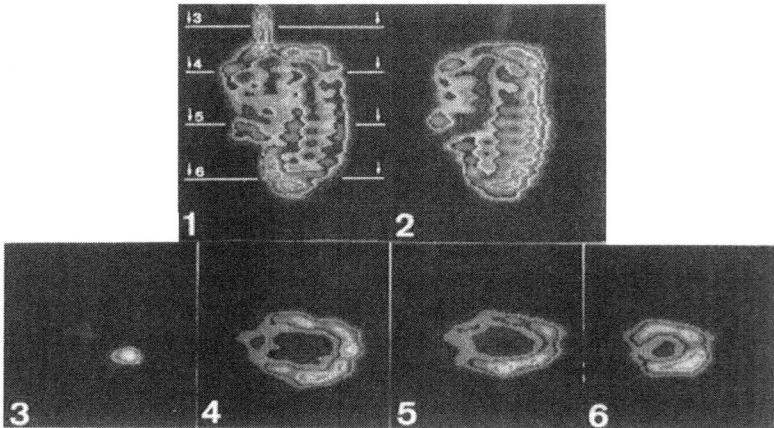

Figure 8.9 Tomographic cross-sectional images of a horse kidney, obtained with the system of Figure 8.8, except that only a linear array (one row of dipoles) was used. Image 1 shows the perfused kidney and image 2 shows the nonperfused kidney. Images 3–6 are cross-sectional views as indicated in image 1. They are calculated from eight different projections obtained by rotating the kidney. (Courtesy of Professor Bolomey, Laboratoire des Signaus et Systémes, Gif-sur-Yvette, France.)

REFERENCES

Anderson, A.P., and S.J. Mawani, "A 'Microwave Eye' Can Be Almost Human," *Proc. 6th European Microwave Conf.*, Rome, September 1976, pp. 105–111.

Arnaud, J.A., *Beam and Fiber Optics*, New York: Academic Press, 1976, 447 p.

Baribaud, M., and M.K. Nguyen, "Maximum Entropy Image Reconstruction from Microwave Scattered Field Distribution," *Proc. 18th European Microwave Conf.*, Stockholm, September 1988, pp. 891–896.

Bolomey, J.C., L. Jofre, and G. Peronnet, "On the Possible Use of Microwave Active Imaging for Remote Thermal Sensing," *IEEE Trans. Microwave Theory and Tech.*, Vol. MTT-31, No. 9, September 1983, pp. 777–781.

Broquetas, A., M. Ferrando, J.M. Rius, L. Jofre, E. de los Reyes, A. Cardama, A. Elias, and J. Ibañez, "Temperature and Permittivity Measurements Using a Cylindrical Microwave Imaging System," *Proc. 17th European Microwave Conf.*, Rome, September 1987, pp. 892–895.

Caorsi, S., G.L. Gragnani, and M. Pastarino, "A Numerical Approach to Microwave Imaging," *Proc. 18th European Microwave Conf.*, Stockholm, September 1988, pp. 897–902.

Chaloupka, H., "Imaging of Objects in a Halfspace with Unknown Permittivity," *Electronics Letters*, Vol. 20, No. 13, June 1984, pp. 570–572.

Chu, T.H., and K.Y. Lee, "Wideband Microwave Diffraction Tomography under Born Approximation," *IEEE AP-S Int. Symp. Digest*, Blacksburg, VA, June 1987, pp. 1042–1045.

Cornbleet, S., *Microwave Optics*, Pure and Applied Physics Series, Vol. 41, London: Academic Press, 1976, 416 p.

Edrich, J., "Centimeter- and Millimeter-Wave Thermography—A Survey on Tumor Detection," *J. Microwave Power*, Vol. 14, No. 2, June 1979, pp. 95–104.

Ermert, H., G. Füller, and D. Hiller, "Microwave Computerized Tomography," *Proc. 11th European Microwave Conf.*, Amsterdam, September 1981, pp. 421–426.

Goldsmith, P.F., "Quasi-Optical Techniques at Millimeter and Submillimeter Wavelengths," *Infrared and Millimeter Waves*, Vol. 6, New York: Academic Press, 1982, pp. 277–343.

Jofre, L., E. de los Reyes, M. Ferrando, A. Elias, J. Romeu, M. Baquero, and J.M. Rius, "A Cylindrical System for Quasi-Real Time Microwave Tomography," *Proc. 16th European Microwave Conf.*, Dublin, September 1986, pp. 599–604.

Junkin, G., and A.P. Anderson, "Limitations in Microwave Holographic Synthetic Aperture Imaging over a Lossy Half-Space," *IEE Proc.*, Vol. 135, Pt. F., No. 4, August 1988, pp. 321–329.

Kraus, J.D., *Radio Astronomy*, 2nd Ed., Powell, OH: Cygnus-Quasar Books, 1986, 690 p.

Mensa, D.L., S. Halevy, and G. Wade, "Coherent Doppler Tomography for Microwave Imaging," *Proc. IEEE*, Vol. 71, No. 2, February 1983, pp. 254–261.

Michiguchi, Y., K. Hiramoto, M. Nishi, T. Ootaka, and M. Okada, "Advanced Subsurface Radar System for Imaging Buried Pipes," *IEEE Trans. Geosci. Remote Sensing*, Vol. GE-26, No. 6, November 1988, pp. 733–740.

Orme, R.D., and A.P. Anderson, "High Resolution Microwave Holographic Technique," *IEE Proc.*, Vol. 120, No. 4, April 1973, pp. 401–406.

Paoloni, F.J., "Implementation of Microwave Diffraction Tomography for Measurement of Dielectric Constant Distribution," *IEE Proc.*, Vol. 134, Pt. H, No. 1, February 1987, pp. 25–29.

Paoloni, F.J., and M.J. Duffy, "Microwave Imaging with Digital Reconstructions," *J. Electrical and Electronics Eng.* (Australia), Vol. 4, No. 1, March 1984, pp. 54–59.

Peronnet, G., C. Pichot, J.C. Bolomey, L. Jofre, A. Izadnegahdar, C. Szeles, Y. Michel, J.L. Guerquin-Kern, and M. Gautherie, "A Microwave Diffraction Tomography System for Biological Applications," *Proc. 13th European Microwave Conf.*, Nürnberg, September 1983, pp. 529–533.

Pichot, C., L. Jofre, G. Peronnet, and J.C. Bolomey, "Active Microwave Imaging of Imhomogeneous Bodies," *IEEE Trans. Antennas Propag.*, Vol. AP-33, No. 4, April 1985, pp. 416–425.

Richards, P.J., and A.P. Anderson, "Microwave Images of Sub-Surface Utilities in an Urban Environment," *Proc. 8th European Microwave Conf.*, Paris, September 1978, pp. 33–37.

Rius, J.M., M. Ferrando, L. Jofre, E. de los Reyes, A. Elias, and A. Broquetas, "Microwave Tomography: An Algorithm for Cylindrical Geometries," *Electronics Letters*, Vol. 23, No. 11, May 1987, pp. 564–565.

Schultz, K.I., and D.L. Jaggard, "Novel Microwave Projection Imaging for Determination of Internal Structure," *Electronics Letters*, Vol. 23, No. 6, March 1987, pp. 267–269.

Solymar, L., "On the Theory and Applications of Volume Holograms," *Proc. 7th European Microwave Conf.*, Copenhagen, September 1977, pp. 175–176.

Tricoles, G., and N.H. Farhat, "Microwave Holography: Applications and Techniques," *Proc. IEEE*, Vol. 65, No. 1, January 1977, pp. 108–121.

Chapter 9
Appendices

9.1 TRANSMISSION LINES

9.1.1 Basic Equations

Input impedance of a line (characteristic impedance $= Z_0$) at distance x from a terminating impedance Z_l:

$$Z_{in}(x) = Z_0 \cdot \frac{Z_l + Z_0 \tanh \gamma x}{Z_0 + Z_c \tanh \gamma x} \tag{9.1}$$

Voltage along line:

$$V(x) = V_0[1 + \Gamma_l \exp(-2\gamma x)] \tag{9.2}$$

Current along line:

$$I(x) = I_0[1 - \Gamma_l \exp(-2\gamma x)] \tag{9.3}$$

Voltage reflection coefficient of the termination, or an interface where the impedance changes from Z_0 to Z_l:

$$\Gamma_l = \frac{Z_l - Z_0}{Z_l + Z_0} \tag{9.4}$$

Voltage standing wave ratio (VSWR):

$$VSWR = \frac{V_{max}}{V_{min}} = \frac{1 + |\Gamma_l|}{1 - |\Gamma_l|} \tag{9.5}$$

9.1.2 TEM Lines

Phase factor:

$$\beta = \frac{\omega \sqrt{\varepsilon_r'}}{c_0} \quad (\varepsilon_r' \gg \varepsilon_r'')$$ (9.6)

Loss factor due to dielectric losses of the insulator:

$$\alpha_d = \frac{\omega}{c_0} \cdot \frac{\varepsilon_r''}{2 \sqrt{\varepsilon_r'}}$$ (9.7)

Coaxial Cable

Electric field (*a* and *b* are the radii of the outer and inner conductors, and $Z_w = \sqrt{\mu_r/\varepsilon_r}$ is the plane wave impedance):

$$E_r = \frac{V_0}{r \log_e(b/a)}$$ (9.8)

Magnetic field:

$$H_\phi = \frac{V_0}{Z_w r \log_e(b/a)}$$ (9.9)

Transmission line characteristic impedance:

$$Z_0 = \frac{Z_w}{2\pi} \log_e(b/a) \approx \frac{60\ \Omega}{\sqrt{\varepsilon_r'}} \log_e(b/a)$$ (9.10)

Loss factor due to finite conductivity σ of the conductors:

$$\alpha_c = \frac{1}{Z_0} \left(\frac{\omega\mu_0}{8\sigma}\right)^{1/2} \left(\frac{1}{a} + \frac{1}{b}\right)$$ (9.11)

Total loss factor:

$$\alpha = \alpha_c + \alpha_d$$ (9.12)

Two-Conductor Cable

The distance between the centers of the conductors is D and the radius of the conductors is a:

$$Z_0 = \frac{Z_w}{\pi \cosh(D/2a)} \approx \frac{120 \ \Omega}{\sqrt{\varepsilon_r'} \cosh(D/2a)} \qquad (9.13)$$

$$\alpha_c = \frac{2}{aZ} \left(\frac{\omega\mu_0}{8\sigma} \right)^{1/2} \frac{D/2a}{[(D/2a)^2 - 1]^{1/2}} \qquad (9.14)$$

Analogy between TEM Lines and Plane Waves

The real parts of the propagation factor of a plane wave and TEM line are equal. The imaginary part of the propagation factor of the plane wave is equal to the loss factor due to dielectric losses of the TEM line. The field reflection coefficient with perpendicular incidence can be calculated from (9.4) by replacing the line and termination impedances with the respective wave impedances.

9.1.3 Waveguides

Rectangular Waveguides

Circular Waveguides

Table 9.1
Properties of Modes in Rectangular Waveguides ($\varepsilon_{0i} = 1$ for $i = 0$ and 2 for $i > 0$)

	TE Modes	TM Modes
H_z	$H_{0,nm} \cos \dfrac{n\pi x}{a} \cos \dfrac{m\pi y}{b} \exp\{-j\beta_{nm}z\}$	0
E_z	0	$E_{0,nm} \sin \dfrac{n\pi x}{a} \cos \dfrac{m\pi y}{b} \exp\{-j\beta_{nm}z\}$
E_x	$Z_{h,nm}H_y$	$\dfrac{-j\beta_{nm}n\pi}{ak_{c,nm}^2} E_{0,nm} \cos \dfrac{n\pi x}{a} \sin \dfrac{m\pi y}{b} \exp\{-j\beta_{nm}z\}$
E_y	$-z_{h,nm}H_z$	$\dfrac{-j\beta_{nm}m\pi}{bk_{c,nm}^2} E_{0,nm} \sin \dfrac{n\pi x}{a} \cos \dfrac{m\pi y}{b} \exp\{-j\beta_{nm}z\}$
H_x	$\dfrac{j\beta_{nm}n\pi}{ak_{c,nm}^2} H_{0,nm} \sin \dfrac{n\pi x}{a} \cos \dfrac{m\pi y}{b} \exp\{-j\beta_{nm}z\}$	$-\dfrac{E_y}{Z_{e,nm}}$

Table 9.1

	TE Modes	TM Modes
H_y	$\dfrac{j\beta_{nm}m\pi}{bk_{c,nm}^2} H_{0,nm} \cos\dfrac{n\pi x}{a} \sin\dfrac{m\pi y}{b} \exp\{-j\beta_{nm}z\}$	$\dfrac{E_x}{Z_{e,nm}}$
$Z_{h,nm}$	$\dfrac{k}{\beta_{nm}} Z_w \ (k = \omega\sqrt{\mu\varepsilon})$	
$Z_{e,nm}$		$\dfrac{\beta_{nm}}{k} Z_w$
$k_{c,nm}$		$\left[\left(\dfrac{n\pi}{a}\right)^2 + \left(\dfrac{m\pi}{b}\right)^2\right]^{1/2}$
β_{nm}		$(k^2 - k_{c,nm}^2)^{1/2}$
$\lambda_{c,nm}$		$\dfrac{2ab}{(n^2b^2 + m^2a^2)^{1/2}}$
Power	$\dfrac{|H_{0,nm}|^2\, ab}{2\,\varepsilon_{0n}\varepsilon_{0m}} \left(\dfrac{\beta_{nm}}{k_{c,nm}}\right)^2 Z_{h,nm}$	$\dfrac{|E_{0,nm}|^2 ab}{8} \left(\dfrac{\beta_{nm}}{k_{c,nm}}\right)^2 \dfrac{1}{Z_{e,nm}}$
$\alpha_{c,nm}$	$\dfrac{2R_m}{bZ_w\,(1 - (k_{c,nm}/k)^2)^{1/2}}$ $\cdot \left[\left(1 + \dfrac{b}{a}\right)\dfrac{k_{c,nm}^2}{k^2} + \dfrac{b}{a}\left(\dfrac{\varepsilon_{0m}}{2} - \dfrac{k_{c,nm}^2}{k^2}\right)\dfrac{n^2ab + m^2a^2}{n^2b^2 + m^2a^2}\right]$	$\dfrac{2R_m}{bZ_w\,(1 - (k_{c,nm}/k)^2)^{1/2}} \dfrac{n^2b^3 + m^2a^3}{n^2b^2a + m^2a^3}$

Table 9.2
Properties of Modes in Circular Waveguides
(See Tables 9.3 and 9.4 for p_{nm} and p'_{nm}; $\varepsilon_{0n} = 1$ for $n = 0$ and 2 for $n > 0$)

	TE Modes	TM Modes
H_z	$H_{0,nm} J_n\left(\dfrac{p'_{nm}r}{a}\right) \exp\{-j\beta_{nm}z\} \begin{Bmatrix} \cos n\phi \\ \sin n\phi \end{Bmatrix}$	0
E_z	0	$E_{0,nm} J_n\left(\dfrac{p_{nm}r}{a}\right) \exp\{-j\beta_{nm}z\} \begin{Bmatrix} \cos n\phi \\ \sin n\phi \end{Bmatrix}$
E_r	$Z_{h,nm}H_\phi$	$\dfrac{-j\beta_{nm}p_{nm}}{ak_{c,nm}^2} E_{0,nm} J'_n\left(\dfrac{p_{nm}r}{a}\right) \exp\{-j\beta_{nm}z\} \begin{Bmatrix} \cos n\phi \\ \sin n\phi \end{Bmatrix}$
E_ϕ	$-Z_{h,nm}H_r$	$\dfrac{-jn\beta_{nm}}{rk_{c,nm}^2} E_{0,nm} J_n\left(\dfrac{p_{nm}r}{a}\right) \exp\{-j\beta_{nm}z\} \begin{Bmatrix} -\sin n\phi \\ \cos n\phi \end{Bmatrix}$
H_r	$\dfrac{-j\beta_{nm}p'_{nm}}{ak_{c,nm}^2} H_{0,nm} J'_n\left(\dfrac{p'_{nm}r}{a}\right) \exp\{-j\beta_{nm}z\} \begin{Bmatrix} \cos n\phi \\ \sin n\phi \end{Bmatrix}$	$-\dfrac{E_\phi}{Z_{e,nm}}$
H_ϕ	$\dfrac{-jn\beta_{nm}}{rk_{c,nm}^2} H_{0,nm} J_n\left(\dfrac{p'_{nm}r}{a}\right) \exp\{-j\beta_{nm}z\} \begin{Bmatrix} -\sin n\phi \\ \cos n\phi \end{Bmatrix}$	$\dfrac{E_r}{Z_{e,nm}}$

Table 9.2

	TE Modes	TM Modes
$Z_{h,nm}$	$\dfrac{k}{\beta_{nm}} Z_w \ (k = \omega \sqrt{\mu\varepsilon})$	
$Z_{e,nm}$		$\dfrac{\beta_{nm}}{k} Z_w$
$k_{c,nm}$	$\dfrac{p'_{nm}}{a}$	$\dfrac{p_{nm}}{a}$
β_{nm}	$\left[k^2 - \left(\dfrac{p'_{nm}}{a} \right)^2 \right]^{1/2}$	$\left[k^2 - \left(\dfrac{p_{nm}}{a} \right)^2 \right]^{1/2}$
$\lambda_{c,nm}$	$\dfrac{2\pi a}{p'_{nm}}$	$\dfrac{2\pi a}{p_{nm}}$
Power	$|H_{0,nm}|^2 \dfrac{Z_w k \beta_{nm} \pi}{4k_{c,nm}^4} (p_{nm}'^2 - n^2) J_n^2 (p'_{nm}) \varepsilon_{0n}$	$|E_{0,nm}|^2 \dfrac{k \beta_{nm} \pi}{Z_w 4 k_{c,nm}^4} p_{nm}^2 \, [J'_n(k_{c,nm}a)]^2 \, \varepsilon_{0n}$
$\alpha_{c,nm}$	$\dfrac{R_m}{aZ_w} \left(1 - \dfrac{k_{c,nm}^2}{k^2} \right)^{-1/2} \cdot \left[\dfrac{k_{c,nm}^2}{k^2} + \dfrac{n^2}{(p'_{nm})^2 - n^2} \right]$	$\dfrac{R_m}{aZ_w} \left(1 - \dfrac{k_{c,nm}^2}{k^2} \right)^{-1/2}$

9.2 RESONANT CAVITIES

9.2.1 Rectangular Cavities

Resonant frequency of TE_{nml} or TM_{nml} mode (cavity size $= a \times b \times d$):

$$f_{r,nml} = \frac{1}{2\sqrt{\mu\varepsilon}} \left[\left(\frac{n}{a} \right)^2 + \left(\frac{m}{b} \right)^2 + \left(\frac{l}{d} \right)^2 \right]^{1/2} \tag{9.15}$$

Fields of TE_{nml} mode in a lossless cavity:

$$\bar{\mathbf{E}} = j\omega\mu A_{nm} \left(-\bar{\mathbf{u}}_x \frac{n\pi}{b} \cos\frac{n\pi x}{a} \sin\frac{m\pi y}{b} \sin\frac{l\pi z}{d} \right.$$

$$\left. + \bar{\mathbf{u}}_y \frac{m\pi}{a} \sin\frac{n\pi x}{a} \cos\frac{m\pi y}{b} \sin\frac{l\pi z}{d} \right) \tag{9.16}$$

$$\bar{\mathbf{H}} = A_{nm} \left\{ \bar{\mathbf{u}}_x \frac{l\pi}{d} \frac{m\pi}{a} \sin\frac{n\pi x}{a} \cos\frac{m\pi y}{b} \cos\frac{l\pi z}{d} \right.$$

$$+ \bar{\mathbf{u}}_y \frac{l\pi}{d}\frac{n\pi}{b} \cos \frac{n\pi x}{a} \sin \frac{m\pi y}{b} \cos \frac{l\pi z}{d}$$

$$\left. - \bar{\mathbf{u}}_z \left[\left(\frac{n\pi}{a}\right)^2 + \left(\frac{m\pi}{b}\right)^2 \right] \cos \frac{n\pi x}{a} \cos \frac{m\pi y}{b} \cos \frac{l\pi z}{d} \right\} \tag{9.17}$$

Fields of TM_{nml} mode in a lossless cavity:

$$\bar{\mathbf{E}} = B_{nm} \left\{ - \bar{\mathbf{u}}_x \frac{l\pi}{d}\frac{n\pi}{a} \cos \frac{n\pi x}{a} \sin \frac{m\pi y}{b} \sin \frac{l\pi z}{d} \right.$$

$$+ \bar{\mathbf{u}}_y \frac{l\pi}{d}\frac{m\pi}{b} \sin \frac{n\pi x}{a} \cos \frac{m\pi y}{b} \sin \frac{l\pi z}{d}$$

$$\left. - \bar{\mathbf{u}}_z \left[\left(\frac{n\pi}{a}\right)^2 + \left(\frac{m\pi}{b}\right)^2 \right] \sin \frac{n\pi x}{a} \sin \frac{m\pi y}{b} \cos \frac{l\pi z}{d} \right\} \tag{9.18}$$

$$\bar{\mathbf{H}} = j\omega\varepsilon B_{nm} \left(\bar{\mathbf{u}}_x \frac{m\pi}{b} \sin \frac{n\pi x}{a} \cos \frac{m\pi y}{b} \cos \frac{l\pi z}{d} \right.$$

$$\left. - \bar{\mathbf{u}}_y \frac{n\pi}{a} \cos \frac{n\pi x}{a} \sin \frac{m\pi y}{b} \cos \frac{l\pi z}{d} \right) \tag{9.19}$$

9.2.2 Circular Cavities

Resonant frequency of TE_{nml} ($x_{nm} = p'_{nm}$, Table 9.3) or TM_{nml} ($x_{nm} = p_{nm}$, Table 9.4) mode (cavity radius a and length d):

$$f_{r,nml} = \frac{1}{2\sqrt{\mu\varepsilon}} \left[\left(\frac{x_{nm}}{\pi a}\right)^2 + \left(\frac{l}{d}\right)^2 \right]^{1/2} \tag{9.20}$$

Fields of TE_{nml} mode in a lossless cavity:

$$\bar{\mathbf{E}} = j\omega\mu A_{nm} \left[\bar{\mathbf{u}}_r \frac{n}{r} J_n\left(\frac{p'_{nm}}{a} r\right) \left\{ \begin{matrix} -\sin n\phi \\ \cos n\phi \end{matrix} \right\} \sin \frac{l\pi z}{d} \right.$$

$$\left. - \bar{\mathbf{u}}_\phi \frac{p'_{nm}}{a} J'_n\left(\frac{p'_{nm}}{a} r\right) \left\{ \begin{matrix} \cos n\phi \\ \sin n\phi \end{matrix} \right\} \sin \frac{l\pi z}{d} \right] \tag{9.21}$$

$$\bar{\mathbf{H}} = -A_{nm} \left[\bar{\mathbf{u}}_r \frac{l\pi}{d}\frac{p'_{nm}}{a} J'_n\left(\frac{p'_{nm}}{a} r\right) \left\{ \begin{matrix} \cos n\phi \\ \sin n\phi \end{matrix} \right\} \cos \frac{l\pi z}{d} \right.$$

$$+ \bar{u}_\phi \frac{l\pi}{d} \frac{n}{r} J_n \left(\frac{p'_{nm}}{a} r \right) \begin{Bmatrix} -\sin n\phi \\ \cos n\phi \end{Bmatrix} \cos \frac{l\pi z}{d}$$

$$+ \bar{u}_z \left(\frac{p'_{nm}}{a} \right)^2 J_n \left(\frac{p'_{nm}}{a} r \right) \begin{Bmatrix} \cos n\phi \\ \sin n\phi \end{Bmatrix} \sin \frac{l\pi z}{d} \Bigg] \tag{9.22}$$

Fields of TM_{nml} mode in a lossless cavity:

$$\bar{E} = B_{nm} \Bigg[-\bar{u}_r \frac{l\pi}{d} \frac{p_{nm}}{a} J_n' \left(\frac{p_{nm}}{a} r \right) \begin{Bmatrix} \cos n\phi \\ \sin n\phi \end{Bmatrix} \sin \frac{l\pi z}{d}$$

$$- \bar{u}_\phi \frac{l\pi}{d} \frac{n}{r} J_n \left(\frac{p_{nm}}{a} r \right) \begin{Bmatrix} -\sin n\phi \\ \cos n\phi \end{Bmatrix} \sin \frac{l\pi z}{d}$$

$$+ \bar{u}_z \left(\frac{p_{nm}}{a} \right)^2 J_n \left(\frac{p_{nm}}{a} r \right) \begin{Bmatrix} \cos n\phi \\ \sin n\phi \end{Bmatrix} \sin \frac{l\pi z}{d} \Bigg] \tag{9.23}$$

$$\bar{H} = j\omega\varepsilon B_{nm} \Bigg[\bar{u}_r \frac{n}{r} J_n \left(\frac{p_{nm}}{a} r \right) \begin{Bmatrix} -\sin n\phi \\ \cos n\phi \end{Bmatrix} \cos \frac{l\pi z}{d}$$

$$- \bar{u}_\phi \frac{p_{nm}}{a} J_n' \left(\frac{p_{nm}}{a} r \right) \begin{Bmatrix} \cos n\phi \\ \sin n\phi \end{Bmatrix} \cos \frac{l\pi z}{d} \Bigg] \tag{9.24}$$

Table 9.3
Values of p'_{nm} for TE Modes in Circular Waveguide (Values of p'_{nm} are zeros of the first derivative of the mth-order Bessel function of the first kind, see Section 9.3.)

n	p'_{n1}	p'_{n2}	p'_{n3}
0	3.832	7.016	10.174
1	1.841	5.331	8.536
2	3.054	6.706	9.970

Table 9.4
Values of p_{nm} for TM Modes in Circular Waveguide (Values of p_{nm} are zeros of the mth-order Bessel function of the first kind.)

n	p_{n1}	p_{n2}	p_{n3}
0	2.405	5.520	8.654
1	3.832	7.016	10.174
2	5.135	8.417	11.620

9.3 BESSEL FUNCTIONS

The mth-order (ordinary) Bessel functions of the first and second kinds are independent solutions of the Bessel differential equation:

$$x^2 \frac{d^2f}{dx} + x \frac{df}{dx} + (x^2 - m^2)f = 0 \qquad (9.25)$$

where $m^2 \geq 0$.

The series expansions of the solutions $J_m(x)$ (first kind) and $N_m(x)$ (second kind) are

$$J_m(x) = \sum_{n=0}^{\infty} \frac{(-1)^n (x/2)^{m+2n}}{n!(m+n)!} \qquad (9.26)$$

$$N_m(x) = \frac{2}{\pi} \left(\gamma + \log_e \frac{x}{2} \right) J_m(x) - \frac{1}{\pi} \sum_{n=0}^{m-1} \frac{(m-n-1)!}{n!} \left(\frac{2}{x} \right)^{m-2n}$$

$$- \frac{1}{\pi} \sum_{n=0}^{\infty} \frac{(-1)^m (x/2)^{m+2n}}{n!(m+n)!}$$

$$\cdot \left(1 + \frac{1}{2} + \frac{1}{3} + \ldots + \frac{1}{n} + 1 + \frac{1}{2} + \frac{1}{3} + \ldots + \frac{1}{m+n} \right) \qquad (9.27)$$

where $\gamma = 0.5772$ is Euler's constant. In some literature, Y_m is used instead of N_m. The four lowest-order functions are presented in Figure 9.1. For large values of x:

$$\lim_{x \to \infty} J_m(x) = \sqrt{\frac{2}{\pi x}} \cos \left(x - \frac{\pi}{4} - \frac{m\pi}{2} \right) \qquad (9.28)$$

$$\lim_{x \to \infty} N_m(x) = \sqrt{\frac{2}{\pi x}} \sin \left(x - \frac{\pi}{4} - \frac{m\pi}{2} \right) \qquad (9.29)$$

For small values of x:

$$J_m(x) \approx \frac{1}{m!} \left(\frac{x}{2} \right)^m \qquad (9.30)$$

$$N_0(x) \approx \frac{2}{\pi} \log_e x \qquad (9.31)$$

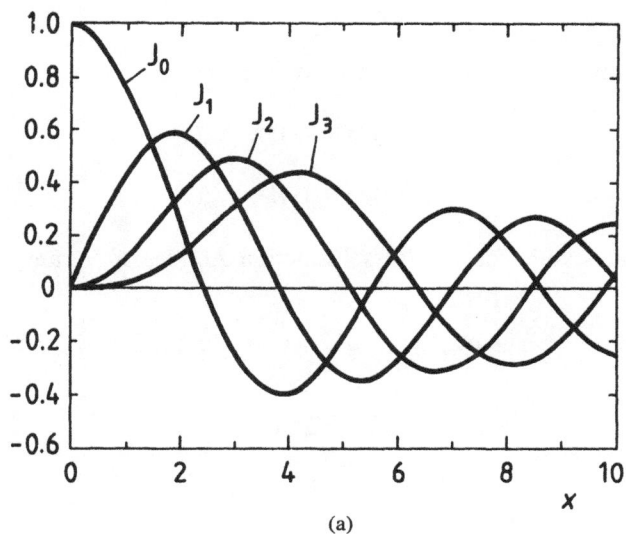

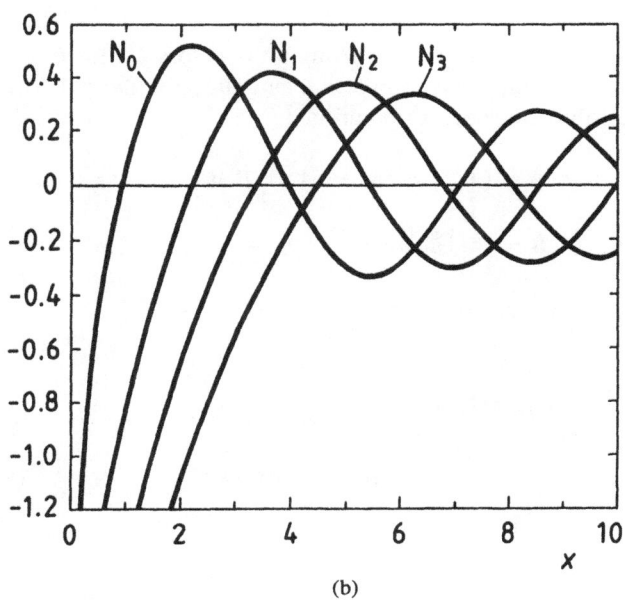

Figure 9.1 The four lowest-order Bessel functions: (a) of the first kind; (b) of the second kind.

$$N_m(x) \approx \frac{1}{\pi} (m - 1)! \left(\frac{x}{2}\right)^{-m} \tag{9.32}$$

Hankel functions of the first and second kinds are

$$H_m^{(1)}(x) = J_m(x) + jN_m(x) \tag{9.33}$$

$$H_m^{(2)}(x) = J_m(x) - jN_m(x) \tag{9.34}$$

Recurrence formulas valid for any Bessel function J_m, N_m, or H_m are

$$x \frac{dZ_m(x)}{dx} = mZ_m(x) - xZ_{m+1}(x) = -mZ_m(x) + xZ_{m-1}(x)$$

$$= \frac{x}{2} [Z_{m-1}(x) - Z_{m+1}(x)] \tag{9.35}$$

9.4 VECTOR RELATIONS

9.4.1 Basic Relations

For vectors $\bar{\mathbf{A}}$ and $\bar{\mathbf{B}}$ having components along unit vectors $\bar{\mathbf{u}}_1$, $\bar{\mathbf{u}}_2$, $\bar{\mathbf{u}}_3$ in a right-hand orthogonal coordinate system (for example, rectangular, circular cylindrical or spherical), we have the following vector operations:

$$\bar{\mathbf{A}} \pm \bar{\mathbf{B}} = (A_1 \pm B_1)\bar{\mathbf{u}}_1 + (A_2 \pm B_2)\bar{\mathbf{u}}_2 + (A_3 \pm B_3)\bar{\mathbf{u}}_3 \tag{9.36}$$

$$\bar{\mathbf{A}} \cdot \bar{\mathbf{B}} = |\bar{\mathbf{A}}| \, |\bar{\mathbf{B}}| \cos \theta = A_1B_1 + A_2B_2 + A_3B_3 \tag{9.37}$$

where θ is the angle between $\bar{\mathbf{A}}$ and $\bar{\mathbf{B}}$.

$$\bar{\mathbf{A}} \times \bar{\mathbf{B}} = -\bar{\mathbf{B}} \times \bar{\mathbf{A}}$$

$$= \bar{\mathbf{u}}_1(A_2B_3 - A_3B_2) + \bar{\mathbf{u}}_2(A_3B_1 - A_1B_3) + \bar{\mathbf{u}}_3(A_1B_2 - A_2B_1) \tag{9.38}$$

$$|\bar{\mathbf{A}} \times \bar{\mathbf{B}}| = |\bar{\mathbf{A}}| \, |\bar{\mathbf{B}}| \sin \theta \tag{9.39}$$

9.4.2 Differential del Operator

Rectangular Coordinates

(f is a scalar function):

$$\nabla f = \text{grad} f = \bar{\mathbf{u}}_x \frac{\partial f}{\partial x} + \bar{\mathbf{u}}_y \frac{\partial f}{\partial y} + \bar{\mathbf{u}}_z \frac{\partial f}{\partial z} \tag{9.40}$$

$$\nabla \cdot \bar{\mathbf{A}} = \text{div} \, \bar{\mathbf{A}} = \frac{\partial A_x}{\partial x} + \frac{\partial A_y}{\partial y} + \frac{\partial A_z}{\partial z} \tag{9.41}$$

$$\nabla \times \bar{\mathbf{A}} = \text{curl} \, \bar{\mathbf{A}}$$
$$= \bar{\mathbf{u}}_x\left(\frac{\partial A_z}{\partial y} - \frac{\partial A_y}{\partial z}\right) + \bar{\mathbf{u}}_y\left(\frac{\partial A_x}{\partial z} - \frac{\partial A_z}{\partial x}\right) + \bar{\mathbf{u}}_z\left(\frac{\partial A_y}{\partial x} - \frac{\partial A_x}{\partial y}\right) \tag{9.42}$$

$$\nabla^2 f = \frac{\partial^2 f}{\partial x^2} + \frac{\partial^2 f}{\partial y^2} + \frac{\partial^2 f}{\partial z^2} \tag{9.43}$$

$$\nabla^2 \bar{\mathbf{A}} = \bar{\mathbf{u}}_x \nabla^2 A_x + \bar{\mathbf{u}}_y \nabla^2 A_y + \bar{\mathbf{u}}_z \nabla^2 A_z \tag{9.44}$$

Circular Cylindrical Coordinates

$$\nabla f = \bar{\mathbf{u}}_r \frac{\partial f}{\partial r} + \bar{\mathbf{u}}_\phi \frac{1}{r} \frac{\partial f}{\partial \phi} + \bar{\mathbf{u}}_z \frac{\partial f}{\partial z} \tag{9.45}$$

$$\nabla \cdot \bar{\mathbf{A}} = \frac{1}{r} \frac{\partial}{\partial r}(rA_r) + \frac{1}{r} \frac{\partial A_\phi}{\partial \phi} + \frac{\partial A_z}{\partial z} \tag{9.46}$$

$$\nabla \times \bar{\mathbf{A}} = \bar{\mathbf{u}}_r\left(\frac{1}{r}\frac{\partial A_z}{\partial \phi} - \frac{\partial A_\phi}{\partial z}\right) + \bar{\mathbf{u}}_\phi\left(\frac{\partial A_r}{\partial z} - \frac{\partial A_z}{\partial r}\right)$$
$$+ \bar{\mathbf{u}}_z\left[\frac{1}{r}\frac{\partial(rA_\phi)}{\partial r} - \frac{1}{r}\frac{\partial A_r}{\partial \phi}\right] \tag{9.47}$$

$$\nabla^2 f = \frac{1}{r}\frac{\partial}{\partial r}\left(r\frac{\partial f}{\partial r}\right) + \frac{1}{r^2}\frac{\partial^2 f}{\partial \phi^2} + \frac{\partial^2 f}{\partial z^2} \tag{9.48}$$

$$\nabla^2 \bar{\mathbf{A}} = \nabla\nabla \cdot \bar{\mathbf{A}} - \nabla \times \nabla \times \bar{\mathbf{A}} \tag{9.49}$$

Spherical Coordinates

$$\nabla f = \bar{\mathbf{u}}_r \frac{\partial f}{\partial r} + \bar{\mathbf{u}}_\theta \frac{1}{r} \frac{\partial f}{\partial \theta} + \frac{\bar{\mathbf{u}}_\phi}{r \sin \theta} \frac{\partial f}{\partial \phi} \tag{9.50}$$

$$\nabla \cdot \bar{\mathbf{A}} = \frac{1}{r^2} \frac{\partial}{\partial r} (r^2 A_r) + \frac{1}{r \sin \theta} \frac{\partial}{\partial \theta} (\sin \theta \, A_\theta)$$

$$+ \frac{1}{r \sin \theta} \frac{\partial A_\phi}{\partial \phi} \tag{9.51}$$

$$\nabla \times \bar{\mathbf{A}} = \frac{\bar{\mathbf{u}}_r}{r \sin \theta} \left[\frac{\partial}{\partial \theta} (A_\phi \sin \phi) - \frac{\partial A_\theta}{\partial \phi} \right]$$

$$+ \frac{\bar{\mathbf{u}}_\theta}{r} \left[\frac{1}{\sin \theta} \frac{\partial A_r}{\partial \phi} - \frac{\partial}{\partial r} (rA_\phi) \right]$$

$$+ \frac{\bar{\mathbf{u}}_\phi}{r} \left[\frac{\partial}{\partial r} (rA_\theta) - \frac{\partial A_r}{\partial \theta} \right] \tag{9.52}$$

$$\nabla^2 f = \frac{1}{r^2} \frac{\partial}{\partial r} \left(r^2 \frac{\partial f}{\partial r} \right) + \frac{1}{r^2 \sin \theta} \frac{\partial}{\partial \theta} \left(\sin \theta \frac{\partial f}{\partial \theta} \right)$$

$$+ \frac{1}{r^2 \sin^2 \theta} \frac{\partial^2 f}{\partial \phi^2} \tag{9.53}$$

$$\nabla^2 \bar{\mathbf{A}} = \nabla \nabla \cdot \bar{\mathbf{A}} - \nabla \times \nabla \times \bar{\mathbf{A}} \tag{9.54}$$

9.4.3 Vector Identities

(*f* and *g* are scalar functions):

$$\nabla(fg) = g\nabla f + f\nabla g \tag{9.55}$$

$$\nabla \cdot (g\bar{\mathbf{A}}) = \bar{\mathbf{A}} \cdot \nabla g + g\nabla \cdot \bar{\mathbf{A}} \tag{9.56}$$

$$\nabla \cdot (\bar{\mathbf{A}} \times \bar{\mathbf{B}}) = (\nabla \times \bar{\mathbf{A}}) \cdot \bar{\mathbf{B}} - (\nabla \times \bar{\mathbf{B}}) \cdot \bar{\mathbf{A}} \tag{9.57}$$

$$\nabla \times (g\bar{\mathbf{A}}) = (\nabla g) \times \bar{\mathbf{A}} + g\nabla \times \bar{\mathbf{A}} \tag{9.58}$$

$$\nabla \times (\bar{\mathbf{A}} \times \bar{\mathbf{B}}) = \bar{\mathbf{A}}\nabla \cdot \bar{\mathbf{B}} - \bar{\mathbf{B}}\nabla \cdot \bar{\mathbf{A}} + (\bar{\mathbf{B}} \cdot \nabla)\bar{\mathbf{A}} - (\bar{\mathbf{A}} \cdot \nabla)\bar{\mathbf{B}} \tag{9.59}$$

$$\nabla(\bar{\mathbf{A}} \cdot \bar{\mathbf{B}}) = (\bar{\mathbf{A}} \cdot \nabla)\bar{\mathbf{B}} + (\bar{\mathbf{B}} \cdot \nabla)\bar{\mathbf{A}}$$
$$+ \bar{\mathbf{A}} \times (\nabla \times \bar{\mathbf{B}}) + \bar{\mathbf{B}} \times (\nabla \times \bar{\mathbf{A}}) \tag{9.60}$$

$$\nabla \cdot \nabla f = \nabla^2 f \tag{9.61}$$

$$\nabla \cdot \nabla \times \bar{\mathbf{A}} = 0 \tag{9.62}$$

$$\nabla \times \nabla f = 0 \tag{9.63}$$

$$\nabla \times \nabla \times \bar{\mathbf{A}} = \nabla \nabla \cdot \bar{\mathbf{A}} - \nabla^2 \bar{\mathbf{A}} \tag{9.64}$$

9.5 SIGNAL FLOW GRAPHS

9.5.1 Basic Elements

The change of a signal (field, voltage) on the path between two points, here called *nodes*, can be expressed by the value of a *branch* of the signal flow graph [Kuhn, 1963]. The signal at the end of a branch is the one at the input node multiplied by the value of the branch. The flow graphs of some basic elements of microwave signal paths are presented in Figure 9.2. The graph in Figure 9.2(a) presents either a section of a transmission line of length l or a material layer through which a plane wave propagates. In the latter case, γl is replaced by $jkd/\cos \theta$, where d is the thickness of the layer and θ is the direction angle of propagation. In the flow graph of an interface in Figure 9.2(b), the reflection coefficient Γ can be calculated from (1.25) for transmission lines and from (1.46) and (1.47) for plane waves (for perpendicular incidence, see Section 9.1.1).

9.5.2 Reduction of Flow Graphs

The following four rules can be used for simplification of flow graphs:

1. Serial branches can be combined to form one branch, the value of which is the product of the values of the original branches (Figure 9.3a).
2. Parallel branches can be combined to form one branch, the value of which is the sum of the values of the original branches (Figure 9.3b).
3. A feedback branch (one that begins and ends at the same node) with value d can be eliminated by dividing by $1 - d$ the values of all branches entering the node where the feedback branch exists (Figure 9.3c).
4. When a node is split into several new nodes, the resulting flow graph contains once and only once each combination of the input and output branches (ex-

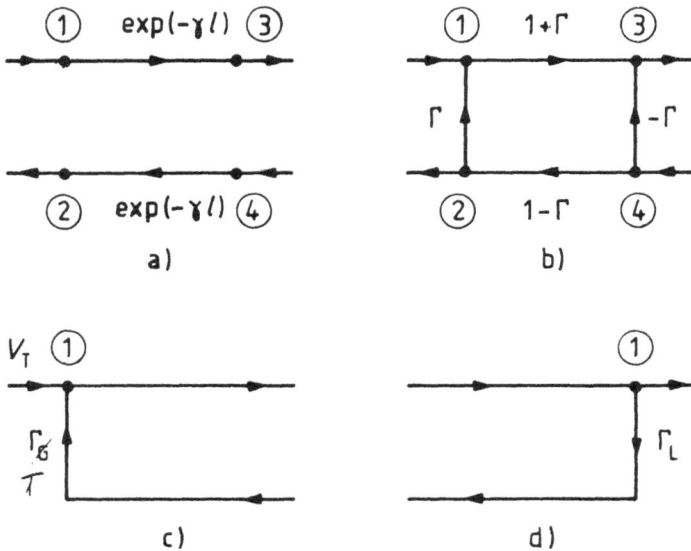

Figure 9.2 Signal flow graphs of some basic elements of microwave signal paths: (a) a transmission line with length l or a material layer (for a plane wave); (b) an interface, where the characteristic impedance of a transmission line or wave impedance of the media changes; (c) a signal transmitter (generator, antenna) with reflection coefficient Γ_T or the output of an "infinite" (lossy or long) transmission line or interface layer; (d) a signal receiver, detector, antenna terminating element (termination, sensor in reflection measurement) or the input of an "infinite" transmission line or material layer.

cluding feedback branches) connected to the original node. Possible feedback branches of the original node are connected to all the new nodes (Figure 9.3d).

9.5.3 Example: Measurement of Reflection Response of a Resonator

In Figure 9.4a, the flow graph is presented for a typical reflection measurement circuit of a resonator with tunable oscillator, circulator, long transmission line (length l), and a phase or power detector at the output. The circulator is assumed to have a forward transmission coefficient c and backward isolation level d, and c and d are complex and $|c| \gg |d|$. The transmission line is assumed to be lossless ($\alpha \approx 0$). When examining the output, we can exclude several, less significant branches, assuming that $|\Gamma_D|$, $|\Gamma_G|$, $|\Gamma_1|$, $|\Gamma_2|$, $|\Gamma_3| \ll 1$, (Figure 9.4b). The remaining flow graph can be simplified by using rules (1) and (4) of Section 9.5.2 to obtain the result in Figure 9.4(c), and further by using rules (1), (2) and (3) to obtain the final result in Figure 9.4(d):

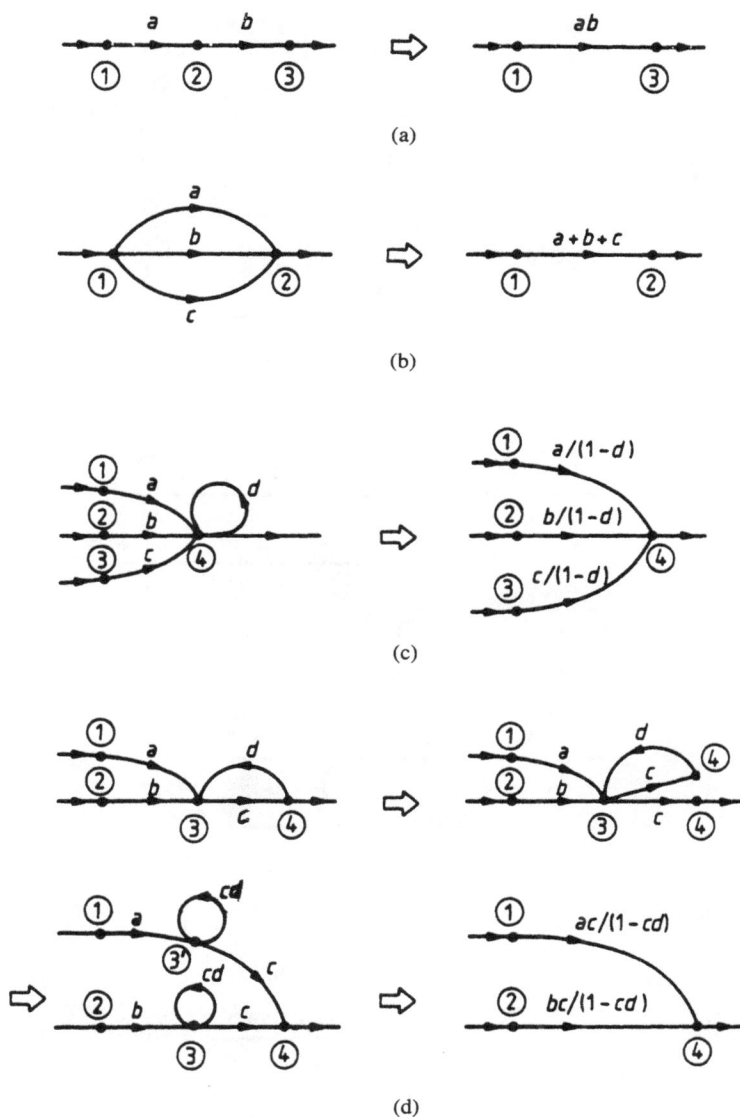

Figure 9.3 Reduction of signal flow graphs: (a) serial branches; (b) parallel branches; (c) feedback branch; (d) splitting of a node.

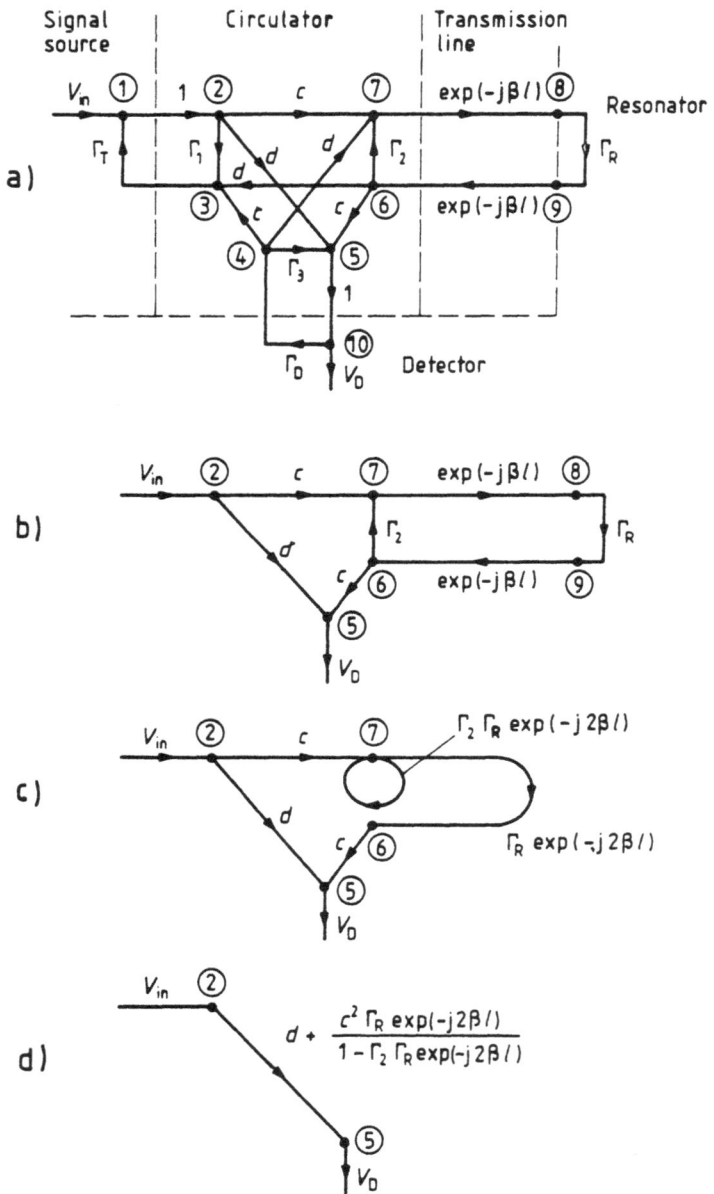

Figure 9.4 An example of a signal flow graph for a typical reflection measurement circuit of a resonator.

$$\frac{V_D}{V_{in}} \approx d + c^2\Gamma_R \exp(-j2\beta l) + c^2\Gamma_2\Gamma_R^2 \exp(-j4\beta l) \qquad (9.65)$$

From the result, we can see the effect of interfering signals—leakage and multiple reflections (first and third terms)—the phase of which changes compared to the desired signal (second term) as frequency (i.e., β) changes. Therefore, interferences cause periodic errors of the measured power (see Section 3.4.3). The effect of multiple reflections can be decreased by adding attenuation to the transmission line section, but then the effect of leakage increases.

REFERENCE

Kuhn, N., "Simplified Signal Flow Graph Analysis," *Microwave Journal,* Vol. 6, No. 11, November 1963, pp. 59–66.

List of Symbols

Symbol	Units	Definition	Section
$A, \|\bar{A}\|$		length of vector $\bar{A}$; $A = \sqrt{\bar{A} \cdot \bar{A}}$ (A may be complex)	
$\|A\|$		absolute value of (phasor) A; $\|A\| = \sqrt{AA^*}$	
$\text{Re}\{A\}$		real part of (phasor) A	
$\text{Im}\{A\}$		imaginary part of (phasor) A	
A	[m²]	area of capacitor plates	2.2.1
A	[1]	density-independent dielectric ratio	2.5.4
A	[dB]	peak insertion loss of a resonance	2.7.4
A	[m³/kg]	constant in the equation of the permittivity of air	2.6.1
A'	[Km²/N]	constant in the equation of the permittivity of air	2.6.1
A,B,C	[*]	results of an independent measurement	2.5.2
A_E	[dB]	field attenuation in decibels	1.4.1
A_e	[m²]	effective area of an antenna	7.2.3
A_{eff}	[m²]	effective area of capacitor plates	2.7.2
A_{max}	[dB]	a_{max} in decibels	2.7.4
A_{nm}	[A/m]	amplitude term of the fields of a TE_{nml} mode of a resonant cavity	9.2.1
A_P	[dB]	power attenuation	1.4.1
A_s	[m²]	area of a sample in a capacitor	2.7.2
A_t	[1]	density-independent dielectric ratio for transmission sensor	4.2.2
A_0	[dB]	peak insertion loss without sample	2.7.4
ΔA	[dB]	attenuation caused by the sample	4.2.2
δA	[dB]	periodic variation of attenuation in a sample caused by multiple reflections	4.2.2
$\bar{a}$	[*]	arbitrary vector	1.3.5
a	[1]	empirical constant in mixing formulas	2.4.2
a	[1]	insertion loss of a resonator	3.2.1
a	[m]	radius of a cylindrical cavity	3.3.2
a	[m]	radius of the aperture of a conical horn antenna	8.2.2
a	[m]	radius of the conductors of a two-conductor line	9.1.2
a	[m]	radius of the outer conductor of coaxial line	2.7.2
a	[m]	width of the broad wall in rectangular waveguide	1.3.4
a_r	[1]	insertion loss at the resonant frequency	3.2.1
a_{rd}	[1]	insertion loss at the resonant frequency with sample	3.4.3
a_{r0}	[1]	insertion loss at the resonant frequency without sample	3.4.3
$\bar{a}_1, \bar{a}_2$	[*]	constants in (8.7) and (8.8)	8.2.1
a_1, a_2, a_3	[m]	axes of an ellipsoid	2.4.3
Δa	[1]	attenuation (change of insertion loss) caused by a sample	4.2.2
B	[W/(Hz·m²·sr)]	brightness	7.2.1
$\bar{B}$	[Vs/m²]	magnetic flux density	1.3.1
B	[m³/kg]	constant in the equation of the permittivity of air	2.6.1

Symbol	Units	Definition	Section
B'	[Km^3/kg]	constant in the equation of the permittivity of air	2.6.1
B_{bb}	[W/(Hz·m^2·sr)]	brightness of a black body	7.2.1
B_{hp}	[Hz]	half-power bandwidth of resonance	3.2.1
B_{hp0}	[Hz]	half-power bandwidth without sample	2.7.4
B_{nm}	[V/m]	amplitude term of the TM$_{nml}$ mode in a resonant cavity	9.2.1
B_0	[S]	susceptance of the field outside an open-ended transmission line sensor	6.2.1
δB_{hp}	[Hz]	error of B_{hp} caused by power measurement error	3.4.3
$(\delta B_{hp})_s$	[Hz]	error of B_{hp} caused by systematic power measurement error	3.4.3
b	[1]	empirical constant for structure-independent calibration factor	2.5.4
b	[m]	radius of the inner conductor of coaxial line	2.7.2
b	[m]	width of the narrow wall of rectangular waveguide	1.3.4
C	[F]	capacitance	2.2.8
C	[Km^3/kg]	constant in the equation of the permittivity of air	2.6.1
C'	[m^3/kg]	constant in the equation of the permittivity of air	2.6.1
C_f	[F]	fringe capacitance	2.7.2
C_{opt}	[F]	optimum end capacitance in permittivity measurements	2.7.4
C_T	[F]	total end capacitance of an open-ended transmission line sensor	6.2.1
C_0	[F]	capacitance without sample	2.7.2
$C(\varepsilon_r)$	[1]	connection between ε_r and perturbations of f_r and Q	2.7.2
c	[m/s]	speed of propagation of electromagnetic wave	1.2
c_0	[m/s]	speed of propagation of electromagnetic wave in vacuum	1.2
D	[m]	distance between the centers of the conductors of a two-conductor line	9.1.2
$\bar{D}$	[C/m^2]	electric flux density	1.3.1
$D, D_\perp$	[1]	reflectivities of a sample	8.2.2
D_c	[m]	diameter of the concave mirror of quasioptical resonator	2.7.2
D_f	[m]	diameter of the flat mirror of quasioptical resonator	2.7.2
D_s	[m]	diameter of the sample in a quasioptical resonator	2.7.2
$D(\theta,\phi)$	[1]	directivity of an antenna	1.3.3
d	[1]	derivative	
d	[m]	diameter of the sample	5.2.7
d	[m]	distance of capacitor plates	2.2.8
d	[m]	length of a cavity resonator	3.3.2
d	[m]	thickness	3.3.4
d_h	[m]	diameter of a helix	2.7.4
d_{minA}	[m]	minimum sample thickness for which attenuation in a sample is monotonic	4.2.2
d_o	[m]	diameter of the outer conductor of a helical resonator	2.7.4
d_s	[m]	diameter of the sample	2.7.4
d_0	[m]	measurement range in holography	8.2.2
d_1, d_2	[m]	distances from a dielectric lens	8.2.2
d_1, d_2	[m]	thicknesses of layers in (2.39)	2.2.8

Symbol	Units	Definition	Section
$d_\perp$	[m]	maximum transverse dimension of the sample in holography	8.2.2
$\hat{\mathbf{E}}$	[V/m]	electric field strength	1.3.1
$\hat{\mathbf{E}}_a$	[V/m]	electric field strength of a coupling aperture	3.3.1
E_a	[V/m]	field applied to an inclusion in a mixture	2.4.3
E_c	[J]	energy content of a capacitor	3.1.1
$\hat{\mathbf{E}}_e$	[V/m]	external electric field strength	2.2.1
$\hat{\mathbf{E}}_i$	[V/m]	internal electric field strength	2.2.1
E_L	[J]	energy content of an inductor	3.1.1
$\hat{\mathbf{E}}_l$	[V/m]	local internal electric field strength	2.2.1
E_{n1},E_{n2}	[V/m]	normal components of electric field strength	2.2.1
$\hat{\mathbf{E}}_p$	[V/m]	electric field strength of a coupling probe	3.3.1
E_q	[J]	energy of electromagnetic quantum	1.2
E_r	[V/m]	reference beam in optical holography	8.2.1
$\hat{\mathbf{E}}_r$	[V/m]	electric field strength in a resonator	3.3.1
E_s	[V/m]	light scattered from a sample in optical holography	8.2.1
$\hat{\mathbf{E}}_0$	[V/m]	electric field amplitude	1.3.2
E_{t1},E_{t2}	[V/m]	tangential components of electric field strength	2.2.1
$E_{0,nm}$	[V/m]	amplitude of the z component of TM_{nm} mode in a waveguide	9.1.3
e	[C]	charge of electron	2.2.4
e	[1]	Euler's constant ≈ 2.718	1.3.4
F	[N]	force	2.2.6
F	[1]	noise figure of a circuit element	7.2.5
f	[Hz]	frequency	1.2
f_c	[Hz]	center frequency of the spectrum of impulse radar	6.3.2
$f_{c,nm}$	[Hz]	cut-off frequency of wave mode nm	1.3.4
f_D	[Hz]	doppler frequency	6.3.2
f_i	[1]	volume fraction of ith constituent or inclusions in a mixture	2.4.2
f_{IF}	[Hz]	intermediate frequency	6.3.2
f_{in}	[Hz]	input frequency of a frequency counter	3.4.2
f_{lhp},f_{uhp}	[Hz]	frequencies of the lower and upper half-power points of a resonance curve	3.4.3
f_{LO}	[Hz]	frequency of the local oscillator	3.4.2
f_{max}	[Hz]	frequency at a maximum of reflection curve	2.7.4
f_r	[Hz]	resonant frequency	3.1.1
f_{rec}	[Hz]	received frequency in the detection of fatigue cracks	5.2.4
f_{rel}	[Hz]	relaxation frequency	2.2.7
f_{r0}	[Hz]	resonant frequency without sample	3.1.1
f_w	[1]	volume fraction of liquid water	2.5.4
f_1,f_2	[Hz]	sweep limits of an FMCW radar	6.3.2
f_1,f_2,f_3	[*]	functions of multivariable measurement	2.5.2

Symbol	Units	Definition	Section
f_∞	[Hz]	resonant frequency of a stripline resonator with infinite suspension height	3.3.4
$f_\perp$	[Hz]	resonant frequency of perpendicular mode	3.2.2
$f_\parallel$	[Hz]	resonant frequency of tangential mode	3.2.2
Δf	[Hz]	bandwidth of radar pulses	6.3.2
Δf_r	[Hz]	change of resonant frequency	3.2.2
$\Delta f_\perp$	[Hz]	change of resonant frequency of perpendicular mode	2.5.4
$\Delta f_\parallel$	[Hz]	change of resonant frequency of tangential mode	2.5.4
δf	[Hz]	frequency measurement error caused by power measurement error	3.4.3
δf	[Hz]	sampling interval in the FM method of holography	8.2.2
δf_D	[Hz]	discretization error of frequency counting	3.4.2
δf_{min}	[Hz]	frequency difference of minima of reflection curve	2.7.4
$(\delta f)_n$	[Hz]	δf caused by noise	3.4.3
$(\delta f_r)_s$	[Hz]	error of f_r caused by systematic power measurement error	3.4.3
G	[1]	gain of an amplifier	3.4.3
G, G_0	[s]	radiation conductance of an open-ended transmission line sensor with and without sample	6.2.1
G_m	[1]	maximum gain of an antenna	7.2.3
$G(\theta, \phi)$	[1]	gain of an antenna	1.3.5
$\bar{\mathbf{H}}$	[A/m]	magnetic field strength	1.3.1
$\bar{\mathbf{H}}_a$	[A/m]	magnetic field strength of a coupling aperture	3.3.1
$\bar{\mathbf{H}}_l$	[A/m]	magnetic field strength of a coupling loop	3.3.1
$\bar{\mathbf{H}}_r$	[A/m]	magnetic field strength of a resonator	3.3.1
$H_{0,nm}$	[A/m]	amplitude of the z component of TE_{nm} mode in a waveguide	9.1.3
h	[m]	height of the sample	5.2.7
h	[Js]	Planck's constant	1.2
h	[m]	suspension height of a stripline resonator	3.3.4
h	[m]	thickness of the substrate of microstrip line	3.3.4
I	[W/m^2]	intensity	1.3.3
I_h	[W/m^2]	intensity of radiation in holography	8.2.1
$I(x)$	[A]	current along a transmission line	9.1.1
$\bar{\mathbf{J}}$	[A/m^2]	current density	1.3.1
$\bar{\mathbf{J}}_{ms}$	[V/m]	magnetic surface current density	1.3.5
$J_m(x)$	[1]	mth-order Bessel function of the first kind	9.3
$\bar{\mathbf{J}}_s$	[A/m]	surface current density	1.3.4
$\bar{\mathbf{J}}_\perp$	[A/m]	transverse component of surface current	8.2.2
j	[1]	imaginary unit	1.3.2
K	[1]	constant depending on the type of the radiometer	7.2.5
K	[1]	correction factor for helical resonator	2.7.4
k	[J/K]	Boltzmann's constant	2.2.6
k	[1/m]	propagation factor = wavenumber	1.3.2

Symbol	Units	Definition	Section
k	[1]	proportionality factor (<1)	3.4.3
$\bar{k}$	[1/m]	wave vector	1.3.2
k'	[1/m]	real part of propagation factor	1.3.2
k''	[1/m]	imaginary part of propagation factor	1.3.2
$\bar{k}',\bar{k}''$	[1/m]	real and imaginary parts of the wave vector	1.4.2
$k_{c,nm}$	[1/m]	cut-off propagation factor of wave mode nm	1.3.4
k_s,k_s',k_s''	[1/m]	propagation factor and its real and imaginary parts in a sample	4.2.2
k_0	[1/m]	propagation factor in vacuum	1.4.1
k_{0x}	[1/m]	propagation factor of surface wave	5.2.1
$\bar{k}_1,\bar{k}_2,\bar{k}_3$	[1/m]	wave vectors in reflection and refraction	1.4.2
L	[H]	inductance	3.1.1
L	[m]	largest diameter of an antenna	1.3.5
L	[m]	location in the waist of Gaussian beam in a corrugated horn	8.2.2
l	[1]	index of a resonant mode	3.3.2
l	[m]	length, distance (radar)	2.7.2
l_r	[m]	length of the reference arm	2.7.4
l_s	[m]	length or thickness of the sample	2.7.2
Δl	[m]	distance resolution (radar)	6.3.2
M	[1]	auxiliary variable in (8.4)	8.2.2
M	[kg]	molecular weight	2.2.2
m	[%]	dry basis moisture	2.6.2
m	[1]	index of a wave or resonance mode	1.3.4
m_s	[kg]	mass of an electron	2.2.4
m_s	[%]	wet basis moisture (by weight) of snow	2.5.4
N	[1/m²]	number of charged particles per volume	2.2.1
N	[1]	prescaling ratio in frequency counting	3.4.2
N	[1]	visibility of a hot spot in radiometry	7.3.2
$N_m(x)$	[1]	mth-order Bessel function of the second kind	9.3
N_s	[1/m²]	number of electrons per volume	2.2.4
N_0	[1]	Avogadro's number $= 6.023 \cdot 10^{23}$	2.2.2
N_0	[1/m²]	number of charged particles at a certain temperature	2.2.6
N_1,N_2,N_3	[1]	depolarization factors of an ellipsoid	2.4.3
n	[1]	index of a wave or resonance mode	1.3.4
$\bar{n}$	[1]	normal unit vector	1.3.4
n	[1]	refractive index	1.4.2
$\bar{P}$	[C/m²]	electrical dipole moment per unit volume	2.2.1
P_{ml}	[W]	power emitted by a matched load	7.2.3
$\bar{P}_r$	[*]	projection operator	1.3.5
P_r	[W]	received power (antenna)	7.2.3
P_t	[W]	transmitted power	1.3.3
$P_n(\theta,\phi)$	[1]	antenna directional power pattern	1.3.5

Symbol	Units	Definition	Section
p_i	[C/m^2]	polarization in an inclusion of a mixture	2.4.3
p_l	[W/m^2]	loss power per unit area	1.3.4
p_m	[C/m^2]	mean polarization of a mixture	2.4.3
p_{nm}	[1]	zero of mth-order Bessel function of the first kind	9.1.3
p'_{nm}	[1]	zero of the first derivative of mth-order Bessel function of the first kind	9.1.3
Q	[C]	charge	2.2.1
Q	[1]	quality factor	3.2.1
Q	[J]	activation energy	2.3.4
Q_l	[1]	loaded quality factor	3.2.1
Q_{l0}	[1]	loaded quality factor without sample	3.4.3
Q_d	[1]	dielectric quality factor	3.2.1
Q_{ext}	[1]	external quality factor	3.2.1
Q_m	[1]	metal quality factor	3.2.1
Q_{m0}	[1]	metal quality factor without sample	3.2.1
Q_{rad}	[1]	radiation quality factor	3.2.1
$Q_{rad,0}$	[1]	radiation quality factor without sample	3.2.1
Q_u	[1]	unloaded quality factor	3.2.1
Q_{u0}	[1]	unloaded quality factor without sample	3.2.1
$Q_\parallel$	[1]	loaded quality factor of tangential resonant mode	2.5.4
q	[C]	charge of a small particle	2.2.1
R	[m]	length of a corrugated horn	8.2.2
R	[m]	curvature radius of concave mirror of quasioptical resonator	2.7.2
R	[J/(mole $\cdot$ K)]	gas constant	2.3.4
R	[1]	power reflection coefficient	1.4.2
R	[1]	ratio between the imaginary and real parts of propagation factor	2.5.4
R_{fz}	[m]	boundary between far and near zones of an antenna	1.3.5
R_m	[Ω]	surface resistance	1.3.4
R_1	[1]	ratio between resonant frequency shifts in orthogonality technique	2.5.4
R_2	[1]	ratio between the shifts of resonant frequency and the inverse of quality factor in orthogonality technique	2.5.4
r	[m]	distance in spherical and cylindrical coordinates	1.3.3
$\bar{r}$	[m]	position vector	1.3.2
r	[m]	radius	2.7.2
$\bar{r}'$	[m]	position vector in the volume of the source or the sample	1.3.5
r_c	[m]	radius of a cylindrical cavity	2.7.2
r_s	[m]	radius of a cylindrical sample	2.7.2
$\bar{r}_t$	[m]	position vector of the transmitter	8.2.2
$\bar{r}'_z$	[m]	position vector along z'-axis in the sample	8.2.2
r_0	[m]	constant distance	1.3.3
$\bar{r}'_\perp$	[m]	position vector in the transverse (x',y') plane in the sample	8.2.2
$\delta r_\perp$	[m]	sampling interval at the aperture in holography	8.2.2

Symbol	Units	Definition	Section
S	[m^2]	area	1.3.5
S	[1]	filling factor of a resonator	3.2.2
S	[%]	salinity	2.3.2
S	[1]	slope in (5.4)	5.2.1
$S_{\parallel}, S_{\perp}$	[1]	filling factors of tangential and perpendicular resonant mode	2.5.4
s	[1]	number of oscillators in electronic polarization	2.2.4
T	[1]	power transmission coefficient	1.4.2
T	[s]	sweep time of an FMCW radar	6.3.2
T	[K]	temperature	
T	[Nm]	torque	2.2.6
T_A	[K]	antenna temperature	7.2.3
T_b	[K]	brightness temperature	7.2.2
T_{bg}	[K]	brightness temperature of the background	7.2.3
T_0	[K]	reference temperature in (2.30)	2.2.6
T_p	[K]	physical temperature of a material	1.5.1
T_s	[K]	system noise temperature of a radiometer	7.2.5
T_s	[s]	usable portion of the sweep time of an FMCW radar	6.3.2
T_N	[K]	noise temperature of a circuit element	7.2.5
t	[1]	field or voltage transmission coefficient	4.2.2
t	[°C]	temperature	2.3.2
t	[s]	time	1.3.1
t_n, t_v	[1]	field transmission coefficients of horizontally and vertically polarized waves	1.4.2
t_s	[1]	transmission coefficient through a sample	2.7.2
ΔT_A	[K]	sensitivity of a radiometer or change in antenna temperature	7.2.5
Δt	[s]	radar pulselength	6.3.2
$\bar{\mathbf{u}}_x$	[1]	unit vector along x-coordinate axis	9.2.1
V	[m^3]	volume	1.3.5
V_{hs}	[m^3]	volume of a hot spot in radiometry	7.3.2
V_m	[V]	output voltage of a mixer	3.4.2
V_{m0}	[V]	constant output voltage of a mixer	3.4.2
V_{out}	[V]	output voltage of a frequency meter	3.4.2
V_T	[V]	tuning voltage of a VCO	3.4.3
V_1, V_2	[V]	input signal voltages of a mixer	3.4.2
$V(x)$	[V]	voltage along a transmission line	9.1.1
v	[m/s]	radial speed of the target of a doppler radar	6.3.2
v_g	[m/s]	group velocity in a waveguide	1.3.4
v_p	[m/s]	phase velocity in a waveguide	1.3.4
W	[m]	maximum width of the scanning aperture	8.2.2
W	[m]	width of the strip of microstrip line	3.3.4
w	[m]	scale radius of a Gaussian beam	2.7.2

Symbol	Units	Definition	Section
w_{01}, w_{02}	[m]	waist scale radii of Gaussian beams	8.2.2
X	[m]	distance between cavities	3.5.1
x	[m]	distance (in rectangular coordinates)	1.3.2
x	[1]	variable in the equation of μ_d	2.2.6
Δx	[m]	distance difference	1.3.2
x, y, z	[*]	variables of multivariable measurements	2.5.2
$Y_m(x)$	[1]	alternative symbol of $N_m(x)$	9.3
Y_s	[S]	admittance of a plate capacitor filled with sample	2.7.2
Y_s	[S]	load admittance caused by a sample	2.7.2
Y_0	[S]	characteristic admittance of a transmission line ($= 1/Z_0$)	3.4.3
Z	[Ω]	impedance	6.2.1
Z_e	[Ω]	wave impedance of TE modes	1.3.4
Z_h	[Ω]	wave impedance of TM modes	1.3.4
Z_{in}	[Ω]	input impedance	1.3.4
Z_l	[Ω]	terminating impedance of a transmission line	9.1.1
Z_s	[Ω]	impedance with sample	2.7.2
Z_w	[Ω]	wave impedance of plane wave	1.4.2
Z_0	[Ω]	characteristic impedance of a transmission line	1.3.4
α	[1]	exponential constant depending on the shape of the sample	5.2.7
α	[1]	empirical constant in Cole-Cole and Cole-Davidson equations	2.2.7
α	[1/m]	loss factor in transmission lines	1.3.4
α	[Cm^2/V]	polarizability of atoms or molecules	2.2.1
α_a	[Cm^2/V]	atomic polarizability	2.2.6
α_c	[1/m]	loss factor due to conductivity losses	9.1.2
α_d	[1/m]	loss factor due to dielectric losses	9.1.2
α_e	[Cm^2/V]	electronic polarizability	2.2.6
α_s	[1]	damping factor (electronic polarization)	2.2.4
α_w	[Cm^2/V]	polarizability of water molecule	2.3.1
β	[1]	exponential constant depending on the shape of the sample	5.2.7
β	[1/m]	phase factor in transmission lines	1.3.4
Γ	[1]	field or voltage reflection coefficient	1.3.4
Γ_B	[1]	reflection coefficient at interface B	2.7.2
Γ_D	[1]	reflection coefficient of a detector	9.5.3
Γ_G	[1]	reflection coefficient of a signal source	9.5.3
Γ_h, Γ_v	[1]	Fresnel field reflection coefficients of horizontally and vertically polarized waves	1.4.2
Γ_l	[1]	load reflection coefficient	9.1.1
Γ_R	[1]	reflection coefficient of a resonator	9.5.3
$|\Gamma|_r$	[1]	magnitude of reflection coefficient at the resonant frequency	3.2.1
$|\Gamma|_{rd}$	[1]	magnitude of reflection coefficient at the resonant	

Symbol	Units	Definition	Section
		frequency with sample	3.4.3
$\|\Gamma\|_{r0}$	[1]	magnitude of reflection coefficient at the resonant frequency without sample	3.4.3
Γ_s	[1]	reflection coefficient of a sample	2.7.2
$\|\Gamma_s\|_{max}$	[1]	local maximum magnitude of the reflection coefficient of a sample	2.7.4
Γ_1, Γ_2	[1]	reflection coefficients of the ends of a resonator	3.2.1
$\Gamma_1, \Gamma_2, \Gamma_3$	[1]	reflection coefficients of a circulator	9.5.3
γ	[1]	Euler's constant	9.3
γ	[1/m]	propagation factor in a transmission line	1.3.4
γ_0	[1/m]	propagation factor without sample	4.3.2
Δ	[1]	calibration factor for structure independent moisture measurements	2.5.4
Δ_z	[m]	depth resolution in holography	8.2.2
$\Delta_\perp$	[m]	transverse resolution in holography	8.2.2
Δx	[*]	change of variable x	
δ	[m]	displacement of charges due to electric field	2.2.1
$\tan\delta$	[1]	loss tangent	2.2.1
δ_s	[m]	skin depth	1.3.4
ε	[F/m]	electric permittivity = dielectric constant	1.3.1
ε_d'	[1]	permittivity of dry material	2.6.1
ε_{eff}	[1]	effective relative permittivity	2.2.8
ε_r	[1]	relative permittivity = dielectric constant	1.3.2
ε_r'	[1]	real part of relative permittivity	1.4.1
ε_r''	[1]	imaginary part of relative permittivity	1.4.1
ε_{ra}	[F/m]	apparent permittivity	2.4.3
ε_{ra}'	[1]	permittivity of air (real part)	2.6.1
ε_{rc}	[1]	permittivity of the insulator of a coaxial cable	6.2.1
ε_{rd}''	[1]	ε_r'' due to dielectric losses	2.2.8
ε_{rh}	[F/m]	permittivity of host material	2.4.3
$\|\varepsilon_{rm}\|$		permittivity tensor of an anisotropic mixture	2.4.3
ε_{rmax}''	[1]	maximum of ε_r'' (Debye relation)	2.2.7
ε_{rms}	[F/m]	static permittivity of a mixture	2.4.3
$\varepsilon_{rm\infty}$	[F/m]	high-frequency permittivity of a mixture	2.4.3
ε_{rp}	[1]	permittivity of a prism	5.2.1
ε_{rs}'	[1]	static relative permittivity	2.2.7
ε_{rsh}	[1]	permittivity of the sample holder	2.7.4
ε_{rw}'	[1]	permittivity (real part) of water vapor	2.6.1
$\varepsilon_{r\infty}'$	[1]	relative permittivity at "infinite" frequency	2.2.7
ε_0	[F/m]	permittivity in vacuum	1.3.1
ε_{0i}	[1]	Neumann factor	9.1.3
η	[1]	emissivity	7.2.2

Symbol	Units	Definition	Section
η	[1]	mismatch loss in the sample to contacting antenna interface in radiometry	7.3.2
θ	[rad]	angle in spherical coordinates	1.3.3
θ	[rad]	optimum coupling angle to a surface wave	5.2.1
θ_c	[rad]	coupling angle at cut-off	5.2.1
θ_{lim}	[rad]	limit angle of total reflection	1.4.2
θ_s	[rad]	phase angle of Γ_s	2.7.4
θ_0	[rad]	opening angle of a conical horn antenna	8.2.2
θ_1, θ_2	[rad]	angles of incidence in refraction	1.4.1
θ_{3dB}	[rad]	half-power beamwidth (HPBW) of an antenna	1.3.5
λ	[m]	wavelength	1.2
$\lambda_{c,nm}$	[m]	cut-off wavelength of wave mode nm	1.3.4
λ_g	[m]	guide wavelength	1.3.4
λ_{g0}	[m]	guide wavelength without sample	4.3.2
λ_s	[m]	wavelength in a sample	2.7.2
λ_0	[m]	wavelength in free space	2.7.4
μ	[H/m]	magnetic permeability	1.3.1
μ	[Cm]	permanent dipole moment of molecule	2.2.6
μ_d	[Cm]	apparent mean dipole moment of molecule	2.2.6
μ_r	[1]	relative permeability	1.3.1
μ_r'	[1]	real part of relative permeability	1.4.1
μ_r''	[1]	imaginary part of relative permeability	1.4.1
μ_w	[Cm]	dipole moment of water molecule	2.3.1
μ_0	[H/m]	permeability in vacuum	1.3.1
Π	[m^3]	polarizability per mole	2.2.2
π	[1]	pi = 3.14159 ...	1.3.3
ρ	[C/m^3]	charge density	1.3.1
ρ	[kg/m^3]	density	2.2.2
ρ_d	[kg/m^3]	density of the dry portion of material	2.5.4
ρ_w	[kg/m^3]	density of water vapor	2.6.1
ρ_0	[kg/m^3]	empirical constant in the Clausius-Mossotti equation	2.2.2
σ	[S/m]	conductivity	3.2.1
τ	[s]	counting period of frequency counting	3.4.2
τ	[s]	integration time of a radiometer	7.2.5
τ	[s]	propagation time of a radar pulse	6.3.2
τ	[s]	relaxation time in dielectrics	2.2.7
τ_m	[s]	relaxation time of a mixture	2.4.3
τ_0	[s]	proportionality factor in (2.46)	2.3.4
ϕ	[rad]	angle in spherical coordinates	1.3.3
ϕ	[rad]	phase angle of the reflection or transmission coefficient of a resonator	3.2.1

Symbol	Units	Definition	Section
$\Delta\phi$	[rad]	phase difference or shift	3.4.2
$\delta\phi$	[rad]	periodic variation of phase shift in a sample caused by multiple reflections	4.2.2
ϕ_a	[rad]	phase lag of an amplifier	3.4.3
ϕ_l	[rad]	phase lag of transmission lines	3.4.3
ϕ_m	[rad]	measured phase	3.4.3
ϕ_m	[rad]	phase lag of the measurement channel	4.4.2
ϕ_{ref}	[rad]	phase of the reference channel	3.4.3
ϕ_s	[rad]	phase lag of the standard load	4.4.2
ϕ_t	[rad]	phase angle of transmission coefficient	4.2.2
ϕ_0	[rad]	constant phase term	1.3.2
ϕ_1, ϕ_2	[rad]	phase angles of reflection coefficients of the terminations in resonator	3.1.2
Ω_A	[sr]	solid angle of an antenna	7.2.3
Ω_s	[sr]	solid angle of a sample	7.2.3
ω	[1/s]	angular frequency	1.3.2
ω'	[1/s]	real part of complex angular frequency	1.4.1
ω''	[1/s]	imaginary part of complex angular frequency	1.4.1
ω_m	[1/s]	angular modulation frequency	4.4.2
ω_n, ω_s	[1/s]	resonant frequency (electronic polarization)	2.2.4
ω_r	[1/s]	angular resonant frequency	3.2.2
ω_{rel}	[1/s]	angular relaxation frequency	2.2.7
$\Delta\omega_r$	[1/s]	(complex) shift of ω_r	3.2.2
∇		del operator (nabla)	1.3.1

Index

Aarholt, E., 260
Academy of Sciences, Minsk (USSR), 253
Activation energy of molecules, 67
    ice, 68
Active imaging, 34
    see also Holographic and Tomographic
    sensors
Adams, M.J., 231, 233, 234
Adsorption (of water), 66–67
    ability of a material, 86–87, 110
Agdur, B., 100, 186
Akhmetshin, A.M., 253
Akyel, C., 173, 190
ALC, see Automatic level control
Altman, J.L., 105, 213
Amato, J.C., 100
Amplification, gain, 21
Amplitude, 6–9
Amplitude measurement, 170, 173–181
    see also Power measurement
Ammonia, 239
ANA, see Network analyzer
Anderson, A.P., 294, 295, 297, 298, 303
Anderson, J.G., 216, 226
Angular frequency, 6
    complex, 22
Antenna
    boundary between near and far zone, 18
    directional (power) pattern, 18, 270, 271
    effective area, 270
    gain, 18, 270
    horn, 29, 202, 222, 224, 231
    reciprocity, 19
    solid angle, 270

Applied Microwave Corp. (US), 250
Arai, I., 260
Arenata, J.C., 190
Arnaud, J.A., 298
Atek, K.A., 260
Athey, T.W., 248
Attenuation of waves, 21–22, 45, 82
    in decibels, 21
Auld, B.A., 196
Automatic level control (ALC), 173, 174, 180, 182
Autronica (Norway), 259, 260
Averaging of measurement results, 176, 202, 211, 218, 272–273

Bach Andersen, J., 287
Backhouse, P., 99
Bahar, E., 227
Bahr, A.J., 238
Bailey, S.J., 260
Baker, J.M., 260
Bardati, F., 286
Baribaud, M., 305
Barlow, H.E.M., 121
Bartashevskii, E.L., 253
Bartlesville Energy Research Center, 191
Barton, D.K., 255
Bastida, E.M., 252
Baumann, S.B., 191
Bellarbi, L., 283
Bennett, P.G., 186
Berteaud, A.J., 101
Bertholdt (company) (FRG), 223
Besada, J.L., 192

Bessel functions, 101, 316–318
  series expansions, 316
Biological substances, 192, 249–250, 286
  in vivo measurement, 250
Birchak, J.R., 70
Black-body radiation, *see* Radiometer sensors
Blast furnace, 225
Bliot, F., 250
Bolomey, J.C., 305
Boltzmann's constant, 268
Boltzmann's statistics, 268
Born approximation, 303, 304–305
Bose, T.K., 260
Boundary conditions
  conducting surfaces, 10
  dielectric interface, 45, 142
Brightness (temperature), *see* Radiometer
  sensors
Brodwin, M., 222
Broquetas, A., 305
Brunfeldt, D.R., 249, 250
Building walls, *see* Moisture measurement
Burdette, E.C., 249
Bussey, H.E., 252

Calibration, 29, 39, 81, 94, 98, 99, 138, 140,
  161, 167–168, 169, 221, 241,
  242–243, 249–250
Caorsi, S., 305
Capacitance (plate capacitor), 128–130
Capacitive sensors, 38, 130
Carleton University (Canada), 227
Cellophane, 226
Cellulose, 66, 67, 76, 90, 186
Chaloupka, H., 87, 302, 303
Chan, W.F.P., 102, 163, 196
Chao, S.H., 101
Characteristic impedance of a transmission
  line, 14–15, 310
Chu, T.H., 302
Chudobiak, W. J., 227
Circulator, 168, 217, 221, 322
Clapeau, M. 196
Clarricoats, P.J.B., 260
Clausius-Mossotti equation, 47–48, 75
Coal, *see* Moisture measurement
Coating (dielectric), thickness, 231–234,
  253–254
Coaxial line, 15, 111–116, 118–119
  attenuation, 110

characteristic impedance, 110
cut-off frequency, 111
fields, 110
semirigid, 185–186
Coke and ore layers, 225
Cole-Cole equation, 56–57
Cole-Davidson equation, 57, 67
Collin, R.E., 8, 14, 15, 18, 139
Compensation, 39, 158, 188, 194, 217, 218,
  222, 226
Concrete, prestressed, 192
Contacting sensor
  reflection, 245–251
  resonant cavity, 151–153, 186–191
  transmission, 213–214, 225–229
Conveyor (measurement on), 201, 212, 223
Cook, R.J., 121
Cornbleet, S., 298
Corona, P., 254
Cotton fibers, *see* Moisture measurement
Crank angle (CA), 191
Cullen, A.L., 102

Dalton, B.L., 180
Debye relation (relaxation), 48, 55–57, 63, 67
Decibel, 21
  relative to 1 mW (dBm), 167
Decréton, M.C., 98
Delaney, A.J., 260
de Loor, G.P., 75
del operator (nabla), 4–5, 319–320
Density, 109, 188–189
  independent measurement, *see*
    Multiparameter measurement
  measurement by gamma-ray transmission,
    29, 81, 222, 237
Depolarization factor, 72
Dielectric constant, *see* Permittivity
Dielectric materials (dielectrics), 20–23
Directional coupler, 168, 179, 221
Distance and displacement measurement, *see*
    Radar sensors
Dnepropetrovsk State University (USSR), 253
Doppler frequency, *see* Radar sensors
Doughty, D.A., 191
Dyson, J.D., 168

Eddy current, 196
Edrich, J., 297
Edvardsson, O., 260

Electric dipole moment, 42
  water, 63
Electric field direction at sample location, 101,
    105, 150
Electric field strength, 5
  external and internal in matter, 42
  inside of an ellipsoid, 72
Electromagnetic fields
  static and time-harmonic, 45
Electromagnetic waves, 3
Emissivity, 32–34
  see also Radiometer sensors
Energy conservation principle, 26
Energy content of peat, 195
Environmental Protection Agency (US), 191
Epstein, B.R., 248
Ermert, H., 302
Esselle, K.P.A.P., 249
Exposure limits of microwaves, 36–38
Extensometer, 192

Faraday rotation, 225
Fibrous materials, 74, 92, 109
Fischer, M., 159, 194
Fish meal, *see* Moisture measurement
Foodstuffs, *see* Moisture measurement
Fourier transform, 55, 253, 294–295, 299
Freon gas, 239
Frequency, 164–166
Frequency measurement, 164–167
  cavity meter, 100–101
  control voltage of VCO, 166, 181
  counter and counting, 164–166, 169, 181,
    190, 193
  counting time, 164
  crystal oscillator, 165
  discretization error, 166, 181
  error sources, 166, 178–181
  frequency-to-voltage converter, 166
  prescaler, 166
  short-term errors, 166
  with network analyzer, 169
Fresnel's field reflection coefficients
  see Reflection coefficients
Fringe capacitance, 96
Fringing field, 130, 157, 186

Gabriel, C., 261
Gajda, G., 248

Gamma (γ) rays, 3–4
  see also Density and Mass measurement,
    Safety
Gardiol, F.E., 8, 14, 15
Garg, S.K., 107
Gas
  analysis by spectrometry, 238–239
  permittivity measurement, 99, 106
Gaussian beam, *see* Quasioptics
Geophysical Survey Systems (US), 262
Giaux, G., 287
Goldsmith, P. F., 298, 303
Grain, *see* Moisture measurement
Granular materials, 92, 109, 160, 222, 226,
    243
Gudmandsen, P., 68
Gypsum, *see* Moisture measurement

Half-power width of resonance peak, 95
Hallikainen, M.T., 107, 119, 288
Hamid, A., 260
Hammerstad, E.O., 15
Harrington, R.F., 8, 141, 145
Hasegawa, S., 90
Haslam, N.C., 283
Hasted, J.B., 53, 58, 62, 64
Heating (treatment) and measurement
    simultaneously with microwaves,
    109–110
Heikkilä, S., 234
Helical coaxial line, 99
Helical resonator, 116, 155, 191–192
Helical waveguide, 121, 239
Helsinki University of Technology
  Laboratory of Applied Electronics, 262
  Radio Laboratory, 190, 193, 194, 234,
    241, 261, 262, 287, 288
Hobson, G.S., 260
Hollow cavities, *see* Resonant cavities
Holographic sensors, 34, 291
  applications, 303–304
    detection of concealed objects, 303
    measurement of reflector antennas,
      303–304
  focusing, 295–297, 302
  images of metallic samples, 300
  imaging of rotating sample (reflection
    tomography), 302–303
  optical holography, 291–292
  projection of the sample, 302–303

quasiholography (amplitude and phase
 recording), 294–300
 depth resolution with FMCW technique,
  299–300
 transverse plane resolution, 300
 speckle, 297
 synthetic aperture imaging, 302
 synthetic aperture radar (SAR), 304
 three-dimensional image, 292, 299, 302,
  303
 true microwave holography, 293–294
  resolution, 294
 two-dimensional imaging of half-space,
  300–302
Hoppe, W., 188
Humidity of air
 effect on resonator sensors, 153, 160, 186,
  194
 relative, 66
 sensor (microwave), 87–90
 for dryers, 190
Hyperthermia (microwave), 33, 34, 286

Ice, 68, 70, 287–288
 permittivity, 66–68
Imaginary unit, 8
Imatran Voima (Finland), 261
Impedance, *see* Characteristic impedance, Input
 impedance
Infrared (range), 3, 4
 absorption lines, 52
 dryer, 194
 heating, 109
 sensor, 38
 thermal radiation, 268, 270
 thermometer, *see* Temperature measurement
Innotec (Finland), 234, 237
Input impedance of a transmission line, 15,
 309
Institute of Physics, Warsaw (Poland), 190
Intensity, 9
 of radiation, 32, 268
 see also Radiometer sensors
Interface detection, *see* Impulse radar
Interference (RFI, EMI), 36–38, 158, 256, 258
Intrusive (*in vivo*) measurements, 30, 195, 250
Ion conductivity, 48, 58–62, 90
IRPA, 38
ISM frequency, 37–38
Isolator, 216–217, 221, 226

Ivoinfra (Finland), 194
Izatt, J.R., 99

Jackson, J.D., 82
Jacobsen, R., 224
Jakkula, P., 227, 260
Jofre, L., 305
Jow, J., 109, 190
Junkin, G., 302

Kajfez, D., 163
Kaliński, J., 221, 225
Kay-Ray (US), 222
Keltronics (Sweden), 261
Kemira (Finland), 227
Kent, M., 67, 86–87, 106, 110, 118, 214,
 219, 224, 226
Khalid, K.B., 105, 214
King, R.S., 192
Klein, A., 81, 212, 221, 222
Kobayashi, S., 176, 185, 188, 260
Kohler, W.E., 75
Konev, V.A., 253
Konopka, J., 190–191
Korneta, A., 196
Kramers-Kronig relations, 82
Kraszewski, A., 225
Kraus, J.D., 272, 303
Kuhn, N., 321

Laboratoire des Signaux et Systemes (France),
 305
Laboratory meter (industrial), 36, 226,
 241–243
Lakshminarayana, M.R., 189
Laminated material, 74, 92, 109, 253
Land, D.V., 287
Leroy, Y., 283, 287
Level sensing, *see* Radar sensors
Ligthart, L.P., 105
Limestone, *see* Moisture measurement
Lindell, I.V., 76
Linzer, M., 182
Liquids, 160, 194, 196, 213, 214, 226, 227,
 243, 249
Looyenga, H., 70
Loss factor, 20
 conductive, 310
 dielectric, 310
Loss tangent, 45, 140

Lossy materials, 95–96, 99, 106, 140, 185, 201, 254
Low-loss materials, 20–23, 97, 102, 106, 173, 207, 253
Lynch, A.C., 102
Lüdeke, K., 279
Lääperi, A., 288

Ma, C.H., 225
Magnetic
  field strength, 5
  flux density, 5
  materials, 21, 86
  surface current, 17
Mamouni, A., 279, 281, 283
Marcuvitz, N., 246
Markowski, J., 176
Marsland, T.P., 249–250
Material
  flow (rate, speed), 185
  stream-forming, 29, 35, 80
Martinson, T., 252
Mass (per area) measurement
  belt weigher, 81
  gamma-ray attenuation, 81, 201, 222, 223
  microwaves, 158
Maxwell's equations, 4–5
Mega System Design (Canada), 226
Measurement frequency, 80, 86
Medical applications, 286–287, 305–306
  see also Biological substances
Mensa, D.L., 302
Merlo, A.L., 191
Metal
  fatigue cracks, 238
  plate thickness, 188
  surface flaws, 196
Metaxas, A.C., 213
Meyer, W., 92, 116
Michiguchi, Y., 260, 302
Microwave ellipsometry, 254
Microwave meter, 35–36
  see also Laboratory, On-line, Portable meters
  price, 80, 164, 186, 216
Microwave resonator, 130–133
  see also Resonator
Microwave sensor
  advantages and disadvantages, 38–39
  environment (of a sensor)

  dirty, 190
  harsh, 190, 221
  measurement system, 35–36
  stabilization of measurement situation, 79–80
  type selection, 36, 79
  working principle, 20–23, 29–34
Microwave spectrometry, 238–239
Microwaves, 3, 4
  guiding, 10–16
  propagation, 9–10
Miller, G.B., 105
Millimeter waves, 3, 98–99, 106, 107, 119–121, 202, 234, 275, 294
Misra, D.K., 250
Mixer, 165, 168–169
  double-balanced, 169
  effect in metal junctions, 238
  harmonic, 165
Mixing formulas, 69–70, 252
  general, 70–71
  inclusion, 71–79
    aligned disks, 72, 73
    aligned needles, 72, 73, 74
    layered spherical, 76
  losses, 77
  material and air, 72–73
  Rayleigh formula, 75
  structure-dependent, 70–79
    small volume fractions, 72–74, 75
    high volume fractions, 74
  structure-independent (exponential model), 70
  two-component mixtures, 74–77
    randomly oriented disks, 76
    randomly oriented needles, 76
    spheres, 76
Miyahara, S., 166, 196
Moisture measurement, 63–65, 222–225
  based on microwave drying, 239–241
  building walls, 192
  coal, 81, 222
  density-independent, 86–87, 143, 194
    cotton fibers, 188–189
    fish meal, 87
    foodstuffs, 86, 222, 223, 224
  grain, 222, 224
  LC-resonator, 128–130
  limestone, 223

moisture profile, 194, 214
paper industry, 186, 193, 224
powders, 222, 226
sand, 223
simultaneous laboratory measurement of
    density and moisture, 241–243
soil, 227
structure-independent
    (coal dust, gypsum, polyfoam, quartz-
    sand, sand), 87
thickness-independent, 85
    veneer sheets, 86, 194
timber, 234
tobacco, 207, 223, 225–226
Morris, H.M., 260
Moschüring, H., 192, 248
Multiparameter (two or more) measurement,
    79–87, 158, 194, 201, 212
auxiliary measurements, 81–82
multifrequency technique, 84
number of independent measurements,
    80–81
orthogonality technique, 84–85, 145, 159,
    189, 191
Multiple reflections, see Standing waves,
    Transmission sensors
Myers, P.C., 278
Mätzler, C., 68

National Research Council of Canada, 226
Nelson, S.O., 92, 105
Network analyzer, 169–170, 245, 246,
    248–249, 250
scalar, 97, 98, 112, 169
vector (VNA), 106, 114, 118, 169
Ney, M., 166
Nguyen, D.D., 278
Nippon Steel Corporation (Japan), 188, 225
Noncontacting measurement, 38
Nondestructive measurement, 192, 233, 238,
    245–246
Nonmagnetic material, 21
Northrop Services, Inc. (US), 191
Nowogrodzki, M., 260
Nyfors, E.G., 70, 159

Oak Ridge National Laboratory (US), 225
Odam (France), 286
Ohno, J., 225
Oil detection (pollution), 287, 288
Oil emulsion, water content, 191, 226

Okamura, S., 219
On-line meters, 36
Optical band, 3, 50
color, 52
thermal radiation, 267–268, 269–270
Orme, R.D., 294
Osterrieder, S., 279
Ostwald, O., 260
Ou, W., 231, 233
Ozamiz, J.M., 216

Pandrangi, R.K., 176
Paoloni, F.J., 300, 305
Paper, cardboard, 159, 186, 193, 212, 213,
    223, 224, 226
Particle board, 193, 223
Particle size measurement, 225
Particulate materials, 187–188
Peat probe, 195
Permeability, magnetic, 5–6, 29, 86
relative, 6
vacuum, 5
Permittivity, electrical (dielectric constant),
    5–6, 29, 41
vacuum, 5, 42
Permittivity measurements in the laboratory,
    93–94
accuracy, 94
free-space methods, 107–108
lumped circuits (plate capacitor), 95–97
resonator method, 97–102
    frequency range, 99
    infinite sample method, 118
    partially filled resonator, 98, 100, 115
    resonator size, 99, 101
    totally filled resonator, 98, 99, 100, 114
sample shape, 99, 106, 109
    arbitrary, 99
summary of measurement ranges, 110
transmission line methods, 102–107
    frequency sweep, 106, 119
    frequency range, 106
    infinite sample method, 106
    open-ended line, 106
    partially filled line, 105
    sample at shorted end, 106, 111
    sample length tuning, 106
wideband measurement, 111
Permittivity, relative, 6, 41–46
air, 46

humid, 66, 87–90
anisotropy, 74, 91, 92, 109, 191, 237
apparent, 71, 74–75
aqueous solutions, 67
complex, 21
dry solids, 50, 92
effective, 69, 70, 71–79, 96, 140, 156–157
frequency dependence, 48, 55–56, 60–62
imaginary part (losses),
    by (ion) conductivity, 59
    dielectric, 59
    frequency dependence (1/f behavior),
        48, 59, 68, 90
    margarine, 59
    maximum with Debye relation, 55–56
    mixtures, 77
mathematical model, 69, 80–81
Maxwell-Wagner effect, 48, 60–62, 90
methacrylate, 121
mixtures, see Mixing formulas
moisture effects, 64, 92
organic materials, 60
polyethylene, 121
PTFE, 121
references for tabulated values, 92
relaxation, see Relaxation frequency
static, 55, 67
structure dependence, 70–79
summary, 92
tobacco, 67
water, 63–65
water vapor, 66
wood, 59, 66, 74
    moist, 59, 90–92
Peronnet, G., 305
Perturbation
    conditions, 101, 118, 141
    formulas, 85, 141–145
    technique, 98, 107
    theory, 141, 213–214, 242
Phase
    angle, 9
    shift, 136–137, 204–211
Phase factor, 214, 310
Phase-locked oscillator, see Synthesizer
Phase measurement, 168–169
    detector, 169, 181, 182, 216, 218, 220
    hybrid junction, 169
Phase modulator, 221
Phase (FM) noise, 180

Phase shifter, 173, 220, 221
Phasor, 8
Plane wave, see Waves
Philip Morris Research Center (US), 240
Philips Research Laboratories (FRG), 188,
    191, 224, 239
Photographic film, 226
Pichot, C., 305
pin diode switch, 156, 193, 217, 219, 305
Pipe (measurement in)
    dielectric, 151, 153, 155, 187, 201, 212,
        222, 226
    metal, 227
Planck's constant, 4, 268
Planck's law, 268
Plasma, 22, 134
    electron density in a Tokamak, 225
Plate-like materials, see Sheet-, Slab-,
    or plate-like materials
Plessey (UK), 225–226
Plywood, see Moisture measurement, veneer
Pneumatic transportation, 187
Polar molecules, 20, 48, 52, 53–54, 55–57,
    63–64
Polarization of matter
    atomic (vibration), 51–52
    dense materials, 47–48
    electronic, 49–51
    orientation
        Debye relation, 55–58
        gases, 52–54
        saturation, 53
        solids and liquids, 55–58
Polarization of waves, 225, 237–238, 254, 270
Polarizability of matter, 42–46, 49–50
    per mole, 47
    water, 63
Polder, D., 75
Polder–van Santen formula, 75–77
Polytechnic University of Madrid (Spain), 192
Portable meter for field use, 36, 82–83, 192,
    194, 250, 252, 262
Powders, 194, 214, 222, 243
    see also Moisture measurement
Power emitted by a matched load, 271
Power measurement (of microwaves), 167–168
    detector (crystal, Schottky-diode, square-
        law), 167–168
    temperature drift, 167
Praxmarer, W., 226

Propagation factor (wavenumber), 6
  in a sample, 203, 213
  in dielectric medium (complex), 20–23
  transmission lines, 11
  vacuum, 23
Pulp, 186, 224

Quality factor (*Q*-factor) of a resonator, 22,
    32, 127–128, 134–140
  dielectric, 114, 139
  external, 95, 134
  loaded, 95, 134, 135
  metal, 95, 134
  radiation, 134, 138, 156, 161
  unloaded, 95, 134, 136
  very high, 95–96
Quasioptics, 295–298
  Gaussian beams, 102, 295–296

Radar sensors, *see also* Holographic sensors,
    29–30, 255
  bandwidth *versus* range resolution, 255,
    256, 258
  doppler radar, 30, 259
  frequency-modulated continuous wave
    (FMCW) radar, 257–258, 302
    single sweep, 261
  impulse radar, 256–257, 302
    applications, 262
  interferometer, 259
  level sensing, 30, 259, 261
  list and references of practical systems and
    applications, 259, 260–261
  pulse radar, 255–256
  subsurface radar, 302
  time-domain reflectometry (TDR) (cable
    radar), 257
  vibration measurement, 261–262
    power lines, 261
    in the industry, 261
Radio waves, 3
Radiometer, 32–34, 272–274
  bandwidth, 272
  correlation, 34, 279–283
  Dicke, 272
  integration time, 272–273
  noise figure and temperature, 274
  null balancing technique, 273–274
  sensitivity, 275
Radiometer sensors, 32–34, 267
  antenna temperature, 270–272

aperture synthesis thermography, 283–285
  depth profile measurement, 283
  resolution, 284–285
background radiation, 272
black-body radiation, 267–269
brightness, 268
brightness temperature, 269–270
  effect of scattering, 288
contacting probes, 277, 279
  arrays, 279
  open-ended waveguide, 279
emissivity, 269, 278
industrial applications, 287
  temperature inside a cement oven, 287
  temperature of a textile web, 287
intensity per bandwidth, 268
medical applications, 286–287
  detection of inflammated areas of
    cancerous tumors, 286
  monitoring during hyperthermia, 286
multifrequency technique, 285–286
  depth profile measurement, 285
  three-dimensional temperature
    distribution, 286
remote sensing, 287–288
subsurface radiometry, 276
  equation of radiation transfer, 276, 287
  penetration depth, 283
  spatial resolution, 279, 281
thermography, 33–34, 277–278, 279
  electrical scanning, 282
  focused (millimeter-wave), 297
visibility of hot spots, 278
  reciprocity principle, 278
Rank Industrial Controls (UK), 226
Raute (Finland), 193
Rayleigh-Jeans law, 268
Reflection coefficient, 25–26
  ellipsometry, 254
  field (Fresnel), 15, 25, 311
  lossy media, 26
  open end of a transmission line, 245–251
  power, 15, 26
    Brewster angle, 26
  resonator, 135
  voltage, 309
Reflection sensors, 29–30, 245–254
  free-space sensors, 252–254
    dielectric coating, 253

thickness and permittivity of laminated
samples, 253–254
open-ended transmission line, 245–251
coaxial line, 30, 246–250
coaxial line resonator, 154–155, 248
connection between reflection coefficient
and permittivity, 247–249
equivalent circuit, 246
reflectometer using FMCW
measurement, 250
waveguide, 251
waveguide sample thickness, 251
slotted waveguide, 252
transmission line as a sample holder, 252
Refraction, 23–25
Refractive index, 25
Reinschlüssel, R., 239
Relaxation frequency of permittivity, 55–57
distribution, 57, 62, 67
ice, 68
Maxwell-Wagner effect, 60–62
shift in mixtures, 78–79
wood, 90–91
Remote Measurements Laboratory of SRI
International (US), 238
Remote sensing, 32–33, 287–288, 304
Resistive sensors, 38
Resonance peak, 135
height, 177, 178, 180
Resonance phenomenon, 127–130
Resonant frequency, 32, 127–130
complex, 142
electrons in an atom, 50
resonator filled with dielectric material,
129, 133–140
Resonant modes
cavities, 150–153, 313–314
degenerate, 150, 153, 191
disturbing, 153
high-order, 101, 116
orthogonal, 84–85
reference, 186
stripline (even, odd), 87, 155–160,
193–194
Resonator, *see also* Perturbation, Resonator
sensors
closed and open, 138
coupling of resonators, 145–149
aperture, 147
capacitive, 148, 157, 170

inductive, 135, 170
location and strength, 149
loop, 146–147
probe, 146, 157
energy stored in a resonator, 134
filling factor, 85, 90, 98, 101, 159
insertion loss, 135, 136–138, 172
length in wavelengths, 132
loss power, 132, 134
lumped circuit, 95–97, 130
mechanical, 127
microwave, 21, 31–32
nonconducting, 163
overcoupled and undercoupled, 135
partially filled, 115, 141, 150
Q-factor, *see* Quality factor
reflection coefficient, 135–136
terminations, 132
totally filled, 114–115, 133–134, 151
untuned ("stirred") cavity, 99
Resonator parameter measurement techniques,
164
accuracy, 173, 176, 184, 188
active measurement method
(resonator as a part of an oscillator),
171–172
effect of filling factor, 184–185
insertion loss method, 135
measurement electronics, *see also*
Frequency, Phase, and Power
measurement
calculation of final results, 186
digital control logic, 186
power consumption, 186
passive phase response measurement
methods, 181–182
derivative measurement, 182–183
errors, 182
errors, effects of interference, 182, 325
tracking systems, 182
passive power (amplitude) response
measurement methods, 173–174
derivative measurement, 179
dynamic range, 177
errors, 178–179
errors, effects of interference, 180, 325
seeking of half-power points, 174–176
stability and repeatability, 181
swept measurement, 184, 190

tracking (lock-in) systems, 176, 184, 188, 195
perturbated resonator, 175–176, 184
reflection coefficient method, 135–136
response time, 173, 185
several resonant modes or resonators, 186
totally filled resonator, 174
Resonator sensors, *see also* Thermal expansion
cavity resonator, 99–101, 150–153, 186–187
cylindrical, 150–151, 188, 190
fields, 313–314
long, 118
rectangular, 118–119, 150, 313
resonant frequency, 150, 313
spherical, 191
split, 188–189
untuned, 99
coaxial resonator, 31, 154–155, 241
open-ended, 192, 245–247
re-entrant cavity, 99–100, 114–115
correlation technique in speed measurement, 187–188
dielectric resonator, 163, 196
effect of air temperature, 153
ferromagnetic resonator, 163, 196
fork type ("snow fork"), 83, 194
microstripline resonator, 155–156, 238
for microwave spectrometry, 239
for permittivity measurements in laboratory, 97–98
quasioptical resonator, 103, 121, 162–163, 196
slotline resonator, 161–162, 194–195
strip resonators, 155–156
arrays, 193
strip shape, 156
stripline resonator, 84, 138, 155–156, 193–194
spatial resolution, 156
two-conductor stripline, 159–160
two-conductor line resonator, 160–161, 194–195
Richards, P. J., 302–303
Rius, J.M., 305
Rod or thread (cylinder-shaped samples), 143–144, 150, 153, 188
Rose, G.C., 107
Rueggeberg, W., 99
Rzepecka, M.A., 110, 118, 176

Saad, T.S., 15
Safety
gamma rays, 81
microwaves, 37–38
Sample
cylindrical, 99, 100, 142
holder, 94, 97, 120
lossy, 95
preparation unit, 35, 80
spherical, 144
Sand, *see* Moisture measurement
SAR, *see* Holographic sensors
Scattering, 28, 70, 211, 212, 286
Schafer, G.E., 260
Schaller, G., 285
Schiek, B., 239
Schilz, W., 210, 260, 279
Schultz, K.I., 302
Semiconductors, 190–191
Semiconductor sensor, 38
Sheet-, slab-, or plate-like materials, 85, 95, 145, 156, 201, 213, 222
Shiraiwa, T., 219, 223, 260
Sihvola, A., 68, 70, 72, 74, 76, 83, 139, 174, 194
Skandinaviska Processinstrument AB (Scanpro, Sweden), 186–187
Skin depth, 10–11
Skolnik, M. I., 219, 255
Skou, N., 288
Slab-like materials, *see* Sheet-, slab-, or plate-like materials
Snell's law, 24
Snow, 68
dry, 70
water equivalent, 287
wet, 82
measurement of wetness and density, 82
Soft materials, 160, 194
Soga, H., 188
Solid samples, 243
Solymar, L., 303
Spalla, M., 251
Spatial resolution of microwave sensors, 39, 162
Speed of light, 3, 46
Speed of propagation
(electromagnetic waves), 3, 46, 82
Sphicopoulos, T., 251
Standing wave ratio (VSWR), voltage, 309

Standing waves,
    in a lumped circuit, 96
    in a resonator, 32, 130, 132
    along a transmission line, 11
    in the transverse plane of a waveguide, 11
Steffens, D., 186
Stiles, W.H., 69
Strain sensitive materials, 190
Stuchly, M.A., 92, 105, 113, 246–248
Stuchly, S.S., 105, 113, 246–248, 260
Stutzman, W.L., 20
Submillimeter waves, 4
    interferometer-polarimeter, 225
Substrates for microwave integrated circuits, 196
Sucher, M., 109
Sugar-water solution, 226
Sumimoto Metal Industries, Ltd. (Japan), 188, 196
Surface current (density), 10, 139
Surface resistance, 11, 139
Surface measurement, 155, 160–161, 192, 195
Surface wave measurement method, 231
Synchronous detector, 176, 272
Synthesized signal source, 166, 169
Synthetic aperture radar, *see* Holographic sensors

Taylor, L., 75
TDR, *see* Radar sensors
Technical University of Denmark, 288
Technical University of Leuna-Merseburg (GDR), 226
TEM lines, 15, 148, 160
    see also Coaxial line, Transmission lines
Temperature measurement
    batteryless miniature transponder, 191
    contacting sensors, 81
    noncontacting (IR) sensors, 81, 109, 236, 287
    with microwaves, *see* Radiometer sensors
Textile, 213
Thansandote, A., 260
Thermal
    energy, 268
    expansion, 153, 160, 186, 191, 192, 194
    radiation, 32, 267
Thickness, elimination of variations, 85, 158
    see also Moisture measurement

    measurement, 29, 158, 231, 254
    see also Metal plate
Thiebaut, J.M., 109
Thomas, C.E., 240
Thorn, R., 260
Thrane, L., 64
Timber
    edging, 234
    knot detection, 34, 234, 235, 237
    moisture measurement, 237
    strength grading, 238
Time-harmonic signal, 6
Tinga, W.R., 76, 92, 105
Tinga formula, 76
Tiuri, M.E., 176, 195, 237, 260–261, 275
To, E.C., 92
Tobacco, *see* Moisture measurement, Permittivity
Toikka, M., 195, 260
Tomographic sensors, 34, 291
    diffraction tomography, 304
        reconstruction procedures, 305
        measurement of isolated organs in water, 305
        medical applications, 305
        real-time imaging, 305
        resolution, 305
    transmission tomography, 304
Top dead center (TDC) of a piston, 191
Toropainen, A.P., 90, 190
Torry Research Station (UK), 226
Total reflection, 25
Toutain, S., 196
Transmission coefficient
    field (voltage), 26, 102
    power, 26, 269
Transmission lines,
    see also Coaxial line, TEM lines, Waveguides
    dispersive, 14
    quasi-TEM lines, 15, 99, 148, 155
        microstrip line, 15–16
        stripline, 15–16
    slow-wave structures, 99
    temperature stabilization, 217
    two-conductor cable, 311
        characteristic impedance, 310
Transmission sensor measurement techniques, 215–216

see also Frequency, Phase, and Power
  measurement
additional reflection measurement, 221, 224
balancing method, 219
  response time, 219
  tracking, 219
direct phase shift or attenuation
  measurement, 216–217
  standard device, 217
  frequency averaging, 211
  response time, 218–219
double transmission, 221
frequency modulation (FM) method, 212,
  224
maximum attenuation, 220
phase shift only, 216
simultaneous measurement of phase shift
  and attenuation, 219–220
temperature stabilization, 217
Transmission sensors, 29, 81, 84, 201
compensation of density variations, 207,
  222
  with gamma attenuation, 207, 222
free space systems, 202
  attenuation or phase, 212, 221
  conveyor-mounted, 223
  diffraction, 211
  minimum attenuation, 207
  minimum sample thickness, 207–208
  pipe-mounted, 202–203, 222–223
  portable moisture meter for grain, 224
  sensor arrays, 221–222
  thin samples, 211–212
guided wave methods, 201, 213, 225–226
dielectric waveguide in a metal pipe, 226
  microstrip sensors, 214, 225–226
  waveguide sensors, 213
multiple reflections and mismatch loss, 206,
  211–212, 213, 222
Trapp, W., 90–92
Tricoles, G., 294, 303
Tsang, L., 28, 75
Two-parameter measurement, *see*
  Multiparameter measurement

Ulaby, F.T., 28, 32, 255, 272, 276, 287
Ultrasonic waves, 259
University of Cambridge (UK), 249
University of Delaware (US), 189
University of Duisburg (FRG), 191

University of Lille (France), 283, 287
University of Nebraska (US), 227
University of Palermo (Italy), 251
University of Sheffield (UK), 196
University of Wollongong (Australia), 300

Vainikainen, P.V., 87, 156, 184, 194
Valmet-Sentrol (Canada), 224, 226
Vector relations, 318
Veneer sheets, *see* Moisture measurement
Vibration measurement, *see* Radar sensors
VNA, *see* Network analyzer
Voltage controlled oscillator (VCO), 163, 166,
  261
von Hippel, A.R., 47, 53, 92, 94

Waldron, R.A., 141
Wang, J. R., 70
Water, *see also* Permittivity
  bound, 57, 64, 66, 90, 92
    degree of binding, 66, 92
  capillary, 67, 90
  drops, 73
  free, 57, 67
  molecule, 62–63
  vapor, 66, 87, 89
Wave, 202
  equation, 5, 6
  impedance, 6, 25, 46
    see also Waveguide
  number, *see* Propagation factor
  vector, 7
Waves
  guided, 11
  inhomogeneous, 26
  plane, 5, 6–7, 9–10, 133, 203
    analogy with TEM lines, 311
  spherical, 5, 9–10, 202
Wavelength, 3–4
  in a dielectric medium, 46
Waveguide, dielectric, 234
Waveguide (hollow), 11
  attenuation,
  characteristic impedance, 14
  circular, 11–12, 311–313
  cut-off frequency, 11
  cut-off wavenumber, 11–12, 311–313
  cut-off wavelength, 12, 311–313
  fields, 11, 15, 311–313
  group and phase velocities, 14

guide wavelength, 14
power, 311–313
rectangular, 12, 14, 311–312
wave impedance, 15, 311–313
wave modes (TE and TM), 12–13
Wavetrap, 226
Weiss, J.A., 98
Wenger, N.C., 100
Williams, R.V., 188
Wolff, I., 191
Wood, *see* Permittivity, relative
Wyslouzil, W., 216, 221, 226

X-rays, 3, 304
Xu, D., 98, 192

Yamanaka, T., 191
YIG (yttrium iron garnet) resonator, 163, 196

Zurcher, J.F., 192

www.ingramcontent.com/pod-product-compliance
Lightning Source LLC
Chambersburg PA
CBHW021428180326
41458CB00001B/182